Ocean
Wave Climate

MARINE SCIENCE

Coordinating Editor: Ronald J. Gibbs, *University of Delaware*

A Continuation Order Plan is available for this series. A continuation order will bring delivery of each new volume immediately upon publication. Volumes are billed only upon actual shipment. For further information please contact the publisher.

Ocean
Wave Climate

Edited by
Marshall D. Earle
and
Alexander Malahoff
National Oceanic and Atmospheric Administration
Rockville, Maryland

PLENUM PRESS • NEW YORK AND LONDON

Library of Congress Cataloging in Publication Data

Ocean Wave Climate Symposium, Herndon, Va., 1977.
  Ocean wave climate.

  (Marine science; v. 8)
  Includes index.
  1. Ocean waves — Congresses. 2. Coasts — Congresses. I. Earle, Marshall Delph,
1913-   II. Malahoff, Alexander, 1939-   III. Title.
GC206.025 1977   551.4'7022   78-24469
ISBN 978-1-4684-3401-9   ISBN 978-1-4684-3399-9 (eBook)
DOI 10.1007/978-1-4684-3399-9

Proceedings of the Ocean Wave Climate Symposium held at Herndon, Virginia,
July 12—14, 1977

© 1979 Plenum Press, New York
Softcover reprint of the hardcover 1st edition 1979

A Division of Plenum Publishing Corporation
227 West 17th Street, New York, N.Y. 10011

PREFACE

Waves critically affect man in coastal regions, including the open coasts and adjacent continental shelves. Preventing beach erosion, designing and building structures, designing and operating ships, providing marine forecasts, and coastal planning are but a few examples of projects for which extensive information about wave conditions is critical. Scientific studies, especially those involving coastal processes and the development of better wave prediction models, also require wave condition information. However, wave conditions along and off the coasts of the United States have not been adequately determined. The main categories of available wave data are visual estimates of wave conditions made from ships at sea, scientific measurements of waves made for short time periods at specific locations, and a small number of long-term measurements made from piers or offshore platforms.

With these considerations in mind, the National Ocean Survey of the National Oceanic and Atmospheric Administration sponsored the *Ocean Wave Climate Symposium* at Herndon, Virginia, July 12-14, 1977. This volume contains papers presented at this symposium. A goal of the symposium was to establish the foundations for a comprehensive and far-sighted wave measurement and analysis program to fully describe the coastal wave climate of the United States. Emphasis was placed on ocean engineering and scientific uses of wave data, existing wave monitoring programs, and modern measurement techniques which may provide currently needed data. The published literature and previous wave symposiums, particularly *Ocean Wave Spectra* in 1961 and *Ocean Wave Measurement and Analysis* in 1974, had drawn together and documented significant advances in wave theories, measurements, and analysis methods. These advances and recent needs for detailed knowledge of wave conditions made it appropriate to hold the symposium, which was attended by a broad spectrum of ocean engineers, scientists, developers of wave measurement systems, and government agencies with a mission involving waves. Information from the symposium has been used by the National Oceanic and Atmospheric Administration (NOAA) and other Federal agencies to plan wave measurement programs. The symposium included eight working groups which engaged in technical discussions and prepared recommendations for NOAA wave measurement

programs. These recommendations are included in this volume. Information from these working groups and from subsequent scientific
reviews is included in many of the contributed papers.

Adequately describing the ocean wave climate involves the
long-term collection of wave data at many locations on an operational basis, which is usually the responsibility of a government
agency. Another purpose of the symposium was to increase knowledge
about important wave programs of various government agencies.
These programs are sometimes not well known but the data and results from these programs are most useful for scientific studies
and practical applications. In this volume, contributors associated
with government agencies describe National Weather Service and U.S.
Navy Fleet Numerical Weather Central wind models which can provide
wind input to wave models, the use of wave data and hindcasts for
ship design by the Naval Ship Research and Development Center,
development of radar wave sensors by NOAA's Wave Propagation
Laboratory, an evaluation of GEOS-3 satellite altimeter wave data
by Wallops Flight Center of the National Aeronautics and Space
Administration (NASA), NOAA's Atlantic Oceanographic and Meteorological Laboratories' work in wave modeling and aircraft wave
measurements, operational wave measurements made by the NOAA Data
Buoy Office, and wave data collected and archived by the United
Kingdom Information and Advisory Service. Summaries of presentations by the U.S. Army Corps of Engineers describing wave modeling
efforts, data collection projects, and Corps' wave data needs are
also provided. The *Ocean Wave Climate Symposium* included discussions of several additional wave related activities of government
agencies. These activities are briefly noted near the end of the
volume.

Other contributors discuss wave data applications for development and use of wave models, mathematical methods for obtaining
directional energy spectra from arrays of wave sensors, nearshore
measurements of the wave climate along the California coast, considerations in applying extreme wave data to ship design, and
recent progress in radar measurements of waves. Several papers
describe various radar wave measurement systems including synthetic
aperture radars, over-the-horizon radars, and ground-wave radars.
Radar systems have a unique potential to provide directional energy
spectra with fine directional resolution at numerous offshore
locations and, without sensors to be installed and maintained at
sea, may offer considerable logistical advantages over *in situ*
sensors.

With the contributed papers and descriptions of other wave
programs, this volume provides an up-to-date summary of wave
measurement methods and wave data applications. Coupled with the
significant advances in wave theory in recent years, this informa-

tion should prove useful to help satisfy a major need for wave data
and verified wave models to specify the ocean wave climate.

We thank the contributors to this volume, the symposium
participants, and the reviewers of the papers. Dr. Ledolph Baer,
Mr. Pat DeLeonibus, Dr. Donald Barrick, Mr. Orville Magoon, and
Dr. Robert Whalin provided valuable advice for planning the
symposium. The Office of Naval Research provided administrative
assistance and the American Institute of Biological Sciences provided logistical support for the symposium. We are particularly
grateful to Mrs. Carol Collom for secretarial assistance and to
Mrs. Mary Lou Lapelosa for her excellent preparation of the manu-
scripts for publication.

Marshall D. Earle
Alexander Malahoff
Rockville, Maryland, 1978

CONTENTS

III.   RECOMMENDATIONS OF SYMPOSIUM WORKING GROUPS

I

WAVE MODELS AND
WAVE DATA APPLICATIONS

Providing Winds for Wave Models

James E. Overland

Pacific Marine Environmental Laboratory

National Oceanic and Atmospheric Administration

ABSTRACT

Wave forecasts and hindcasts are almost universally derived from wind fields. This paper reviews the meteorological computer models used by the National Weather Service and Fleet Numerical Weather Central to analyze and forecast sea level pressure fields. Geostrophic and surface wind relationships are reviewed and it is shown that errors in the input pressure fields are comparable to regional and seasonal thermal corrections to the geostrophic drag coefficient. The meteorological record is also the data source for hindcasting waves during extreme events such as hurricanes and provides frequency information as the basis for climatological assessment studies.

INTRODUCTION

Modern wave models do not make direct use of wave data but consist of relations which deduce waves from the time and space variation of an imposed wind field. The importance of accurate winds on the final wave calculations is demonstrated by a causal investigation of significant wave height tables. An increase in wind speed from 10 to 13 $m\text{-}s^{-1}$ (i.e., 5 knots) results in an increase in significant wave height from 2.5 to 4.0 m. Duration of wind from a constant direction is also important. At a wind speed of 13 $m\text{-}s^{-1}$, increasing the wind duration from 6 to 18 hours results in a significant wave height increase from 2.2 to 3.4 m. Such duration considerations result in quite different sea conditions for hurricanes, fast moving east coast U.S. storms, and steady post cold front winds over the northeast Pacific.

The definition of surface wind is an important concept. The universal parameter is the wind stress on water. As this parameter cannot be easily and routinely measured directly, it is calculated from the wind at a given anemometer height, usually 10 or 20 m, by means of a drag coefficient (e.g. *Garrett*, 1977). As wind speed varies with height, it is essential to specify the anemometer elevation when referring to a "surface wind." The drag coefficient is a measure of the amount of bite which an air mass can exert on the water surface. It depends on the stability which is related to the difference between the density of the air in actual contact with the surface (which may be determined directly from the sea surface temperature), and that of the overflowing air. Cold air outbreaks behind a cold front with moderate wind speeds overflowing warm water can often generate larger and steeper waves in a shorter amount of time than other types of storm situations containing stronger winds.

Wind fields may be specified in two manners. The first involves direct analyses or forecasts of sea level pressure, and inference of the winds from the pressure fields. Spacing and accuracy of routine ship wind observations are usually insufficient to analyze wind fields directly, but pressures are coherent on length scales sufficient for delineating major weather features. Analyses are dependent upon the resolution and methodology of automated analysis techniques (or the skill of a hand analyst), which in turn are dependent upon the spatial distribution of surface observations. Both sea level pressure reports and ship and buoy wind measurements are generally utilized in analyzing sea level pressure charts. Forecasts of sea level pressures are almost exclusively derived from modern numerical weather prediction. Accurate forecasts are dependent upon adequate initialization at all elevation levels of an atmospheric prediction model and the accuracy and resolution of the numerical technique. Thus, the limitations on meteorological forecasting provide a boundary on the ability to forecast the wave field.

The second approach to specifying winds is the use of a storm parameterization in which the surface wind field is entirely specified in terms of a few storm parameters, such as storm size, intensity, and motion. This is a typical approach to treating severe storms such as hurricanes, in which little direct data is available and the storm size is smaller than the typical resolution of standard pressure analyses.

This paper briefly reviews the development of numerical weather prediction and outlines the operation and accuracy of the analysis-forecast cycle as it relates to wave forecasting at the major U.S. forecast centers. In determining wave climatologies, the meteorological record is generally the only sufficiently long data source on which to base return frequencies. An illustration

is given applying the joint probability method to the assignment
of return frequencies to parameters of a hurricane storm model.
Finally, methods of relating surface stress to the sea level
pressure field are discussed.

NUMERICAL WEATHER PREDICTION

A necessary condition for forecasting is that the atmosphere
be orderly to the extent that the events being predicted must not
be overwhelmed by interactions with smaller scales that are not
explicitly accounted for in the predictive system. The predominance
of vorticities in the atmosphere is an important feature from this
standpoint of predictability. Vortices in a fluid that is not
forced (or in which external forcing is small and/or slow acting)
are longer lived, slower to move and develop, and slower to exchange energy with other atmospheric phenomena, relative to other
types of fluid motion. This is a consequence of conservation of
angular momentum on a rotating earth.

The first operational numerical weather predictions depended
on the predominance of vorticity in the atmosphere in the storm
scale. The only physical mechanism contained in the first model
(*Charney et al.*, 1950) was the horizontal rearrangement of the
vertical component of vorticity. The vertical structure of the
atmosphere was treated as one layer. This model is called a
barotropic model because it assumes a high correlation between
the pressure and temperature fields (i.e., isobars parallel to
isotherms) and an exact relation between the wind field and the
pressure field, i.e., a geostrophic balance.

The barotropic model is still in use today as it is a successful method of predicting storm advection. However, it is the slight
imbalance between physical processes assumed to be in equilibrium
by the barotropic model that leads to growth and development of
storm systems. In the middle 1960's, numerical weather prediction
took a more direct approach in which more basic physical laws or
primitive equations were integrated from an initial state of the
atmosphere. These consist of Newton's second law of motion in three
dimensions (the momentum equations), the first law of thermodynamics,
conservation of mass, density, pressure, and temperature. The
system of equations is nonlinear, which has led to a great deal of
applied mathematical work in the field. Fortunately, the terms with
derivatives in time are linear and first order leading to an initial-
value problem. The system has derivatives in three-dimensional space
and time so that one must deal with four-dimensional fields of data.
The continuous equations are transformed into finite difference equations in which derivatives are replaced by difference ratios for
digital computers (*Phillips*, 1959; *Arakawa*, 1966; *Marchuk*, 1967;

Shuman, 1974). At the present time, the National Meteorological
Center of the National Weather Service, NOAA, makes use of three
IBM 360/195 computers for data processing and model forecasting,
each with a capacity of about 10-15 million instructions per second.
Further background is provided in *Haltner* (1971), *Monin* (1972),
Fawcett (1977), and *Shuman* (1978).

FORECAST ANALYSIS-CYCLE

The two major weather centers concerned with wave forecasting
are the National Meteorological Center (NMC) in Camp Springs,
Maryland, and the U.S. Navy Fleet Numerical Weather Central (FNWC)
in Monterey, California.

Current numerical weather prediction centers work on an
analysis-forecast cycle in which either 6- or 12-hour forecasts
of parameter fields are used as first-guess fields for a new
analysis. The first-guess fields are used in data checking, in
providing continuity and in supplying reasonable estimates in
regions lacking observations. Two types of objective analysis
procedures are in use at NMC, referred to as the "Cressman" and
"Hough." The *Cressman* (1959) method begins with a first-guess
value at each grid point of a numerical mesh and adjusts this value
with the observations in the vicinity using a sequence of scans.
On each succeeding scan the area of influence of the weighting
function is decreased. The Cressman analysis is used with the
limited area models. The present global analysis at NMC consists
of fitting Hough functions to observed data (*Flattery,* 1970).
Hough functions are solutions to Laplace's tidal equation, which
imply a near-geostrophic relation between the wind and mass
fields. It contrasts with the Cressman by being spectral in
character as opposed to local domain of influence.

FNWC uses a "Fields by Information Blending" approach to
analyzing sea-level pressure (*Holl and Mendenhall,* 1971). "Fields
by Information Blending" is a variational approach which analyzes
a scalar distribution by weighting and blending information on
the value, gradient, and Laplacian of the field. It accepts a
mixed set of observations and fields such as a first-guess field
along with estimates of reliability of each type of observation.

NMC now has five distinct models in operation as summarized
in Table 1.

The Barotropic-Mesh Model employs the barotropic vorticity
equation for the 500 mb pressure level. It also uses an 800-500 mb
thickness equation to estimate surface drag and topographical influence. It covers the northern hemisphere with a square mesh of

TABLE 1. NMC operational models June 1, 1977 (after
Shuman, 1977). The first column indicates how
many hours after observation time the models
are run. The models are described in the text.
The next three columns indicate the aerial
coverage, number of levels in the vertical, and
grid length of each model. For example, the
6LPE covers the hemisphere at a resolution of
381 km, has 6 levels in the vertical and forecasts for 84 hours from 00GMT and 48 hours from
12GMT data.

Starting Time After 00 & 12 GMT	Model	Area	No. of Levels	Grid Size	Forecast Period	
01:15	Barotropic-Mesh	Hemisphere	2	381 km at 60N	48 hr	
01:30	LFM	North America	6	190.5 km at 60N	48 hr	
04:00	6LPE	Hemisphere	6	381 km at 60N	00 GMT: 84 hr 12 GMT: 48 hr	
07:30	MFM	300X300KM^2	10	60 km	48 hr	
10:00	9L Global	Global	9	2.5 Deg Latitude	00 GMT: 6 hr 06 GMT: 6 hr 12 GMT: 6 hr 18 GMT: 18 hr	

Analysis methods: Hough and Cressman

points on a polar stereographic projection (Figure 1). A major
purpose of the barotropic mesh model is that it is started at H +
01:15 hours (H being one of the synoptic hours 0:00 GMT or 12:00
GMT) and, as it runs very quickly on the computer, provides the
field forecaster with an early prediction from the latest data.

The Limited-Area Fine-Mesh Model (LFM) is the model that
fulfills the main requirement for weather guidance over the continental United States (*Howcroft*, 1971). It covers North American
and nearby waters (Figure 2) and is started after receipt of most
North American upper air sounding reports. It employs the primitive
equations in six layers and includes surface drag, topography,

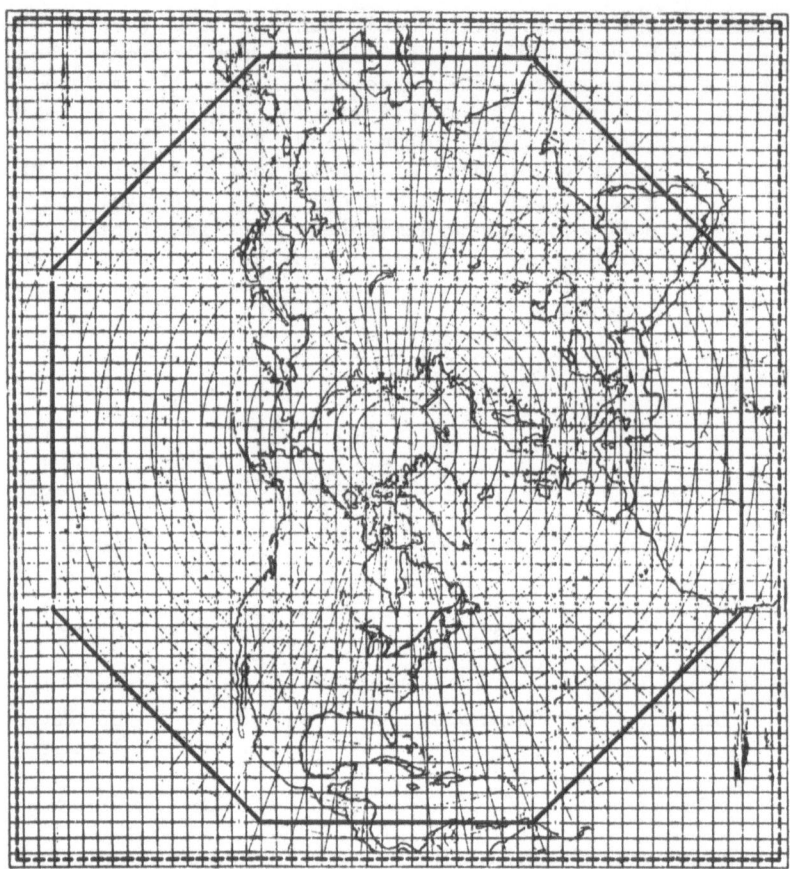

Fig. 1. Location of the grid points used for the NMC coarse
mesh numerical models. The grid overlays a polar stereographic
map projection of the Northern Hemisphere. The mesh separation is
such that the grid points are 381 km apart at 60°N with less separation distance to the south and greater separation to the north.

feedback of latent heat, convective parameterization and long- and
short-wave radiation.

   The Six-Layer Primitive Equation Model (6LPE) has the same
physics as the LFM (*Shuman and Hovermale,* 1968) but it covers the
northern hemisphere, and its grid size is double that in Figure 1.
The 6LPE provides guidance for commercial avaiation and maritime
interests, is used in the NMC 5-day forecast program, and provides
lateral boundary conditions for the LFM forecasts. The northern
hemisphere singular wave forecasting model, produced by the National
Weather Service, is based upon the 1000-mb wind forecasts from the
6LPE. The 1000-mb winds have been extrapolated from the lowest

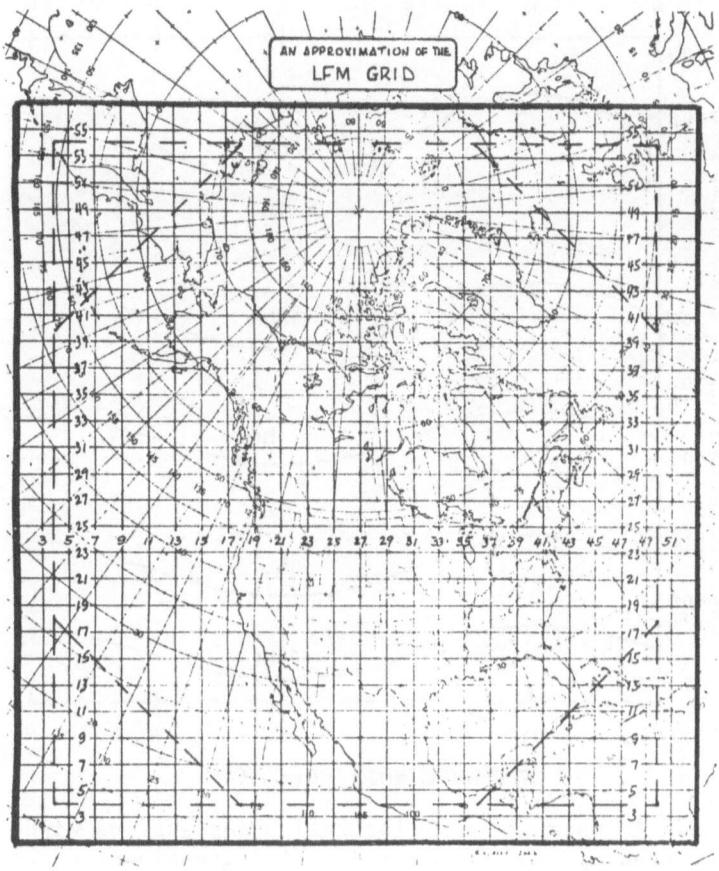

Fig. 2. The NMC limited-area fine mesh (LFM) grid on a polar stereographic grid. The model domain is North America with twice the resolution (190.5 km at 60°N) as the coarse mesh grid. Only the odd numbered grid lines are shown.

layers of PE model to convert from the model's internal vertical coordinate system (the sigma system) to standard pressure levels. For the wave forecast, empirical corrections are made to the 1000-mb winds which are based upon comparison of the 6LPE winds with 1 year of wind observations from ocean station vessels. These corrections (*Pore and Richardson*, 1969) are shown in Figures 3 and 4. Note that there are different corrections for different forecast periods as an attempt to remove the bias introduced by the forecast model. The most significant difference in the corrections is between analyzed winds at +00 and forecast winds. Input to the spectral wave model at FNWC is based upon a five-layer primitive equation atmospheric model which runs on the same 381-km polar stereographic grid as the

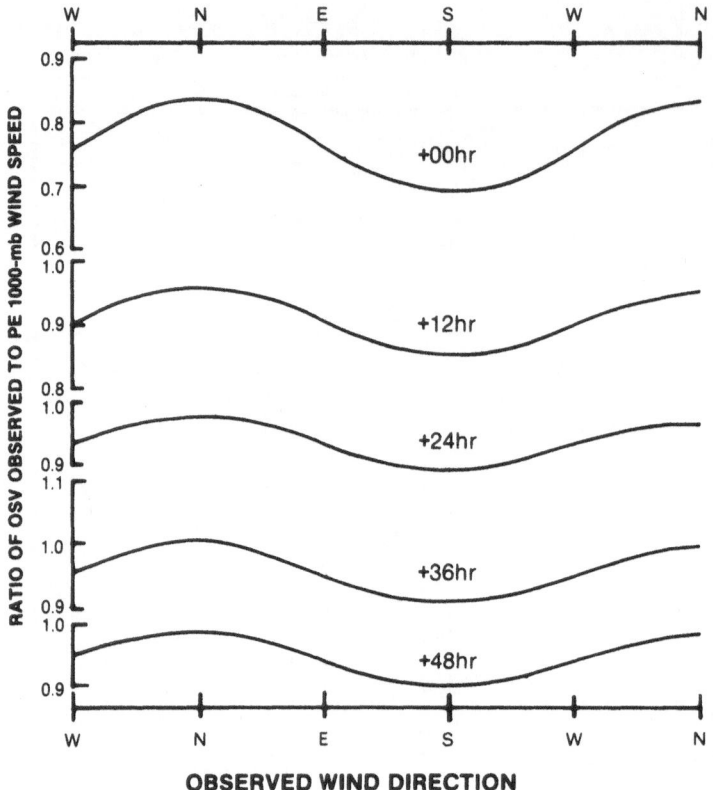

Fig. 3.   Adjustment ratio for estimating surface winds from the
6LPE 1000 mb wind speed.   Wind speed ratios are given on the left
hand side.   The independent parameters are wind direction (bottom)
and forecast hour (different lines).   The latter account for the
fact that the numerical models tend to lose energy with time so
that the wind speed reduction is greater for a 12 hour forecast
than a 36 or 48 hour forecast.

6LPE (*Kesel and Winninghoff*, 1972).   The forecast sea-level pressure
and air temperature, as well as a sea-surface temperature field, are
taken as input to a diagnostic planetary boundary layer model.   This
model infers an equivalent 19.5-m elevation wind which is consistent
with the growth formulas used by the spectral wave model (*Kaitala*,
1976).

The Movable Fine-Mesh Model (MFM) was originally designed to
predict the path of hurricanes.   It is the most highly resolved
model run at NMC, with 10 vertical layers and a 60-km mesh covering
an area about the size of the continental United States.   It is not
run regularly, but is on call when a hurricane or flash flood threat
exists.

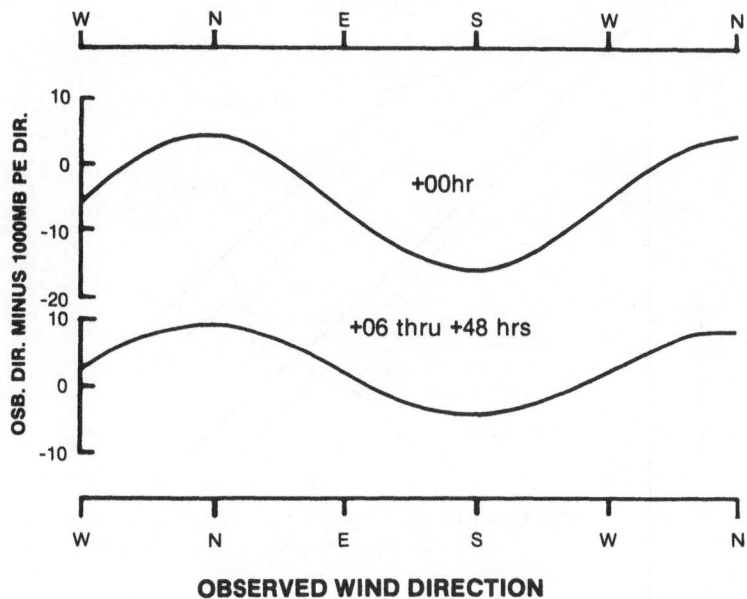

Fig. 4. Adjustment angle for estimating surface wind direction
from the 6LPE wind direction. The corrections between observed (OBS)
and 1000 mb PE wind directions are shown as a function of wind direction and forecast hour.

The Nine-Layer Global (9L GLOBAL) is the only global model run
operationally and principally serves to provide a first-guess for
the global objective analysis. It contains physical processes
similar to the 6LPE but is run on a 2.5° latitude-longitude grid.
The 9L GLOBAL is run at H + 10:00 from +00:00 and +06:00 data. The
late data cutoff of 10 hours after observation times insures that
all available data is included in the initial conditions for the
model run which generates the first-guess field for the next 12-
hour analysis cycle.

MODEL ACCURACY

Two sets of statistics will be shown to indicate the relative
accuracy of the LFM and 6LPE models. Figure 5 shows the skill
score of the LFM and 6LPE averaged over 12 months, December 1975
through November 1976, for the 500-mb (50K pascals) height field
and the sea-level pressure field. The measure of skill is based
upon the S_1 score (*Teweles and Wobus*, 1954) given by the normalized
error in gradients over the grid, i.e.,

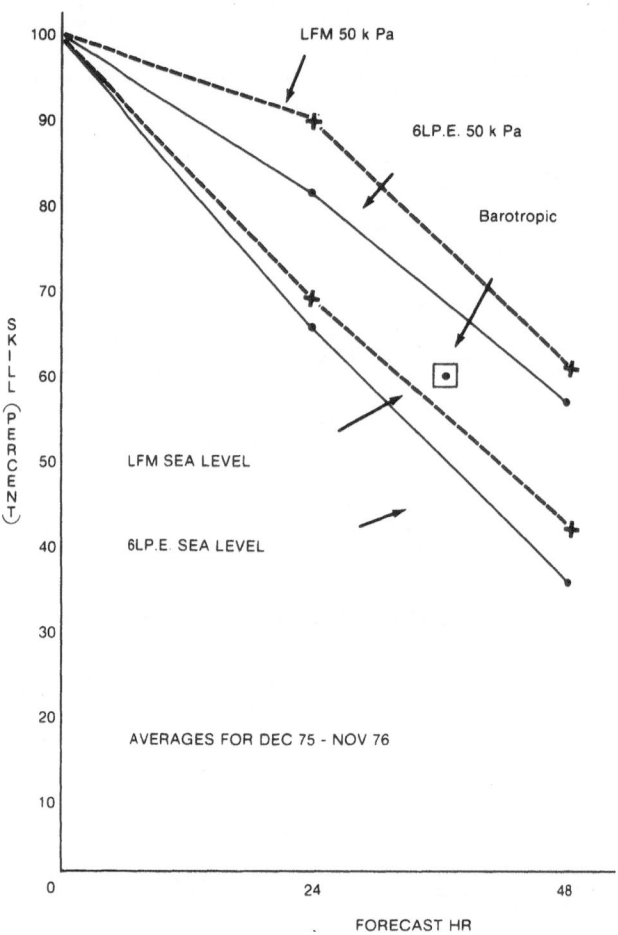

Fig. 5. Skill scores for the LFM and 6LPE models for both the sea level pressure and 500 mb height fields as a function of forecast hour. Skill score is related to normalized errors in gradients as described in the text.

$$100 \ \Sigma |e_a| / \Sigma |G_L| \qquad\qquad (1)$$

where e_a is the error of the forecast pressure difference between adjacent grid points and G_L is the observed or forecast pressure difference between grid points, whichever is larger. As shown skill (percent) is defined as $2X \ (70-S_1)$, which experience has shown to yield 100 for a virtually perfect chart and 0 for a worthless one. By 48 hours there is significant reduction in skill

score, particularly for placement of surface features. These
statistical plots provide information on overall performance, but
of more specific interest to wave forecasting are the motion and
development of individual storm systems. *Brown* (1975) has documented 32 36-hour forecasts of the position and central pressure
for 10 east coast winter storms during January-April 1975. The
storms were between Florida and Newfoundland with 24 of the 32
cases being over the ocean. At 36 hours (Table 2a) the overall
average position error for the LFM is 186 nmi, roughly half of the
6LPE error of 354 nmi. Figure 6 shows the distribution of observed
low positions relative to the forecast position. Both models tend
to move systems too slowly. Table 2b lists the error in central
pressure of the lows. The error of the LFM forecasts was greater
than the PE's in the southern area as a result of a few forecast
central pressures that were much too low. Nevertheless, the
average algebraic error of the LFM was positive. Overall, the
LFM absolute error is 4.1 mb less than the PE. The 6LPE forecast
central pressure is strongly biased toward being too high since
the algebraic error is large. It also suffers from a large standard deviation which implies that it is not possible to remove
central pressure error by simple correction factors.

FUTURE TRENDS IN NUMERICAL FORECASTING

By comparison of the LFM and 6LPE models, NMC has had mounting
evidence that the key to advances in operational numerical weather
prediction is reduction of truncation error. Truncation error is
a purely mathematical error which occurs when finite difference
ratios from grids replace derivatives in the basic meteorological
equations. The smearing of features smaller than about four grid
lengths results in a general reduction of the speed of propagation
of lows and an inability to resolve initial development of baro-
clinically unstable flow that can grow into meteorologically significant features.

The most straightforward way of reducing truncation error is to
reduce grid length, and this is the path chosen by NMC for the
immediate future. Table 3 shows the NMC operations as projected for
implementation during 1978. The LFM will be replaced by essentially
the same model but with the grid size adjusted to approximately two-
thirds of the previous value. NMC is considering three candidates
for replacement of the 6LPE:

1. Keep the same model, but reduce the grid length to half
   the size.

2. The 9L HEM, which is a version of the 9L GLOBAL, but with
   intervals between points reduced to 2° and covering only
   the northern hemisphere.

TABLE 2. This table lists average errors in position and central pressure for ten east coast storms stratified by latitude and model (more than one prognosis are made on each storm). Absolute error averages the error regardless of sign while algebraic error retains sign in computing the average error.

TABLE 2a. Error in position of lows.
January 1, 1975 to April 15, 1975

Area	No. of Progs.	Magnitude of Vector Error In Deg. of Lat.	Standard Deviation	No. of Progs.	Magnitude of Vector Error In Deg. of Lat.	Standard Deviation
		36-Hr. PE (10 lows)			36-Hr. LFM (10 lows)	
25 to 38.9N	10	6.4		10	3.5	
39 to 47N	20	5.6		22	2.8	
25 to 47N	30	5.9 (354)*	2.50	32	3.1 (186)*	1.67

*nautical miles

TABLE 2b. Error in central pressure of lows.

Area	Absolute Error (mb)	Standard Deviation	Algebraic Error (mb)	Absolute Error (mb)	Standard Deviation	Algebraic Error (mb)
	36-Hour PE			36-Hour LFM		
25 to 38.9N	3.9		+ 3.7	4.4		+ 1.8
39 to 47N	12.2		+11.1	5.7		+ 3.5
25 to 47N	9.4	7.31	+ 8.6	5.3	3.95	+ 2.9

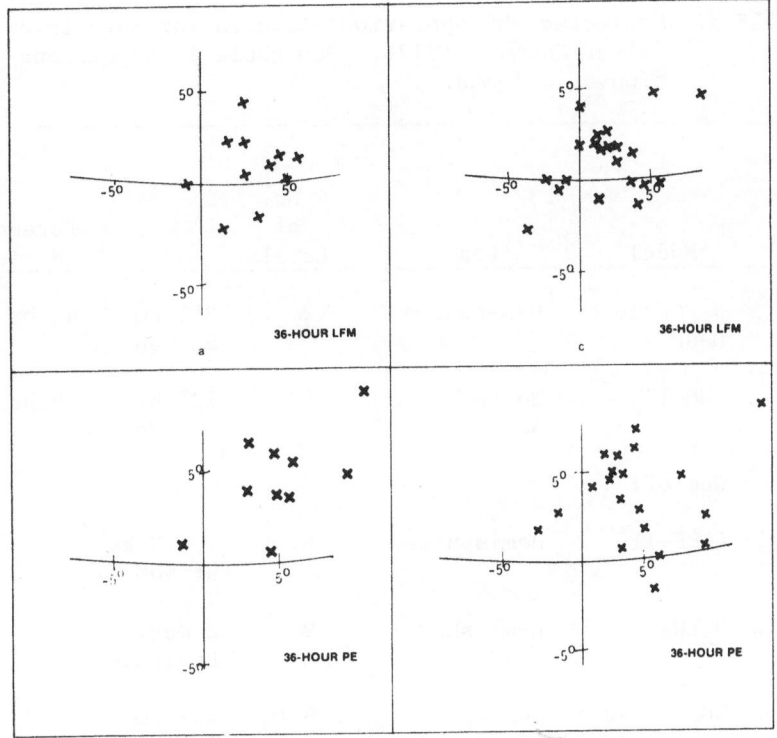

Fig. 6.   Position errors of low pressure centers for east coast winter storms during January-April 1975.   Data is stratified according to model (i.e., LFM or 6LPE) and storm position north or south of 39°N.   Storm positions south of 39°N are on the left and storm positions north of 39°N are on the right.   Small crosses indicate the true position of the storm center relative to the 36 hour predicted position.   The scale is 5 degrees of latitude and longitude. Note that almost all storms have moved farther in a northeast direction than predicted by the models.

   3.   A Nested Grid Model (NGM) which would have two grids,
        a coarse and fine mesh, A and B, which are fully inter-
        active.   In operation the location of grid B could be
        moved to cover the problem of the day.

   FNWC plans to replace its current northern hemisphere five-
level atmospheric model by 1978.   The new model (*Mohok and Kaitala*, 1976) is global and is formulated on a 2.5° latitude-longitude mesh. It makes use of fourth-order spatial finite difference approxima-
tions for improved control of truncation error.   The most substantial change involves an improved parameterization of the planetary bound-
ary layer based upon modern boundary layer similarity theory.

TABLE 3.   Projected NMC operational models for late 1978
(after *Shuman*, 1977). See table 1 for explanation of columns.

Starting Time After 00 & 12 GMT	Model	Area	No. of Levels		Forecast Period
01:15	Barotropic-Mesh	Hemisphere	2	381 km at 60N	48 hr
01:30	LFM-II	North America	6	127 km at 60N	48 hr
04:00	One of:				
a.	6LPE-II	Hemisphere	6	190.5 km at 60N	
b.	9 LHEM	Hemisphere	9	2 deg. Latitude	
c.	NGM: Grid A Grid B	Hemisphere 25000 x 25000 km^2	8-10 8-10	448 km 224 km at 60 N	
07:30	NFM	3000 x 3000 km^2	10	68 km	48 hr
10:00	9L GLOBAL	GLOBAL	9	2.5 deg. Latitude	6 hr. cycle and 'catch-up'

Analysis methods: Optimum interpolation and Cressman

Besides model replacement, NMC also plans to make major
changes to the global analyses system. While the Hough analysis
was ideal for analyzing a mix of wind, temperature, and pressure
reports, it did not properly account for varying accuracy of different sources of measurements. NMC now believes that the latter
is more important and is developing an optimum interpolation system
(*Gandin*, 1963) that will account for the statistics of the errors in
different types of measurements such as radiosonde observations,
winds found by tracking clouds from satellites, indirect temperature
soundings from satellites, constant pressure balloons, and surface
wind estimates derived from SEASAT-A.

WINDS IN THE COASTAL ZONE

The step function in physical parameters at the air/sea/land triple point can lead to sharp and intense variations in the meso-scale wind field along both U.S. coasts. Mesoscale circulations respond to local forcing such as sea breeze and katabatic (density driven) flow of cold air down mountains such as in southern Alaska, or are the result of modification of synoptic systems by topography and land-water contrasts in surface heating and friction.

The existing technique of the National Weather Service for predicting coastal winds is a set of linear regression equations that relate surface winds at selected locations to output variables of the 6LPE model (*Feit*, 1976). The equations were based upon surface data from lightship stations along the east coast, Coast Guard coastal stations and an offshore platform along the west coast, and ships of opportunity divided into 12 regions in the Great Lakes (Figures 7, 8, and 9). A multiple screening technique was used to select predictors from a large set of possible primitive equation parameters.

Figure 10 shows the average explained variance of the dependent data for 12:00 GMT primitive equation origin time for the west coast and east coast for both summer and winter. The explained variance of all parameters is low. The main factors that contribute to uncertainties in the coastal winds are forecast errors in the 6LPE and the inability of the regression model to resolve coastal modification of the large-scale flow, particularly as a function of direction and local circulations such as sea breeze. For example, the east coast equations are better in winter, presumably because of the importance of organized synoptic systems, while summer is dominated by sea breeze. Another feature of the regression forecasts that *Feit* (1976) points out is that the procedure tends to over-forecast the lower wind speeds but under-forecast the higher wind speeds. This is a consequence of treating the 6LPE output as an independent variable in the regression analysis.

There is currently keen interest in coastal meteorology reflected by the establishment of a committee on meteorology of the coastal zone by the American Meteorological Society. While immediate emphasis of coastal wind studies will remain upon case studies of processes, results of these studies will eventually make their way into operational systems.

PARAMETRIC STORM MODELS

Standard synoptic meteorological analysis techniques cannot provide a description of the wind field in hurricanes because of

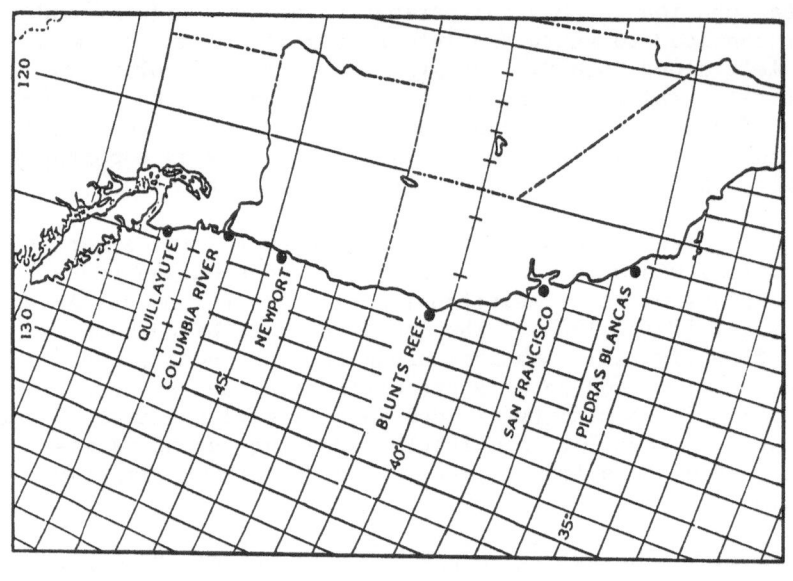

Fig. 8. Forecast locations for west coast coastal wind forecasts.

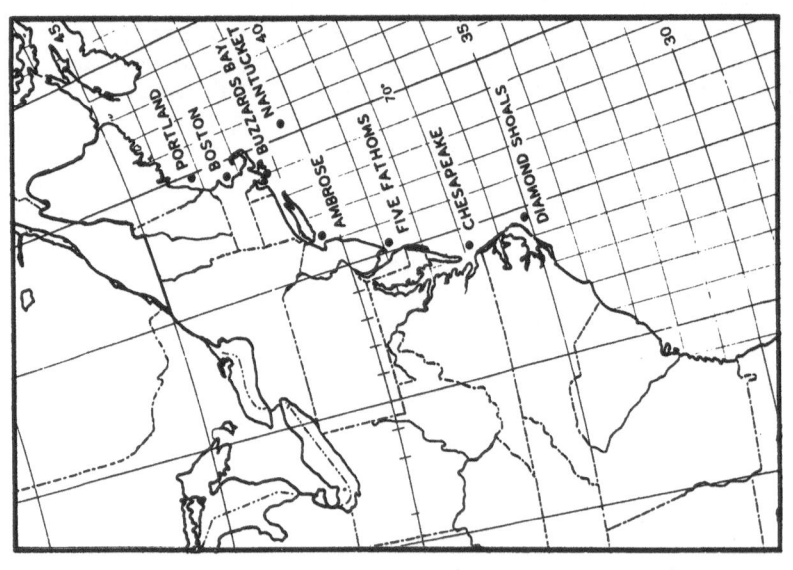

Fig. 7. Location of forecast points (light ships used in the development of offshore wind forecasts).

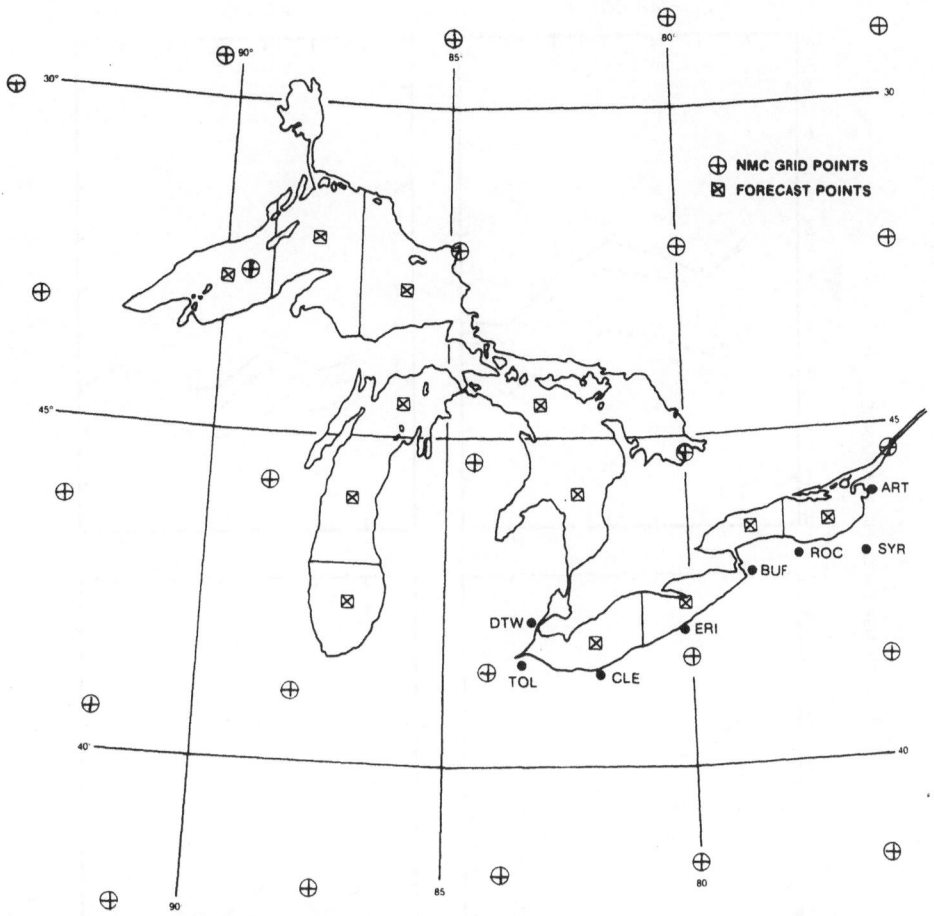

Fig. 9. Forecast locations and closest 6LPE grid points used in derivation of Great Lakes winds from 6LPE forecasts.

their small relative size and the lack of direct measurements. Hurricanes can be characterized by 4 parameters: central pressure deficit, horizontal size, forward speed and direction. The horizontal size parameter is often the radius to maximum wind speed. The pressure is often given by an exponential:

$$P(r) - P_o = (P_n - P_o) \exp (-R/r) \qquad (2)$$

where $P(r)$ is the pressure at a distance r from the center, P_o is the pressure at the storm center, and P_n is the sea level pressure

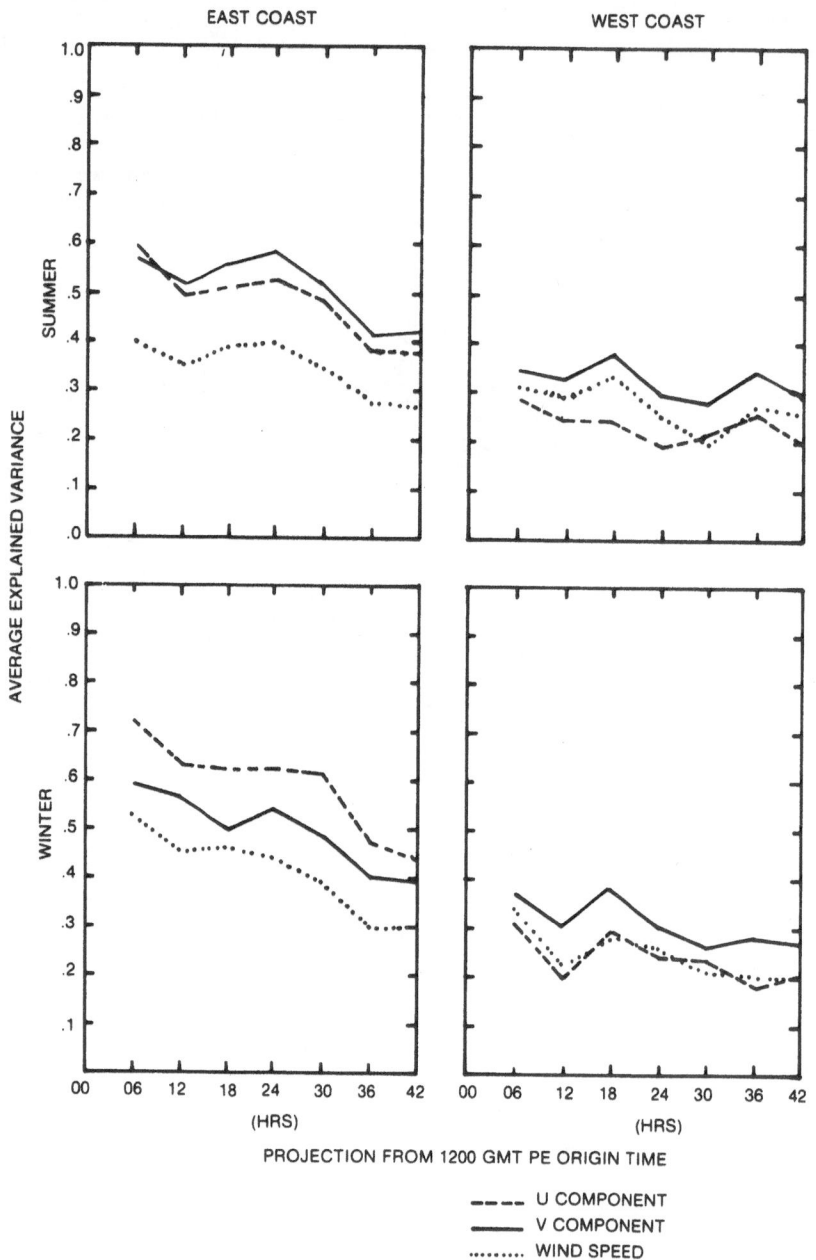

Fig. 10. East and west coast explained variance averaged over all stations for 1200 6LPE origin time as a function of forecast hour stratified for summer and winter. Explained variance is the amount of the wind record (light ship or other) accounted for by linear regression of numerical model output parameters.

at the outer limits of the hurricane. R is a horizontal size parameter and is usually the radius to maximum winds.

This equation can be solved for the pressure gradient:

$$\frac{dP}{dr} = (P_n - P_o) \frac{R}{r^2} \exp (-R/r).$$ (3)

The exponental distribution is documented for a large number of storms (*Myers*, 1954; *Graham and Hudson*, 1960; *Myers and Malkin*, 1961). The Army Corps of Engineers uses as a reference a Standard Project Hurricane, or SPH (*Graham and Nunn*, 1959). The SPH storm defines a maximum surface wind in terms of this pressure gradient:

$$V_x = .865 \ V_g = .865 \left[(\frac{1}{\rho_a e}) \ \sqrt{P_n - P_o} - \frac{Rf}{2} \right]$$ (4)

where V_g is the maximum gradient wind computed from the radial component of the momentum equation. Air density is ρ_a, f is the Coriolis parameter, and e is the base of the natural logarithm. Next, the wind profile $V(r)/V_x$ is specified empirically from hurricane wind patterns observed along the Atlantic and Gulf coasts summarized by the nomogram in Figure 11 (*Schwerdt*, 1972). The angle between the wind direction and a tangent to the circle centered at the middle of the storm (inflow angle) is given in linear pieces (0° - 10°) out to R, (10° - 25°) from R to 1.2R and constant 25° for r greater than 1.2R. The wind speed is adjusted for the forward speed of the storm by

$$V_{xp} = V - \frac{T}{2} \ (1 - \cos\theta)$$ (5)

where V_{xp} is the wind speed at any point in the wind field, V is the stationary storm speed at that point, T is the speed of forward motion of the storm and θ is the angle between the wind vector and the storm motion vector.

The SPH storm fits the limited data sets available but has been criticized because the pressure distribution, wind speeds, and inflow angles are not dynamically consistent. *Cardone et al.* (1976) used a storm model proposed by *Chow* (1971) in which the momentum balance equations in two dimensions are solved numerically based upon the assumption of an exponentially dependent radial pressure distribution. For hindcasting hurricane Camille the radial distribution of surface measurements to determine parameters $P_o - P_n$

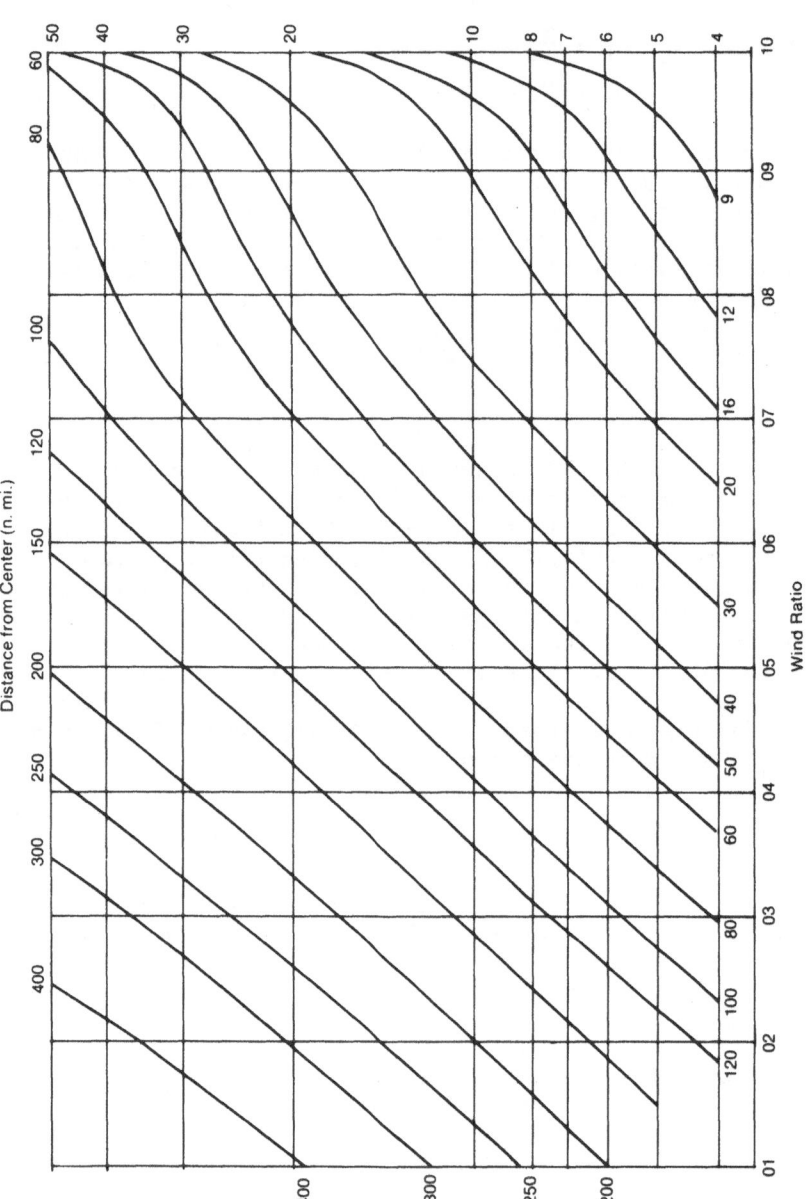

Fig. 11. Empirical curves for estimating wind speed as a function of distance from the center of the storm for the U.S. Army Corps of Engineers Standard Project Hurricane (SPH). Radius to maximum winds is input on the vertical axis, distance is input along diagonal lines with the speed ratio read at the bottom.

and R was produced from coastal station reports by a simple time space transformation as the storm crossed the coastline (Figure 12). Figure 13 shows the wind field produced by the *Chow* model and Figure 14 shows a wind speed comparison for Camille as the storm approached the coast of Louisiana. Another hurricane model is the SPLASH model (*Jelesnianski*, 1965, 1966, 1972) which is used by the National Weather Service in forecasting hurricane storm surge. The SPLASH model is similar in construction to the *Chow* model in that it numerically solves for a dynamically consistent wind field. However, unlike the *Chow* model it specifies the relative wind profile

$$V(r) = V_{max} \frac{2Rr}{R^2+r^2} \qquad\qquad (6)$$

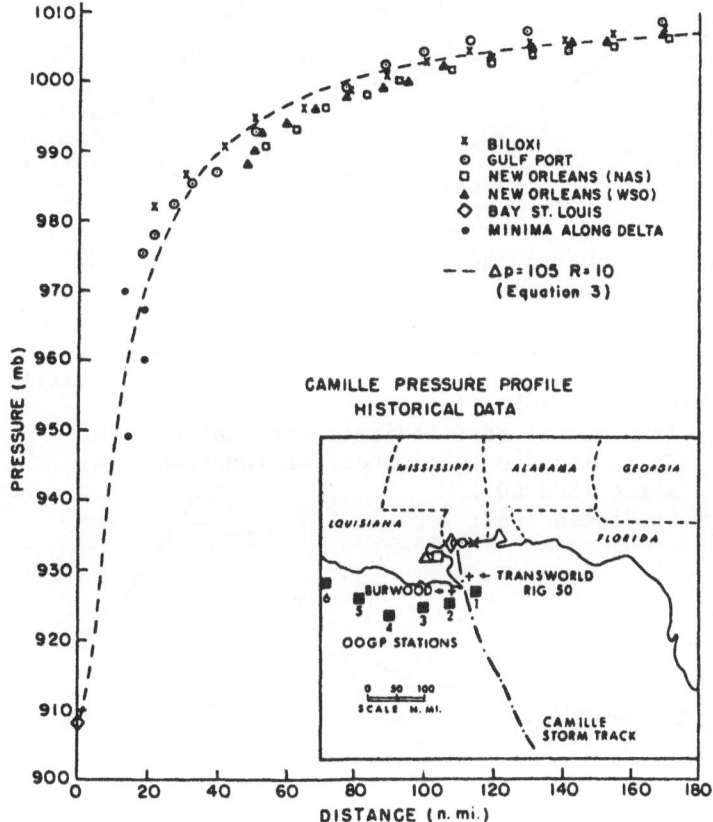

Fig. 12. Pressure profile fitted to composite of coastal station pressure measurements during the approach of Hurricane Camille (after *Cardone et al.*, 1976).

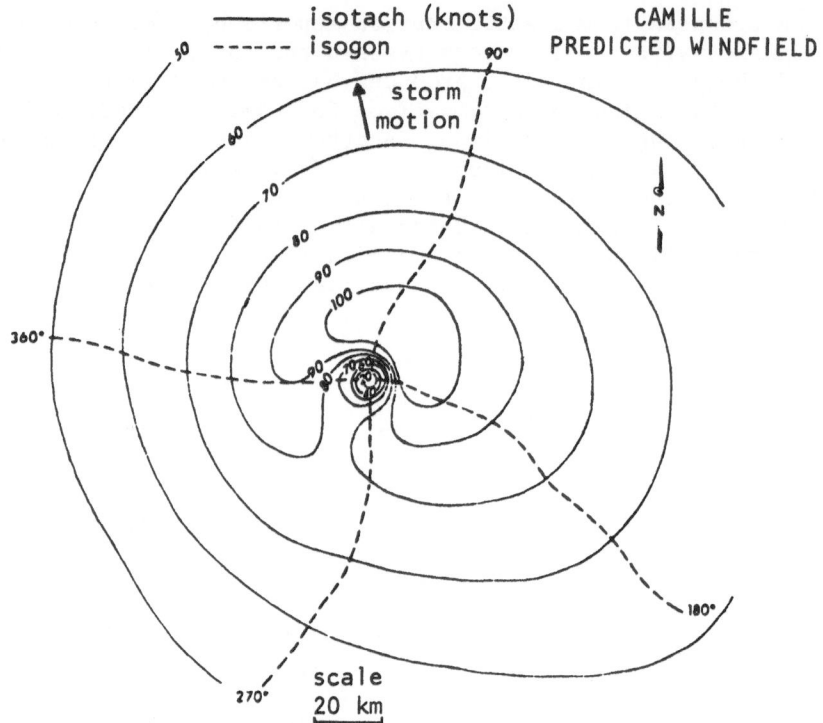

Fig. 13. Hurricane Camille parametric wind field.

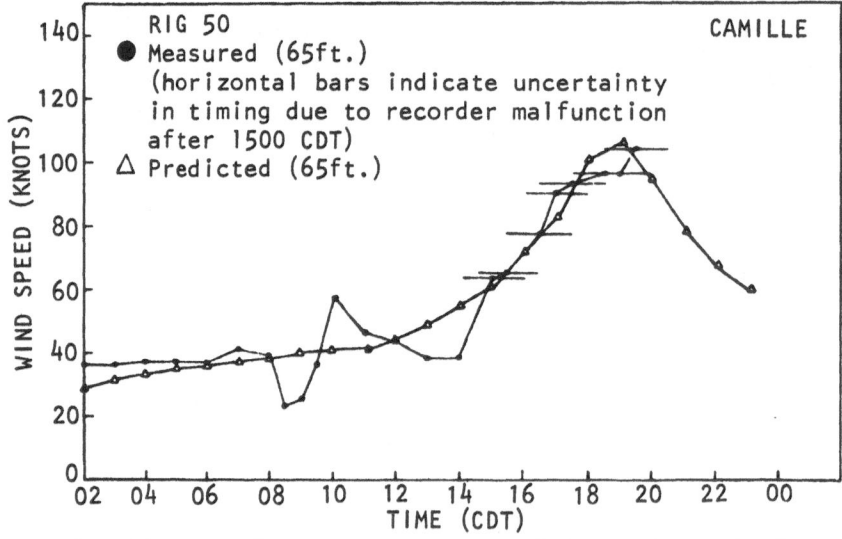

Fig. 14. Predicted and measured wind speed at Rig 50 during
Hurricane Camille. The position of Rig 50 is shown in Fig. 12

which gives a sharper peak to the radial wind distribution than other models. V_{max} is the wind speed at the radius of maximum winds.

Extratropical mid latitude storms are not generally amenable to parameterization. However, one notable exception is a study by Fernandez-Partagas (*Mooers et al.*, 1976) for east coast winter storms. Figure 15 shows the time dependent wind field for a model storm which was based upon 34 storms affecting the New York Bight. Pressure, temperature, humidity, cloud cover, precipitation, momentum, heat flux, wind stress curl and divergence fields were also calculated.

METEOROLOGICAL BASIS FOR WAVE CLIMATOLOGIES

The limited availability of wave data for design applications in an adequate format and resolution with associated return frequencies has been a major problem. Observational data such as ship reports are limited in accuracy by biases in measurement and location which render such data useless for some applications, while long time series of reliably measured data are extremely scarce. The hindcast approach uses wave models as a transfer function between wave histories at a point and the meteorological record. This meteorological record is generally the only sufficiently long time series to establish climatologies for extreme events. The general procedure is to calibrate the model on case studies of major events and extrapolate to an ensemble of events based upon the meteorological storm climatology. However, care must be taken to determine the sensitivity of the model to its tuning parameters when extrapolating to realizations quite different from the case studies.

The most ambitious hindcasting program is currently underway at FNWC to run a spectral wave model from 20 years of atmospheric pressure fields. This program is briefly described by *Bales and Cummins* (1978) in this volume. A spectral wave climatology will be developed including frequency of occurrence, persistence and height information at the 3400 grid points of the model.

Another approach is to determine exceedence frequencies from running a wave model with a parametric storm model in which the range of storm parameters forms a storm climatology. Such a climatology has been developed for hurricanes along the East and Gulf Coast by *Ho et al.* (1975). Each hurricane parameter, central pressure, radius to maximum winds, direction of approach to the coast and forward speed is viewed as a climatological variable that has a probability distribution at each coastal point and exhibits a smooth variation of these probabilities along the coast. The curves for central pressure are shown in Figure 16. The final parameter necessary for the climatology is the frequency of storms per 100 years

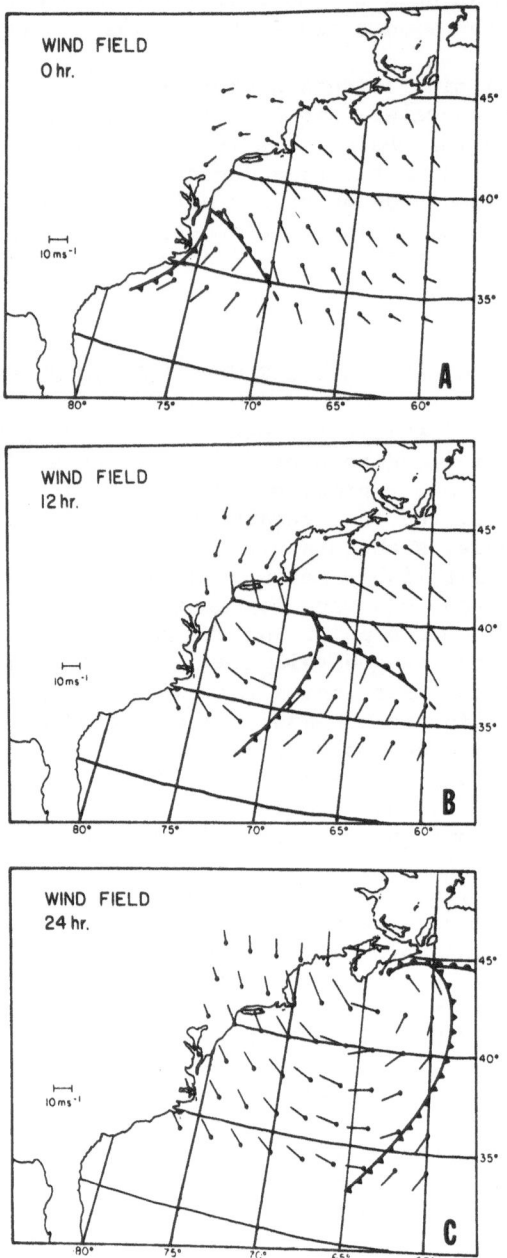

Fig. 15. A sequence of parametric wind fields for an extratropical east coast storm. The frontal system is seen to mature
as it moves on a northeast trajectory. Strong winds behind the
front are indicative of an unstable planetary boundary layer.

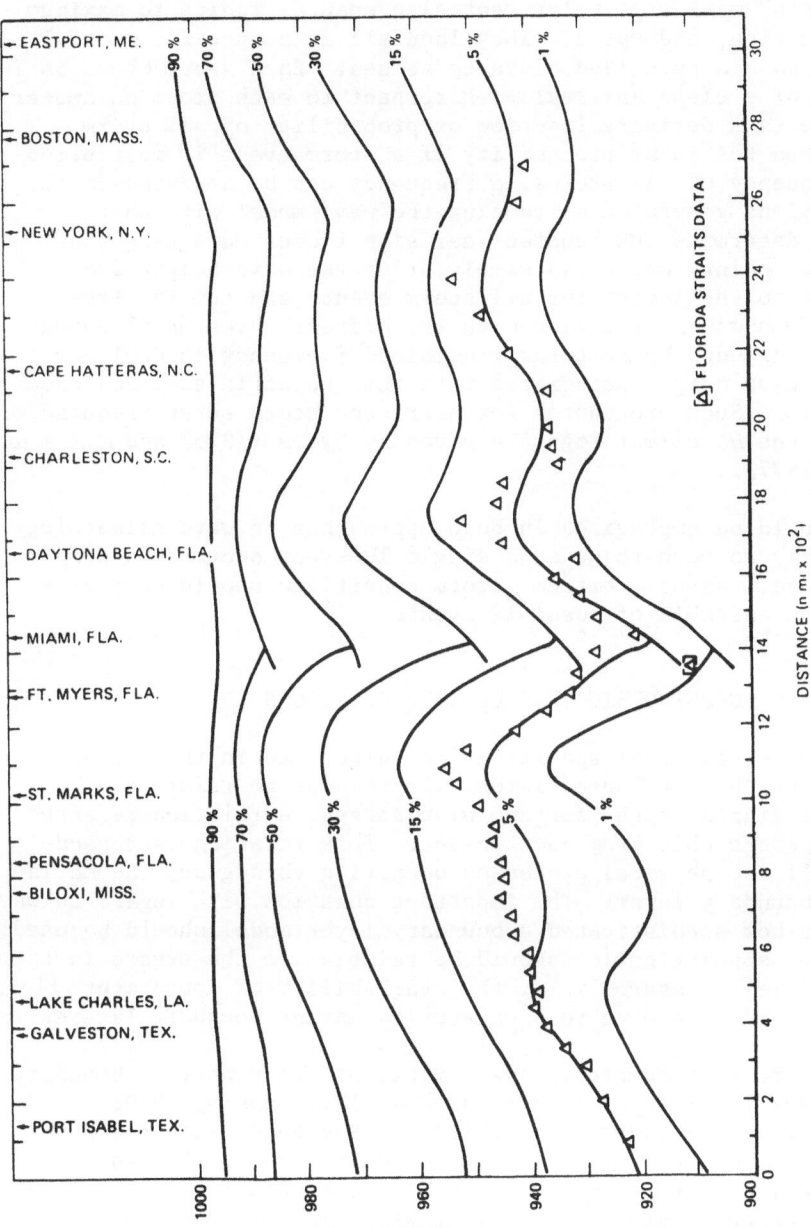

Fig. 16. Probability distribution of central pressure of hurricanes, Gulf and East Coasts (1900–1973). Numbered lines denote the percent of storms with central pressure equal to or lower than the value indicated along the ordinate. The plotted points (Δ) are unsmoothed data for 5 n-mi segments along the coast at the 5% level.

per 10 nautical miles of coast as shown in Figure 17. The curves
from *Ho et al.* are based upon all storms from 1871-1973. One
approach to using this information is to define a sequence of
"storm events" with particular central pressure, radius to maximum
winds, direction, and speed. They landfall at a specific coastal
point or bypass a specified distance at sea. Each storm event is in
the middle of a class interval with respect to each storm parameter
and represents a definite fraction or probability of all storm
events. When the joint probability of a storm event is multiplied
by the frequency of all storms, a frequency can be assigned to the
wave conditions generated by running the wave model with that
storm. To determine the hundred-year significant wave height at
a particular point, one would simply order the wave height from
the highest to the lowest for all storm events and sum the frequencies of occurrence starting with the highest waves until enough
events are included so that their combined frequency is 0.01 per
year. The wave height associated with this point is exceeded once
in 100 years. Such procedures for hurricane storm surge frequencies
based upon the *Ho* climatology are given by *Myers* (1975) and *Myers and
Overland* (1977).

It should be emphasized in both approaches to wave climatology
that there is no such thing as a single 100-year storm or a single
design storm. Rather, extreme storm conditions should be represented as an ensemble of possible events.

GEOSTROPHIC-SURFACE WIND RELATIONS

Most meteorological specifications discussed in this paper
are dependent upon pressure fields. It remains to relate the
pressure gradients to the surface wind stress, a relation referred
to as the geostrophic drag coefficient. This relation is dependent upon all the physical processes occurring throughout the marine
planetary boundary layer. The important question with regard to wave
modeling is how sophisticated a boundary layer model should be used;
the level of sophistication should be relative to the errors in the
input sea level pressure field, the availability of temperature fields
and present understanding to parameterize marine boundary layers.

There are many comprehensive reviews of the planetary boundary
layer (*Brown*, 1974; *Arya*, 1977; *Tennekes*, 1973; *Monin*, 1970;
Deardorff, 1974; *Wyngaard et al.*, 1974). Beginning with the geostrophic wind, corrections are made for curvature of the isobars.
This correction is very important for low pressure systems as winds
corrected for curvature (the gradient wind) may often be less than
70% of the geostrophic wind. *Hasse* (1976) has recently argued
that acceleration caused by convergence of isobars can add another
correction of 10-20%. While his data set was for the North Sea,

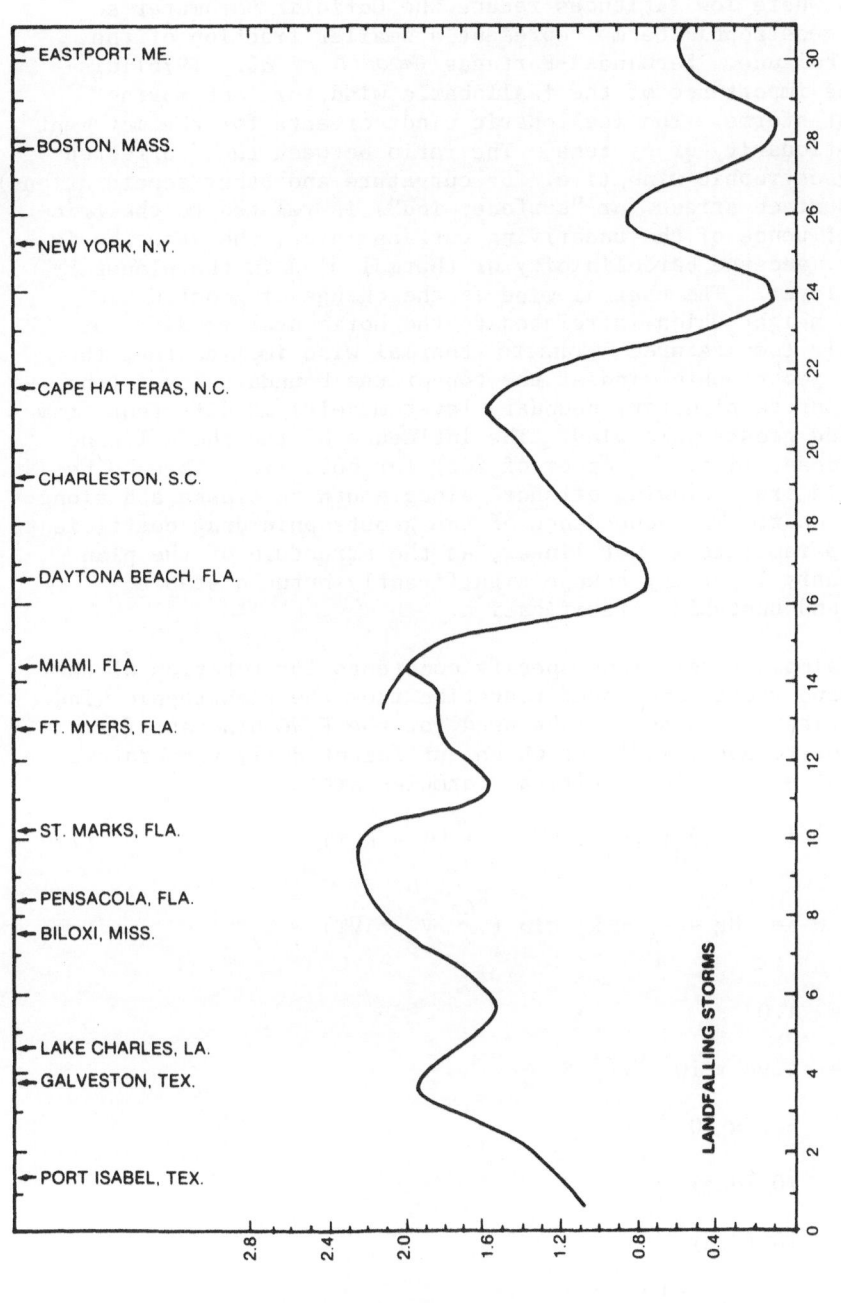

Fig. 17. Frequency of landfalling tropical storms and hurricanes (1871-1973) for the Gulf and East Coasts of the United States.

this correction can be relevant for frontal passages in the Gulf of Mexico where low latitudes reduce the Coriolis parameter so that the geostrophic terms represent a smaller fraction of the momentum balance. Fernandez-Partagas (*Mooers et al.*, 1976) discusses the importance of the isallobaric wind for fast-moving east coast storms. The isallobaric wind corrects for the movement or nonstationarity of systems. The ratio between the "corrected" surface geostrophic wind (i.e. for curvature and other accelerations) and the surface stress (or "surface wind") is related to the frictional influence of the underlying surface layer, the atmospheric stability, and the baroclinicity or thermal wind in the planetary boundary layer. The thermal wind is the change of geostrophic wind with height which is related to the horizontal gradient of surface air temperature. Nonzero thermal wind implies that the corrected geostrophic wind at the top of the boundary layer (the normal input to planetary boundary layer models) is different from the surface geostrophic wind. The influence of the thermal wind is significant (i.e., of order of 20%) for cold air outbreaks behind a cold front blowing offshore along southern Alaska and along the east coast. The dependence of the geostrophic drag coefficient on these parameters is not linear, as the structure of the planetary boundary layer may change significantly between stable, neutral, and unstable stratification.

The simplest relations specify constants for rotation of the surface wind vector and speed reduction from the geostrophic wind. Such empirical formulas will be used for the FNWC hindcast for years prior to about 1974 for which sufficient daily temperature data do not exist. The empirical formulas are:

$$|\vec{v}| = |\vec{v}_g| (K_1 + K_2 \vec{v}_g \cdot \nabla T - K_3 T) \tag{7}$$

$$\theta = \theta g - K_4 + K_5 \sin (\frac{K_6 \pi}{2} \vec{v}_g \cdot \nabla T) - K_7 T$$

K_1 = 1.0

K_2 = 4.94 x 10^2 $^\circ C^{-1}$ s

K_3 = 6.7 x 10^{-3} $^0 C^{-1}$

K_4 = 10 (deg)

K_5 = 10 (deg)

K_6 = 2.47 x 10^3 $^\circ C^{-1}$ s

K_7 = 0.5 $^\circ C^{-1}$ (deg)

where $|\vec{V}|$ is the wind speed, θ is the wind inflow angle, $\vec{V}_g$ is the geostrophic wind, T is the sea surface temperature, and ∇T is the gradient of the sea surface temperature. The dot product between sea surface temperature and the wind direction is a correction for baroclinicity and stability since it is positive for cold air advection and negative for warm air advection.

More ambitious studies, particularly hindcasts, *Isozaki et al.* (1975) make use of more general boundary layer models. *Cardone's* (1969) model separates the boundary layer into two regions, a surface layer and an Ekman layer. Wave development is considered indirectly through the dependence of surface roughness upon surface stress. *Cardone's* model is representative of the class of single point analytical models with dependences of geostrophic drag coefficients and inflow angles on atmosphere stability and thermal wind roughly comparable to more recent theories and data sets (*Arya and Wyngaard*, 1975). These dependences are shown in Figures 18 and 19.

Overland and Gemmill (1977) compared wind velocity measurements at two NOAA buoys in the New York Bight with winds inferred by various approaches based upon the input from the LFM sea level pressure analyses. *Cardone's* model was compared to empirical surface/gradient wind coefficients, stratified by stability, derived from data for the same New York Bight region. *Cardone's* model effectively duplicated the regional climatology by inclusion of air stability and thermal wind. For example, neglect of the strong baroclinicity associated with the temperature gradient established between cold continental air and the Gulf Stream region increased the mean absolute error of *Cardone's* model from 2.9 to 3.8 ms^{-1} for wind speeds greater than 12.5 ms^{-1}. The largest errors for both *Cardone's* and the empirical model were caused by erroneous estimates of the geostrophic wind rather than any inaccuracy in the models. *Overland and Gemmill* concluded that *Cardone's* model is comparable to seasonally and regionally stratified empirical models when both are based upon routine synoptic analyses. For winds in excess of 10 ms^{-1}, the LFM-based analyses provided adequate specification of the offshore wind field 81% of the time, as defined by having their mean absolute vector error < 5 ms^{-1}. The remaining 19% included cases of underestimation of detail structure in the vicinity of fronts.

CONCLUSIONS

a. Deep water (> 100 km offshore). Wave analysis and forecasting in deep water is dependent upon a time history of the wind

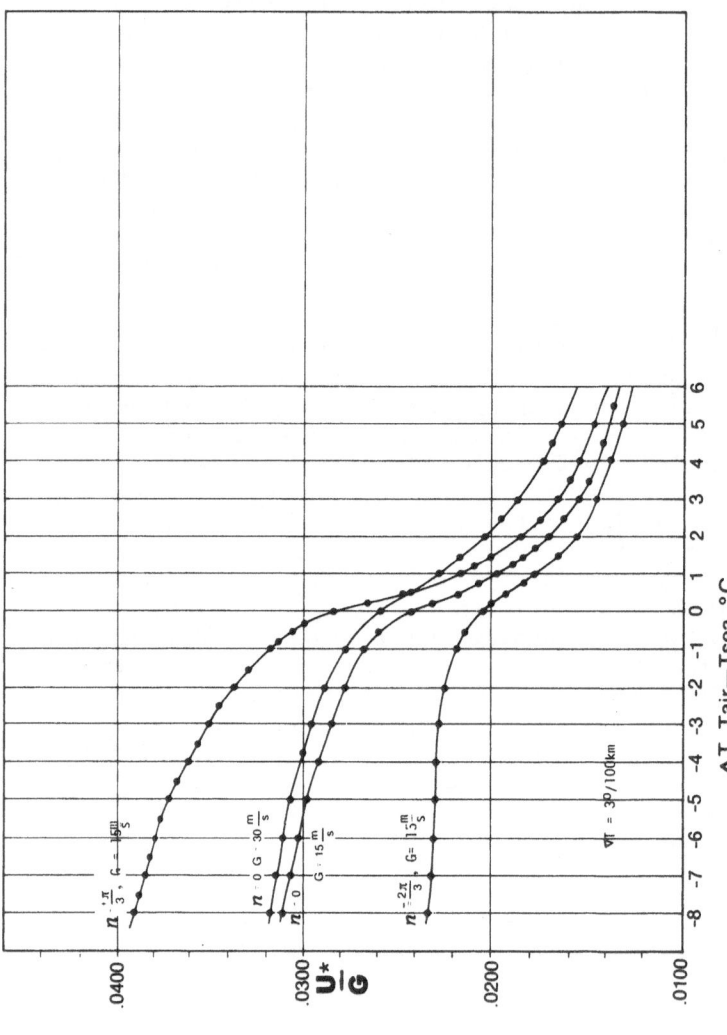

Fig. 18. Predicted geostrophic drag coefficient, U_*/G, as a function of air-sea temperature difference and thermal wind from _Cardone's_ (1969) planetary boundary layer model. The friction velocity, U_*, is defined as $\sqrt{\tau/\rho_a}$ where τ is the surface wind stress and ρ_a is the air density. Three curves show the relation for three values of η the angle between the thermal wind and the geostrophic velocity (measured counterclockwise) for a geostrophic wind speed of 15 ms^{-1}. The fourth curve is for a geostrophic speed of 30 m/s. The strength of the thermal wind is 3°/100km. Note the curve is relatively flat for unstable stratification, water warmer than air.

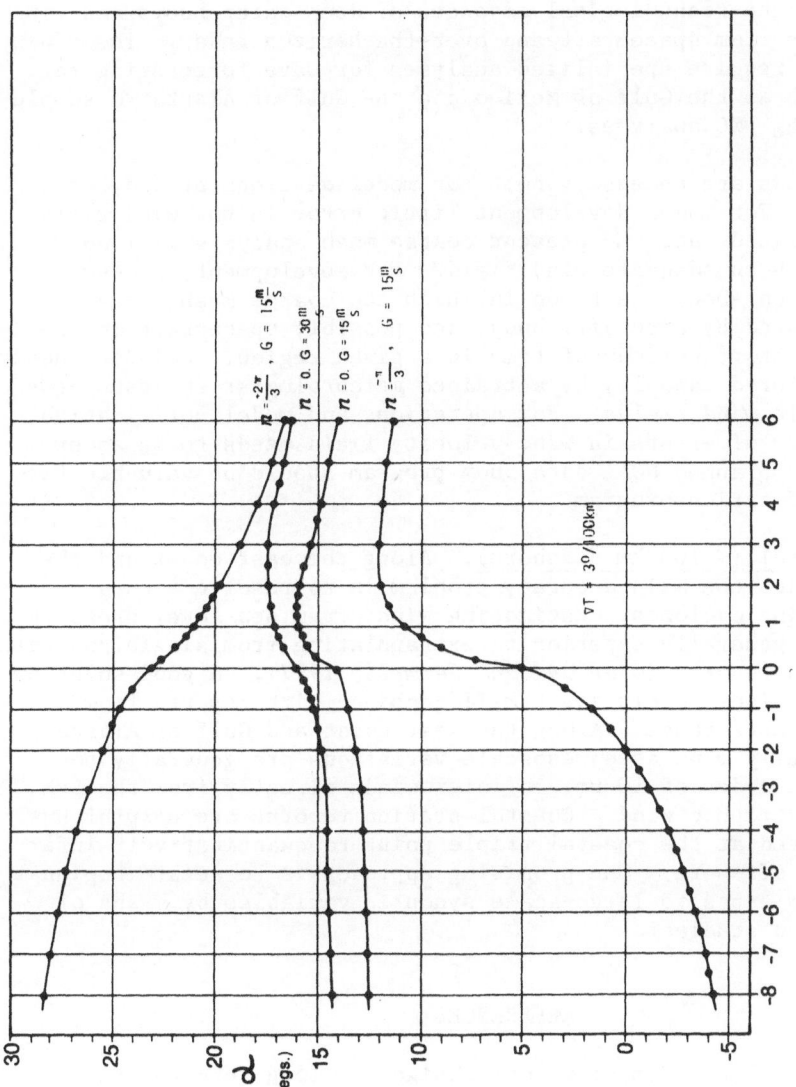

Fig. 19. The inflow for the same set of parameters in Figure 18. The inflow angle is the angle of rotation of the surface wind vector toward low pressure from the direction of the geostrophic wind velocity. Note that both the inflow angle and geostrophic drag coefficient are very sensitive to stability in the near neutral stratification range.

field. Errors in the wave field must be separated into wave modeling errors and errors in the input meteorology. The finer mesh models at NMC should improve first-guess fields over the ocean by providing better continuity of systems moving offshore. The data base for objective analyses should increase in the future with the technological advances of deep water buoys and wind measurements from spacecraft and over-the-horizon radar. These new sources may require specialized analyses for wave forecasting for regions such as the Gulf of Mexico and the Gulf of Alaska to supplement existing NMC analyses.

Wind fields are necessary both for model development and for operations. For model development little error in the wind field can be tolerated, and the present coarse mesh analysis will not always supply an adequate wind field. For development, a case study approach should be taken in which the coarse mesh analysis is supplemented by aircraft, buoy, and possibly spacecraft observations for short periods of time in a given region. All data needs to be considered manually by a trained meteorologist to insure the best possible wind fields. For operations and model verification, the magnitude of errors in wind velocity field needs to be ascertained. The planned NOAA data buoy program should be valuable for such a verification program.

b. <u>Coastal (< 100 km offshore)</u>. Along the east coast and the Gulf of Mexico the main accuracy problem is to resolve accurately rapidly moving cyclones. Estimating winds from sea-level pressure analysis is generally superior to extrapolating from single coastal station wind reports (*Overland and Gemmill*, 1977). A good solution will be the finer resolution LFM-II combined with reports from several offshore buoys. Along the west coast and Gulf of Alaska topographically generated mesoscale variations are generally too fine (on the order of 10 km) to infer winds directly from the LFM surface geostrophic wind. Coastal station reports are helpful but badly situated at the coastal triple point to quantitatively infer wind speeds offshore. One promising approach is to relate regional flow pattern types to large-scale synoptic variables by means of regional wind studies.

REFERENCES

Arakawa, A. 1966. Computational design for long-term numerical integration of the equations of fluid motion: Two-dimensional flow, Part I. *J. Comput. Phys.*, *1*, 119-143.

Arya, S. P. S. 1978. Comparative effects of baroclinicity and rotation on drag laws for the atmospheric boundary layer. Submitted to *J. Atmos. Sci.*

Arya, S. P. S. and J. C. Wyngaard. 1975. Effect of baroclinicity on wind profiles and the geostrophic drag law for the convective planetary boundary layer. *J. Atmos. Sci.*, *32*, 767-778.

Bales, S. L. and W. E. Cummins. 1978. Wave data requirements for ship design and operation, In: *Ocean Wave Climate*, M. D. Earle and A. Malahoff (Eds.), Plenum, New York (this volume).

Brown, R. A. 1974. *Analytical Methods in Planetary Boundary-Layer Modeling*, John Wiley and Sons, New York, 148 pp.

Brown, H. 1975. Comparison of the position and central pressure errors of east coast lows on 36-hour PE and LFM prognoses. NMC Technical Attachment No. 75-4. National Weather Service, Washington, D.C.

Cardone, V. J. 1969. Specification of the wind distribution in the marine boundary layer for wave forecasting, Report GSL-TR-69-1, New York Univ. School of Engineering and Science, Bronx, N.Y., 131 pp.

Cardone, V. J., W. J. Pierson, and E. G. Ward. 1976. Hindcasting the directional spectra of hurricane generated waves, *J. Petrol. Technology, 28,* 385-394.

Charney, J. G., R. Fjortoft, and J. von Neumann. 1950. Numerical integration of the barotropic vorticity equation, *Tellus, 2,* 237-254.

Chow, S. 1971. A study of the wind field in the planetary boundary layer of a moving tropical cyclone, MS Thesis, Dept. of Meteorology and Oceanography, New York University.

Cressman, G. P. 1959. An operational objective analysis system. *Monthly Weather Review, 87,* 367-374.

Deardorff, J. W. 1974. Three-dimensional numerical study of the height and mean structure of a heated planetary boundary layer for use in general circulation models. *Monthly Weather Review, 100,* 93-106.

Fawcett, E. B. 1977. Current capabilities in prediction at the National Weather Services' National Meteorological Center, *Bull. Am. Meteorological Soc., 58,* 143-149.

Feit, D. 1976. Single station marine wind forecasts based on model output statistics. Conference on Coastal Meteorology, Sept. 21-23, 1976, *Am. Meteorological Soc.,* 83-87.

Flattery, T. W. 1970. Spectral models for global analysis and forecasting. *Proc. Sixth AWS Technical Exchange Conference.* AWS Tech. Rept. 242, Scott Air Force Base, Illinois, 42-54.

Gandin, L. S. 1963. *Objective Analysis of Meteorological Fields.* English Translation by Israel Program for Scientific Translations, Jerusalem, 1965, 242 pp.

Garrett, J. R. 1977. Review of drag coefficients over oceans and Continents, *Monthly Weather Review, 105,* 915-929.

Graham, H. E., and Hudson, G. N. 1960. Surface winds near the center of hurricanes, *National Hurricane Research Project, Report 39,* Dept. of Commerce, Washington, D.C. 200 pp.

Graham, H. E. and D. E. Nunn. 1959. Meteorological considerations pertinent to Standard Project Hurricane, Atlantic and Gulf coasts of the United States, *National Hurricane Research Project, Report 33,* U.S. Weather Bureau and Corps of Engineers, Washington, D.C.

Haltner, G. J. 1971. *Numerical Weather Prediction*, John Wiley
and Sons, New York, 317 pp.

Hasse, L. 1976. A resistance law hypothesis for the nonstationary
advective PBL, *Boundary Layer Meteorology*, *10*, 393-407.

Ho, F. P., R. W. Schwerdt, and H. V. Goodyear. 1975. Some
climatological characteristics of hurricanes and tropical storms,
Gulf and East Coasts of the United States, *NOAA Technical Report*,
NWS 15, National Weather Service, U.S. Dept. of Commerce, Silver
Spring, Maryland.

Holl, M. and B. Mendenhall. 1971. FIB - Fields by Information
Blending. Final report to Commanding Officer, Fleet Numerical
Weather Central, Project M-167, 66 pp.

Howcroft, H. 1971. Local forecast model: Present status and
preliminary verification. Office Note No. 50, National
Meteorological Center, National Weather Service, Washington,
D.C. 22 pp.

Isozaki, H., T. Uji, T. Seto, and A. Kaietsu. 1976. Ocean waves
on Sagmi Bay caused by Typhoon 7410. *Oceanographical Magazine*
27, 1-23.

Jelesnianski, C. P. 1965. A numerical computation of storm tides
induced by a tropical storm impinging on a continental shelf,
Monthly Weather Review, *93*, June 1965, 343-358.

Jelesnianski, C. P. 1966. Numerical computations of storm surges
without bottom stress, *Monthly Weather Review*, *94*, 379-394.

Jelesnianski, C. P. 1972. SPLASH (Special program to list
amplitude of surges·from hurricanes) I Landfall Storms, *NOAA
Technical Memorandum*, *NWS TDL-46*, National Weather Service,
U.S. Dept. of Commerce, Silver Spring, Maryland.

Kaitala, H. 1976. A brief description and preliminary evaluation
of an operational planetary boundary layer parameterization
scheme. Naval Oceanic Research and Development Activity, Bay
Saint Louis, Mississippi, 48 pp. (unpublished manuscript).

Kesel, P. G. and F. H. Winninghoff. 1972. The Fleet Numerical
Weather Central operational primitive-equation model, *Monthly
Weather Review*, *100*, 360-373.

Marchuk, G. I. 1967. *Numerical Methods in Weather Prediction*,
Translated by K. N. Trirogoff and V. R. Lamb, edited by A
Arakawa and Y. Mintz, Academic, New York, 1974, 277 pp.

Mihok, W. and Kaitala, H. 1976. U.S. Navy Fleet Numerical
Weather Central operational five-level global fourth-order
primitive-equation model. *Monthly Weather Review*, *104*,
1527-1550.

Monin, A. S. 1970. *Boundary Layers in Planetary Atmospheres in
Dynamic Meteorology*, (P. Morel Editor), Reidel Publishing,
Boston, Mass., 421-468.

Monin, A. S. 1972. *Weather Forecasting as a Problem in Physics*,
MIT Press, Cambridge, Mass., 199 pp.

Mooers, C. N. K., J. Fernandez-Partagas, and J. F. Price. 1976. Meteorological forcing fields of the New York Bight, *Technical Report TR76-8*, Rosenstiel School of Marine and Atmospheric Science, University of Miami, Miami, Florida, 151 pp.

Myers, V. A. 1954. Characteristics of United States hurricanes pertinent to levee design for Lake Okeechobee, Florida, *Hydrometeorological Report No. 32*, U.S. Weather Bureau and Corps of Engineers, Washington, D.C.

Myers, V. A. 1975. Storm tide frequencies on the South Carolina coast, *NOAA Technical Report NWS 16*, National Weather Service, U.S. Department of Commerce, Silver Spring, Maryland, 79 pp.

Myers, V. A. and W. Malkin. 1961. Some properties of hurricane wind fields as deduced from trajectories, *NHRP Report No. 49*, U.S. Weather Bureau, 45 pp.

Myers, V. A. amd J. E. Overland. 1977. Storm tide frequencies for Cape Fear River, *J. of Waterways, Harbors and Coastal Engineering Div., Am. Soc. Civ. Engineers, 103*, WW4, 519-535.

Overland, H. and W. Gemmill. 1977. Specification of marine winds in the New York Bight. *Monthly Weather Review, 105*, 1003-1008.

Phillips, N. A. 1959. An example of nonlinear computational instability. *The Atmosphere and the Sea in Motion*, Rockefeller Institute Press, New York. 501-504.

Pore, N. A. and W. S. Richardson. 1969. Second interim report on sea and swell forecasting, *ESSA Technical Memorandum WBTM TDL 17*, National Weather Service, Silver Spring, Maryland.

Shuman, F. G. 1974. Analysis and experiment in nonlinear computational stability. *Proc. of the Symposium on Difference and Spectral Methods for Atmosphere and Ocean Dynamics Problems*, 17-22 September, Novosibirsk, 1973. Part 1, Computer Center, Siberian Branch, U.S.S.R. Academy of Sciences, 51-81.

Shuman, F. G. 1977. Plans of the National Meteorological Center for numerical weather prediction, Official Note, National Meteorological Center, National Weather Service, NOAA, 14 pp. (unpublished manuscript).

Shuman, F. G. 1978. Numerical weather prediction, *Bull. Am. Met. Soc., 59*, 5-17.

Shuman, F. G. and J. B. Hovermale. 1968. An operational six-layer primitive equation model. *J. Applied Meteorology, 7*, 525-547.

Schwerdt, R. W. 1972. Revised Standard Project Hurricane criteria for the Atlantic and Gulf Coasts of the United States. *Memorandum HUR 7-120*, Office of Hydrology, National Weather Service, Silver Spring, Maryland.

Tennekes, H. 1973. Similarity laws and scale relations in planetary boundary layers, In: *Workshop on Micrometeorology*, Am. Meteorological Soc., Boston, Mass., 177, 216.

Teweles, S. and H. Wobus. 1954. Verification of prognostic charts, *Am. Meteorological Soc., 35*, 455-463.

Wyngaard, J. C., O. R. Cote, and K. S. Rao. 1974. Modeling the atmpsheric boundary layer, In: *Advances in Geophysics, 18A*, Academic, New York, 193-211.

Practical Determinations of Design Wave Conditions

Marshall D. Earle

National Ocean Survey

National Oceanic and Atmospheric Administration

ABSTRACT

Wave models which can be used to determine design wave conditions for coastal and offshore structures are reviewed. Comparisons are made between model results and wave measurements and between results from different models. Overall, the accuracy of the wave models is approximately 1-2 meters but, due to a lack of measured wave data, even this number is an estimate. There is a need to evaluate directional spectra models by comparison with measured wave energy directional spectra.

INTRODUCTION

The extreme wave conditions which a coastal or offshore structure is designed to survive are called design wave conditions. These conditions are usually expressed in terms of wave characteristics as a function of occurrence probability. For example, the wave height associated with a 100-year return period is frequently used to determine the design of offshore structures. The return period for a given extreme wave height is the average time between occurrences of waves with heights equal to or greater than the given wave height. Design wave conditions could be obtained from a statistical analysis of measured wave data covering a very long time period. For nearly all offshore locations, wave data covering a statistically significant time period do not exist. Consequently, wave models are used to hindcast extreme waves for subsequent statistical analysis and extrapolation. The accuracy of the models directly affects the accuracy of the final design criteria.

39

The specification of design wave conditions is an important area of applied oceanography. Oceanographers are sometimes not familiar with the procedures, including the use of wave hindcast models, that are used to determine design wave conditions. The purpose of this paper is to review wave hindcast models which have been used or are suitable to determine design wave conditions. A second paper (*Cardone and Ross*, 1978) in this volume more fully describes wave model physics and recent developments in wave modeling including the use of spectral and parametric wave models.

WAVE MODELS

SMB and PNJ Models

Many wave models have been used to calculate design wave conditions. This paper does not review earlier wave models such as the SMB (Sverdrup, Munk, and Bretschneider) method (e.g. U.S. Army Coastal Engineering Research Center Shore Protection Manual, 1973) and the PNJ (*Pierson, Neumann, and James*, 1955) method which have been described many times in the scientific literature. While useful for providing wave condition estimates, the frequently used graphical forms of these methods have been technically surpassed for providing design wave conditions. The SMB method has been automated and used for wave forecasts by the U.S. Navy Fleet Numerical Weather Central but forecasts are now made with directional spectra models. The National Weather Service of the National Oceanic and Atmospheric Administration presently uses a modified and automated SMB method to prepare wave forecasts (*Pore and Richardson*, 1969). *Enfield* (1973) has devised a semi-automated form of the PNJ method which considers some effects of time and spatially varying fetches in a manner similar to the graphical procedures described by *Wilson* (1955). Automated SMB and PNJ methods have not been used for design wave calculations.

Wilson's Wave Model

Wilson (1955) developed a graphical significant wave height model that considers time and spatially varying wind fields. This model, which is empirical, has been further developed and described by *Wilson* in a series of papers (1961, 1963, 1965, 1966). His wave model with minor variations has been widely used by the offshore oil and construction industries (*Patterson*, 1971, 1972; *Wilson et al.*, 1973; *Bea*, 1974; *Ward et al.*, 1977). The foundations of *Wilson's* model are empirical relationships between the wind speed U, the wind fetch F, the significant wave height H_s, and the deep water phase speed C_s associated with the significant wave height. In terms of the nondimensional parameters C_s/U, gH_s/U^2, and gF/U^2, these relationships are given by

$$\frac{c_s}{U} = f_1 \left(\frac{gF}{U^2}\right) \qquad (1\text{-}a)$$

and

$$\frac{gH_s}{U^2} = f_2 \left(\frac{gF}{U^2}\right) \qquad (1\text{-}b)$$

where f_1 and f_2 are empirical functions of gF/U^2.

Wilson (1965, 1966) recommends the following empirical relationships for f_1 and f_2:

$$\frac{c_s}{U} = 1.37 \left[1 - \left\{1 + 0.008 \left(\frac{gF}{U^2}\right)\right\}^{1/3}\right]^{-5} \qquad (2\text{-}a)$$

$$\frac{gH_s}{U^2} = 0.30 \left[1 - \left\{1 + 0.004 \left(\frac{gF}{U^2}\right)\right\}^{1/2}\right]^{-2} \qquad (2\text{-}b)$$

Most recent applications of *Wilson's* method (*Patterson*, 1971, 1972; *Wilson et al.*, 1973; *Feldhausen et al.*, 1973; *Bea*, 1974; and *Ward et al.*, 1977) use Equations (2-a,b). *Wilson's* earlier descriptions of his model (1961, 1963) and recent use of his model by *Dexter* (1974) rely on a different expression for the empirical functions f_1 and f_2.

Significant wave heights are generated and propagated along a line toward a site where wave forecasts or hindcasts are wanted. For design calculations, the orientation of the line is selected to coincide with the expected propagation path of the largest waves during individual storms or hurricanes. For situations in which directions associated with maximum wave heights cannot be accurately estimated ahead of time, *Wilson's* model is applicable to propagation paths radiating from the site of interest. The prediction equations which are used in *Wilson's* model are obtained by differentiating Equations (2-a,b) to obtain

$$\frac{dc_s}{dx} = (8.013)10^{-2}U^{-1}(z_1)^{-6/5} \left[z_1^{1/5} - 1\right]^{-2} \qquad (3\text{-}a)$$

$$\frac{dH_s}{dx} = (2.918)10^{-2} (Y_1)^{-3/2} \left[Y_1^{1/2} - 1\right]^{-1} \tag{3-b}$$

where

$$U = U(x,t) \qquad C_s, U \text{ in knots}$$

$$Z_1 = \frac{1.37}{(1.37 - C_s/U)} \qquad x \text{ in nautical miles}$$

$$Y_1 = \frac{0.30}{(0.30 - \frac{gH_s}{U^2})} \qquad H_s \text{ in feet}$$

The wind speed component, $U = U(x,t)$, along the wave propagation path is a function of position x and time t. Equations (3-a,b) are numerically integrated as wave energy travels from a starting point to the site at the deep water group velocity, $C_s/2$, associated with the significant wave height. Although the numerical time and space increments are variable, most applications of *Wilson's* model have used a time step Δt of 1 hour and a space increment Δx of 10 nautical miles along the wave propagation path. Initially, the group velocity is too small for wave energy to travel a distance Δx in time Δt and wave energy is propagated to intermediate positions with a constant time step Δt. When the group velocity is large enough for wave energy to travel a distance greater than Δx, wave energy is propagated in constant space increments Δx with times calculated from the group velocity. Wind speed components are computed at any point or time by interpolation between discrete values used to specify $U(x,t)$. Following a comment by *Walden* (in *Wilson*, 1963), *Ward et al.* (1977) propagate energy at the group speed associated with the period of maximum spectral energy. Empirically, the period of maximum spectral energy is about 1.14 T_S where T_S is the significant period (*Darbyshire*, 1959).

    Wave decay is not discussed in several descriptions and applications of *Wilson's* model (*Wilson*, 1961, 1963, 1965, 1966; *Wilson et al.*, 1973; and *Feldhausen et al.*, 1973). Equations (3-a,b) require that

$$\frac{C_s}{U} < 1.37 \tag{4-a}$$

$$\frac{gH_s}{U^2} < 0.30 \tag{4-b}$$

If either of the conditions are not met, *Patterson* (1971, 1972), *Bea* (1974), and *Ward et al.* (1977) calculate wave decay according to empirical equations which approximately agree with the *Sverdrup and Munk* (1947) wave prediction theory. Wave decay in *Dexter* (1974) is computed from *Bretschneider's* (1952) decay curves.

For practical applications, *Wilson's* model is an important advance over previous models because his model considers time varying and spatially changing wind fields. In addition, *Wilson's* model is simple to computer program and very little computer time is required for running the model. A disadvantage is that the model is a significant wave height model which does not provide wave energy frequency spectra or wave energy directional spectra. *Wilson* (1963) and *Wilson et al.* (1973) have fit empirical wave energy directional spectra to significant wave heights computed by the model. Because *Wilson's* model is not a spectral model, users of his model generally refract deep water significant wave heights and periods computed from the model to obtain shallow water wave conditions. This procedure is not as accurate as refracting directional wave energy spectra.

Directional Wave Spectra Models

Since the early 1960's, there have been major advances in numerical models to predict directional wave energy spectra. The basis for these models is the wave energy balance equation which is sometimes called the equation of radiative transfer for waves. This equation is given by

$$\frac{\partial E}{\partial t}(f,\theta,\vec{x},t) = -\vec{V}_G(f,\theta,\vec{x},t)\cdot\nabla E(f,\theta,\vec{x},t) + S(f,\theta,\vec{x},t) \tag{5}$$

where E is the two dimensional directional energy spectrum; f is the wave frequency; θ is the wave direction; $\vec{x}$ is position; t is time; $\vec{V}_G$ is the wave group velocity and S is a source function that represents all processes that add or remove energy from the spectrum. The first term on the right of Equation (5) accounts for wave propagation and the source function S accounts for wave generation and decay. Numerical spectral models solve Equation (5) for E over a grid of points overlying the region of interest. Depending on the model, time steps usually range from 1 to 6 hours.

Cardone (1974) has reviewed the development of spectral models including propagation schemes, incorporation of recent wave generation data, and comparisons to some measurements. These points are also discussed by *Cardone and Ross* (1978) in this volume. The most extensively documented and tested spectral models follow from the work of *Baer* (1962) and *Pierson et al.* (1966). Other spectral models, such as the models developed by *Barnett* (1968) and *Ewing* (1971) which include parameterized nonlinear wave interaction effects, have not been widely used probably because of large computer time requirements. A version of the *Pierson et al.* model is used by the U.S. Navy Fleet Numerical Weather Central for operational wave forecasting (*Lazanoff and Stevenson,* 1975) and the models have been modified for application to hurricanes in the Gulf of Mexico (*Cardone et al.,* 1976). The hurricane model and its calibration with wave data are discussed in this volume by *Cardone and Ross* (1978).

Gelci and his coworkers have developed a series of directional wave spectral models based on Equation (5) since 1956 (*Gelci et al.,* 1956; *Gelci et al.,* 1957). These models are somewhat different than other spectral models in that the source function is approximated by empirical equations for wave growth and decay. *Gelci and Devillaz* (1970) describe a recent version of the model that has been operationally used by the French for wave forecasting. A computer program for this model has been prepared by the U.S. Navy Fleet Numerical Weather Central (*Rabe,* 1970). *Earle* (1975) and *Earle and Burns* (1975) determined design wave conditions off the U.S. Atlantic Coast with a version of the Fleet Numerical Weather Central model.

For the *Gelci* model, the source function is of the form

$$S = P(f,U)D(\theta_{wave}, \theta_{wind}) - (cf^{-4}m_o)E \qquad (6)$$

where P is a wave growth function which depends on wave frequency f and wind speed U; D is a function that spreads wave energy over different directions and which depends on the wave direction θ_{wave} and wind direction θ_{wind}; c is a wave decay coefficient; and m_o is the square root of the total wave energy integrated over all wave frequencies and directions. The first term in Equation (6) represents wave growth and the second term represents wave decay which is exponential with time. The wave growth function is given by

$$P = KT^2\left(\frac{U-2T}{U}\right)^3 \qquad T < \frac{U}{2} \quad \text{(T in seconds, U in knots)}$$

$$P = 0 \qquad T > \frac{U}{2} \qquad (7)$$

where $T = f^{-1}$ is the wave period and K is an empirical constant. The energy spreading function is given by

$$D = \frac{2}{\pi} \cos^2 (\theta_{wave} - \theta_{wind}) \qquad \left| \theta_{wave} - \theta_{wind} \right| < \frac{\pi}{2} \qquad (8)$$

$$D = 0 \qquad \left| \theta_{wave} - \theta_{wind} \right| \geq \frac{\pi}{2}$$

Gelci's model is thus based on the energy balance equation; Equation (5), with relatively simple empirical functions for wave growth, energy spreading, and decay. This model, however, requires consideraly less computer time than other spectral models.

Thom's Statistical Method

Thom investigated the use of the Fréchet probability distribution to determine extreme wind speeds over land (*Thom*, 1967) and over oceans (*Thom*, 1973a) and extreme wave heights at ocean station vessels (*Thom*, 1971). Based on this work, he proposed a method (*Thom*, 1973b) whereby the probability distribution for extreme wave heights at a location can be determined from the maximum mean monthly wind speed at the same location. The method was developed for extratropical storm regions and only a preliminary method was developed for tropical storm regions.

The Fréchet probability distribution is given by

$$G(H_s) = \exp \left[-\left(\frac{H_s}{\beta_s} \right)^{-\gamma_s} \right] \qquad (9)$$

where, in this case, H_s is the significant wave height; $G(H_s)$ is the probability for a significant wave height less than or equal to H_s; β_s is a constant known as the scale parameter; γ_s is a constant known as the shape parameter; and the subscripts s indicate that these values are those for the probability distribution of extreme significant wave heights. The Fréchet distribution can be written as a Fishet-Tippett type 1 distribution which is given by

$$G'(y) = \exp \left[-\exp \left(\frac{\alpha - y}{\lambda} \right) \right] \qquad (10)$$

where

$$y = \ln H_s \quad \text{or} \quad H_s = \exp(y)$$
$$\alpha = \ln \beta_s \quad \text{or} \quad \beta_s = \exp(\alpha)$$
$$\lambda = \gamma_s^{-1} \quad \text{or} \quad \gamma_s = \lambda^{-1}$$

and $G^I(y)$ is the cumulative probability distribution for y. For
twelve ocean station vessels, *Thom* (1971) fitted Equation (10) to
the logarithms of the highest significant wave heights observed
each year (for up to sixteen years). For all ocean station
vessels, the values of β range from 18 ft to 37 ft and the values
of γ range from 4 to 11.

Based on previous use (*Thom*, 1967; *Thom*, 1973a) of the Fréchet
distribution with extreme wind speeds, *Thom* (1973b) proposed that

$$\beta_s = 0.455\, \beta_v \tag{11}$$

and

$$\gamma_s = \frac{2}{3}\, \gamma_v \tag{12}$$

where β_v is the scale distribution and γ_v is the shape parameter
for extreme wind speeds. The numerical value $\gamma_s = 6.0$ is recommended
by *Thom* (1973b) and β_v is calculated from

$$\beta_v = (373.8\, \max(\overline{V}) + 542.4)^{1/2} - 23.3 \tag{13}$$

in which $\max(\overline{V})$ is the maximum mean monthly wind speed (mph) and the
units for β_s are feet when Equations (11) and (13) are used. Thus,
by use of Equations (11), (12), and (13) in Equation (9), the significant wave height probability distribution can be calculated.

There are several problems with *Thom's* method which are mainly
due to the fact that it is a statistical method rather than a physical model of wave generation. Most importantly, wave heights are
calculated solely from the maximum mean monthly wind speed so that
the method does not consider effects of wind duration and fetch.
The effects are significant near coasts where waves may be fetch
limited and in the open ocean where storm-generated waves are a
function of the time and spatial variation of the wind field. In
addition, other probability distributions may fit the observed data
as well as or better than the Fréchet probability distribution. For
example, log-normal probability distributions are widely used to
approximate the probability distribution for extreme wave heights.
Petrauskas and Aagaard (1971) discussed methods for selecting and
evaluating different probability distributions. *Borgman and Resio*
(1976) have recently discussed the use of different distributions
as part of an overall review of extreme wave statistics. The
variability of the scale and shape of parameters at different ocean
station vessels indicates that a Fréchet distribution defined by
constant parameters does not fit the observed data very well.
Finally, the use of this statistical method should be confined to

extratropical regions because of the great difficulty in trans-
forming wind probability distributions to wave probability
distributions for tropical storms and hurricanes. Hurricanes,
however, produce design wave conditions in many regions where
these conditions must be known for ocean engineering applications.

WAVE MODEL COMPARISONS

Comparisons with Wave Measurements

The best way to determine the accuracy of a particular wave
model is to compare wave conditions computed by the model to
measured wave conditions. In particular, knowledge of frequency
spectra, directional spectra, and extreme waves are useful. Unfortunately, little suitable wave data are available and few
detailed comparisons have been made.

Feldhausen et al. (1973) compared wave hindcasts made by
several investigators for a severe winter storm (December 1959)
in the North Atlantic. Significant wave heights up to 13 m were
measured by a ship-borne wave recorder at ocean station J. Significant wave height time histories are shown in Figure 1 which
shows that differences between models are typically 2-3 m.
Feldhausen et al. (1973) computed linear regression lines between
the measured significant wave heights and the significant wave
heights calculated by each investigator. These results are summarized in Table 1. From this table, *Wilson's* model which is
based on Equation (2) provides the best overall results, particularly
because of its small intercept value. Figure 2 shows the linear regression analysis for *Wilson's* method with Equation (2). Except for
the near zero intercept, the scatter shown in this plot is qualitatively similar to scatter for the other models. From Table 1,
intercepts for the other models range approximately between 1 and 3
meters. The importance of accurate wind fields for wave forecasting
and hindcasting is well known. A potential problem with the comparisons made by *Feldhausen et al.* (1973) is the fact that each
investigator prepared his own wind fields from the same synoptic
atmospheric pressure and wind data. Thus, part of the differences
between models may be caused by differences between wind fields.

A directional wave spectra model similar to that used by the
U.S. Navy Fleet Numerical Weather Central was used to hindcast
storms near ocean stations J and K in the North Atlantic ocean
during December 1973 and January 1974 (*Salfi and Pierson,* 1977).
Hindcast significant wave heights were generally within 1.5 meters
of measured significant wave heights. *Salfi and Pierson* (1977)
noted that results could be improved by decreasing the angular

TABLE 1.   Least-squares fit between measured and hindcast
           significant wave heights at ocean station J in
           the North Atlantic Ocean during a storm in
           December 1959 (after *Feldhausen et al.*, 1973).

Wave Model	Linear Regression Equation for Wave Height in Meters (y = hindcast, x = measured)	Correlation Coefficient
Barnett (1968)	y = 0.91x + 1.72	0.866
	y = 0.82x + 2.82	0.769
Bretschneider (1963)	y = 1.20x − 1.76	0.923
Darbyshire (1961)	y = 0.82x + 2.36	0.710
Inoue (1967)	y = 0.89x + 1.57	0.896
	y = 0.95x + 1.10	0.765
Pierson et al. (1966)	y = 0.72x + 3.06	0.813
	y = 0.71x + 2.56	0.817
	y = 0.70x + 2.07	0.808
Wilson (1965)	y = 1.04x + 0.63	0.750
Wilson et al. (1972)	y = 1.02x − 0.09	0.885

resolution (30°), spatial grid size (150 nautical miles), and
time step (3 hours) of the model.

Several wave model comparisons and calibrations with measured
wave data use data that were collected in the Gulf of Mexico as
part of the Ocean Data Gathering Program which was sponsored by
a consortium of oil companies. *Ward* (1974) gives an overview of
this program which included several instrumented platforms and
Hamilton and Ward (1974) discuss the quality and the reduction of
the data. Much of the data is available to the public from NOAA's
Environmental Data Service. The data includes wave height data
as hurricane Camille approached an instrumented offshore oil platform off the Mississippi Delta in water 300 feet deep. Hurricane
Camille was one of the worst hurricanes to occur in the Gulf of
Mexico and the maximum wave heights measured at this platform were
about 23 meters (75 feet). This data set has been frequently used
to calibrate wave models. The most extensive calibration of a
model that has been used to determine design wave criteria has been
for the spectral model described by *Cardone et al.* (1976). These
calibrations are also briefly discussed in this volume (*Cardone
and Ross*, 1978).

Wilson's model as described in this paper and as referenced by
Feldhausen et al. (1973) has been applied to determine hurricane
generated waves by the offshore oil industry (*Patterson*, 1971, 1972;

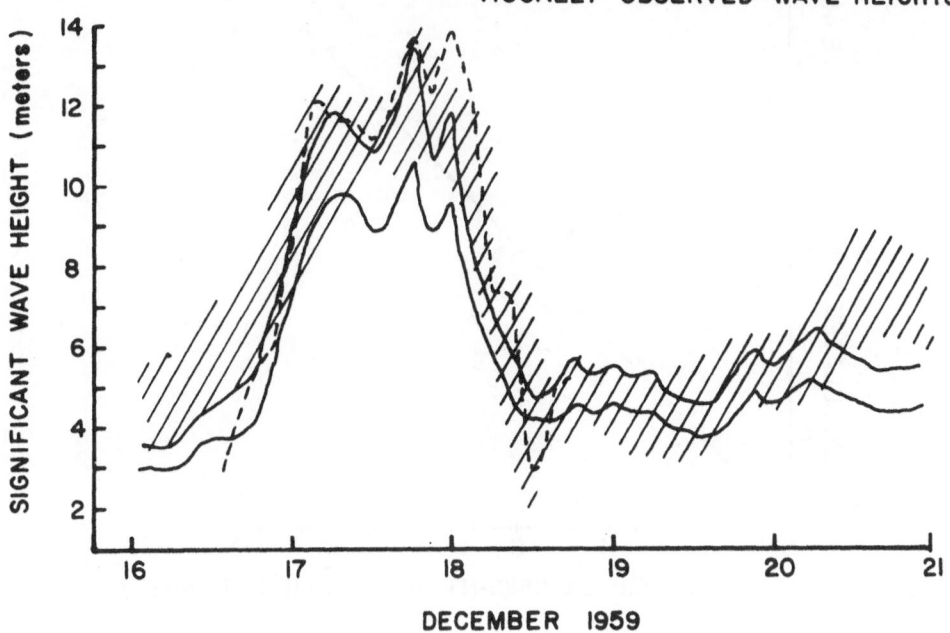

—90% CONFIDENCE LIMITS FOR
MEASURED WAVE HEIGHTS

//////RANGE OF CALCULATED VALUES

— — — VISUALLY OBSERVED WAVE HEIGHTS

Fig. 1.   Comparison of measured significant wave heights and
significant wave heights from different models (after *Feldhausen
et al.*, 1973).   Significant wave heights computed by several
models fill the cross-hatched region.   The agreement with data is
good but differences between models are generally 2-3 meters.   The
comparisons are at ocean station J in the North Atlantic Ocean.

Wilson et al., 1973; *Bea*, 1974; and *Ward et al.*, 1977).   For most
of these hurricane applications, the wind fields were adjusted
(*Patterson*, 1971, 1972) so that hindcast waves agreed, to within
about 2 feet, with maximum wave heights measured during hurricane
Carla in 1961 and hurricane Camille in 1969.   The hurricane Camille
wave data became available to the public in 1974 and the *Wilson*
model calibration with these data is not described in the literature.   *Gelci's* model has not been widely used to determine extreme
waves.   In the applications by *Earle* (1975) and *Earle and Burns*
(1975), the wave decay coefficient was selected to provide agreement with measured wave data during hurricane Camille.   With the

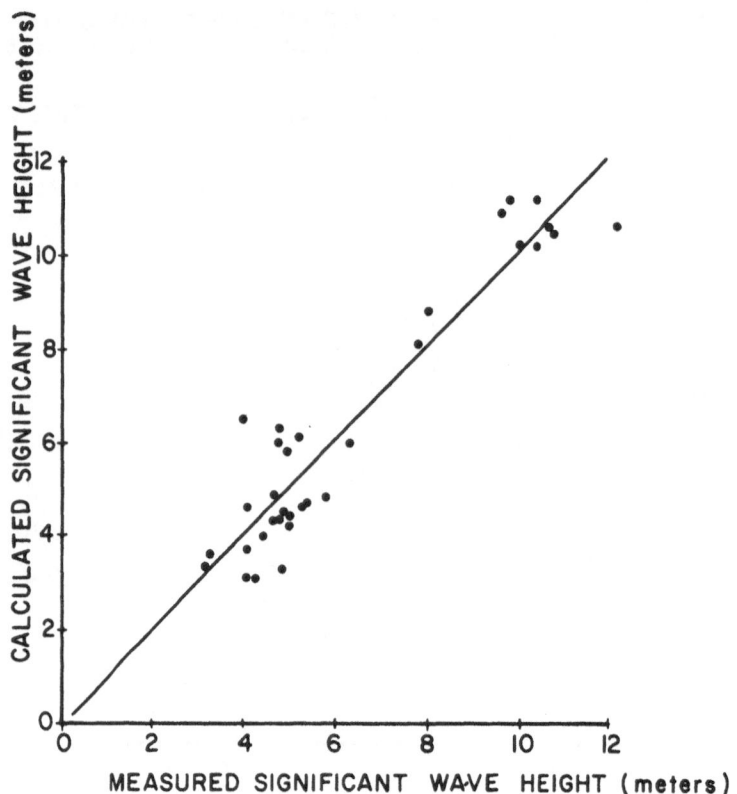

Fig. 2.   Least-squares fit between significant wave heights
calculated by *Wilson's* model and measured significant wave heights
(after *Feldhausen et al.*, 1973).   There is considerable scatter
with greater differences between calculated and measured values
at lower wave heights.   The scatter is similar to that for other
wave models.   The measured data is from a storm at ocean station J
in the North Atlantic Ocean.

same decay coefficient, hindcast and careful estimates of maximum
wave heights off Delaware during the March 1962 storm matched to
within 2 feet.   There is no way to compare results from *Thom's*
method to short-term wave measurements.   Wave data can only be
used with this method by determining long-term wave statistics
from long-term wave data.

Simulated Comparisons

Without adequate wave data, model differences can be studied by comparing model results with simulated wind fields. *Dexter* (1974) tested modified forms of *Wilson's* model which is described earlier in this paper, two versions of the *Pierson et al.* model (1966) and *Barnett's* model (1968). Model wave growth was compared by applying a spatially constant 15 meter-second^{-1} wind to each model for 30 hours and model wave decay was compared by running each model for a subsequent 30 hours with no wind. Such a simple test provides only a general feel for the relative comparability of the models. To compare *Wilson's* significant wave height models to the other models which are spectral models, *Dexter* calculated significant wave heights from the wave energy spectra. *Dexter* also computed significant wave heights with *Bretschneider's* (1970) revised Sverdrup-Munk-Bretschneider (SMB) method. While suitable for this comparison, the latter method is not well suited for determining wave conditions due to time and spatially varying actual wind fields. Figure 3 shows the time history of significant wave heights. The spread in significant wave heights for all models is about 1 meter which is about 20 percent of the mean significant wave height at 30 hours. *Dexter* does not fully discuss the large significant wave height differences during decay and does not compare spectra computed by the spectral models. Figure 4 shows the time history of the frequency of maximum spectral energy and indicates that the models result in reasonably consistent values but the corresponding period differences between models range to about 3 seconds. For the *Wilson* and SMB methods, the significant frequency is plotted in Figure 4. The significant frequency can be approximately converted to the frequency of maximum spectral energy by dividing by 1.14 (*Darbyshire*, 1959) so that the *Wilson* and SMB curves should be moved upward in Figure 4.

Comparison of Design Wave Criteria

The relative validity of different wave models can also be examined by comparing design wave conditions based on the models. Such a comparison has the advantage of smoothing model-to-model differences during individual storms of hurricanes. Such differences may be due to somewhat different wind fields, different spatial positions relative to storm tracks, different model grid size, or a variety of other causes. Systematic differences, however, would still result in design wave criteria differences. In addition, a comparison of design wave criteria is the only way to evaluate *Thom's* method which does not involve hindcasting individual storms and hurricanes.

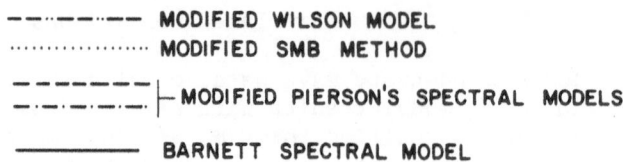

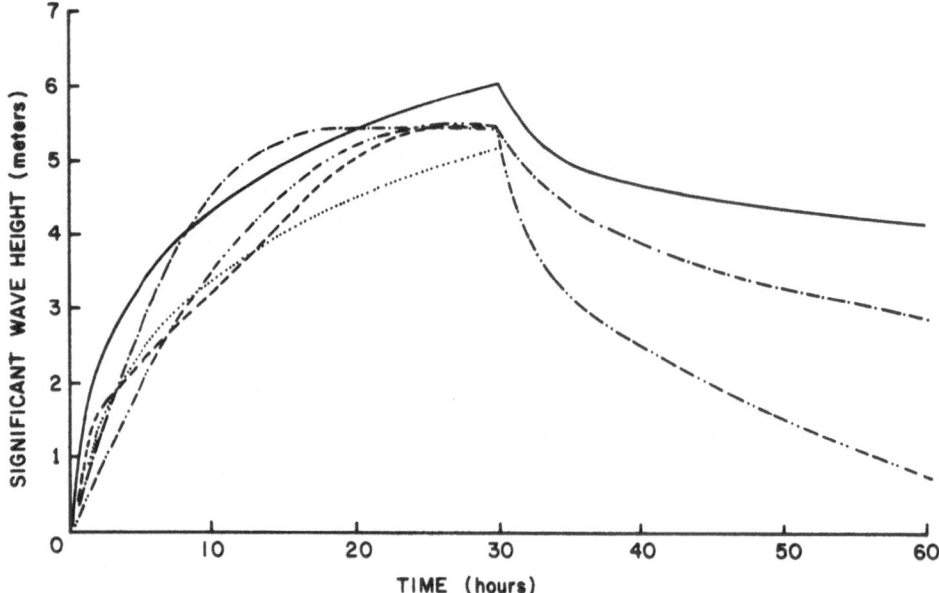

Fig. 3. Time histories of significant wave heights from a simulated comparison (from *Dexter*, 1974). During wave growth for the first thirty hours, the maximum difference between models is approximately 1 meter.

The most comprehensive design wave condition results for the Gulf of Mexico have been summarized by *Bea* (1974). These results are based on the use of *Wilson's* model to determine wave conditions during past hurricanes between 1900 and 1969 at nine sites offshore of Texas and Louisiana and are consistent with several recent proprietary studies of extreme waves in the Gulf of Mexico. Figure 5 is a comparison of the results reported by *Bea* (1974) and by *Quayle and Fulbright* (1975) who used *Thom's* method. The choice of a normal probability abscissa is for convenience and a different type of plotting would be chosen to extrapolate to long return periods from individual wave hindcasts or observations. In Figure 5, *Bea's* results are averages for his nine sites and the results of *Quayle and Fulbright* are averages for their three offshore regions covering Texas and Louisiana.

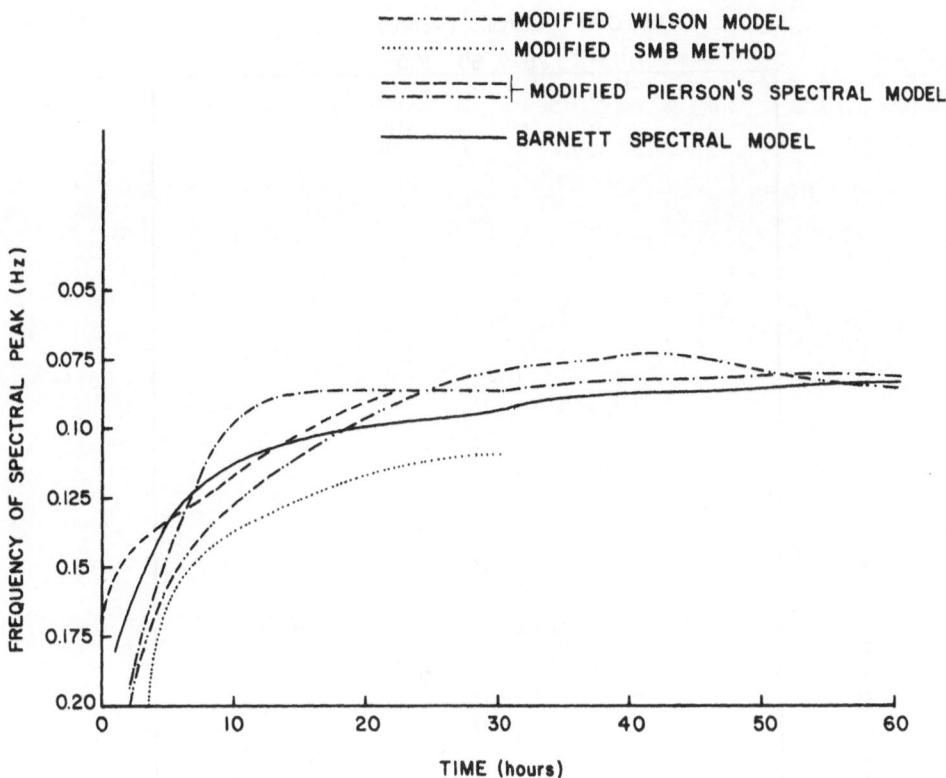

— · — · — · — · — MODIFIED WILSON MODEL
· · · · · · · · · · · · · · · · · MODIFIED SMB METHOD
— · — · — · — · —┤ MODIFIED PIERSON'S SPECTRAL MODELS
————————— BARNETT SPECTRAL MODEL

Fig. 4. Time histories of frequency of maximum spectral energy for three spectral models and significant frequency for the *Wilson* and SMB methods (from *Dexter*, 1974). The values are reasonably consistent but result in period differences up to about 3 seconds. The *Wilson* and SMB curves would be moved upward if empirically converted to the frequency of maximum spectral energy.

Discrepancies between significant wave heights for the two methods range from 10 feet for the 10-year return period to 16 feet for the 100-year return period. For design purposes, the probable maximum wave height is generally used and the discrepancy between probable maximum wave heights for the 100-year return period is approximately 30 feet = 1.86 x 16 feet.

Several recent design wave condition studies have been made for the Atlantic continental shelf. *Yang et al.* (1974) used *Thom's* method and estimated extreme waves from ship observations. *Earle* (1975) determined design wave conditions at Diamond Shoals

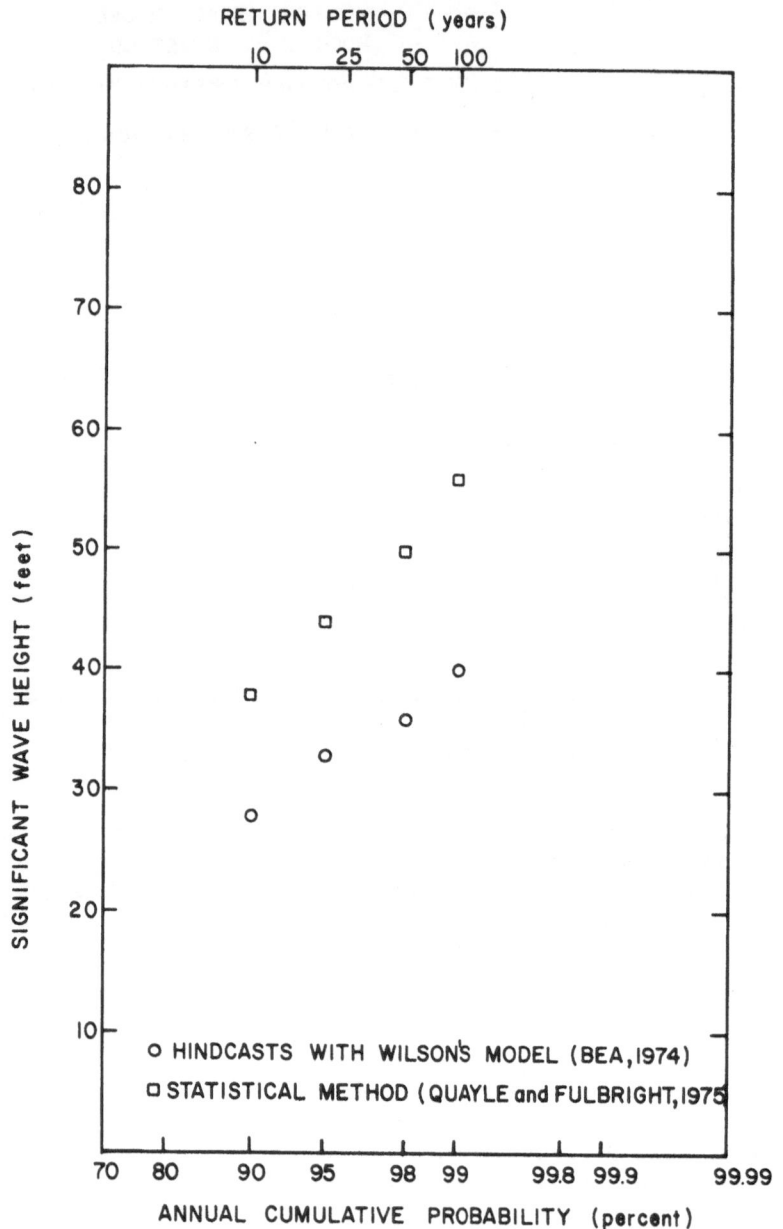

Fig. 5. Comparison of design wave conditions for deep water off Texas and Louisiana. There are major differences between results based on hindcasts with *Wilson's* model and results based on relationships between wave and wind speed statistics.

light tower off Cape Hatteras from hindcasts for storms and hurricanes between 1944 and 1973. These hindcasts were made with
Gelci's model. Design criteria from this study were later provided
for other U.S. Coast Guard light tower locations along the Atlantic
Coast (*Earle and Burns*, 1975). *Quayle and Fulbright* (1975) summarize conditions computed by *Thom's* method for all U.S. coastal
areas. *Ward et al.* (1977) describe design wave conditions off the
Atlantic coast based on hindcasts of hurricanes between 1900 and
1975 with *Wilson's* model in the form described in this paper.
Figure 6 compares design wave conditions from these studies for a
location in deep water off Delaware. *Yang's* (1974) results are
taken directly from his paper which is site specific to Delaware
coastal waters. He utilizes *Thom's* (1973b) mixed distribution for
both tropical and extratropical storms and a modified version of
this method in which the Pierson-Moskowitz spectrum is used to relate wind speeds to wave heights. The results for *Earle* (1975) are
for a site at the shelf break directly off Delaware Bay. These
results are not given in this form in *Earle* (1975) or *Earle and
Burns* (1975) because these reports provide design wave criteria
for shallow water sites where wave conditions are modified by
refraction, shoaling, and bottom friction. The results from
Quayle and Fulbright are for their region 7, which is directly off
Delaware Bay. The results from *Ward et al.* (1977) are based on
hurricane wave hindcasts and are for their site 3, which is near
the shelf break and just north of Delaware Bay.

The three applications of *Thom's* method provide design wave
criteria that are far higher than those provided by applications
involving hindcasts and ship observations. Differences between
significant wave height results for the statistics method and
results for the other methods are roughly 10 feet for the 10-year
return period and greater than 17 feet for the 100-year return
period. These differences are consistent with the differences
for the Gulf of Mexico. As for the Gulf of Mexico, the probable
maximum wave height discrepancy for the 100-year return period
between *Thom's* method and hindcasting methods is at least 30 feet
for the 100-year return period. The two hindcasting studies were
entirely independent; involved separate determinations of wind
fields; and used different wave models. Moreover, the hindcasting
studies and the ship observations, which cover a 24-year (1949-
1972) time period, result in significant wave height design
criteria differences of about 5 feet. The fact that the ship
observations are lower is reasonable because ships try to avoid
high seas particularly during hurricanes.

Dattatri (1974) compared results computed by *Thom's* method to
extrapolations based on 1 year of measured wave data off India.
The 100-year return period maximum wave height based on the statistical method was 91 feet and that based on the data was 33 feet.

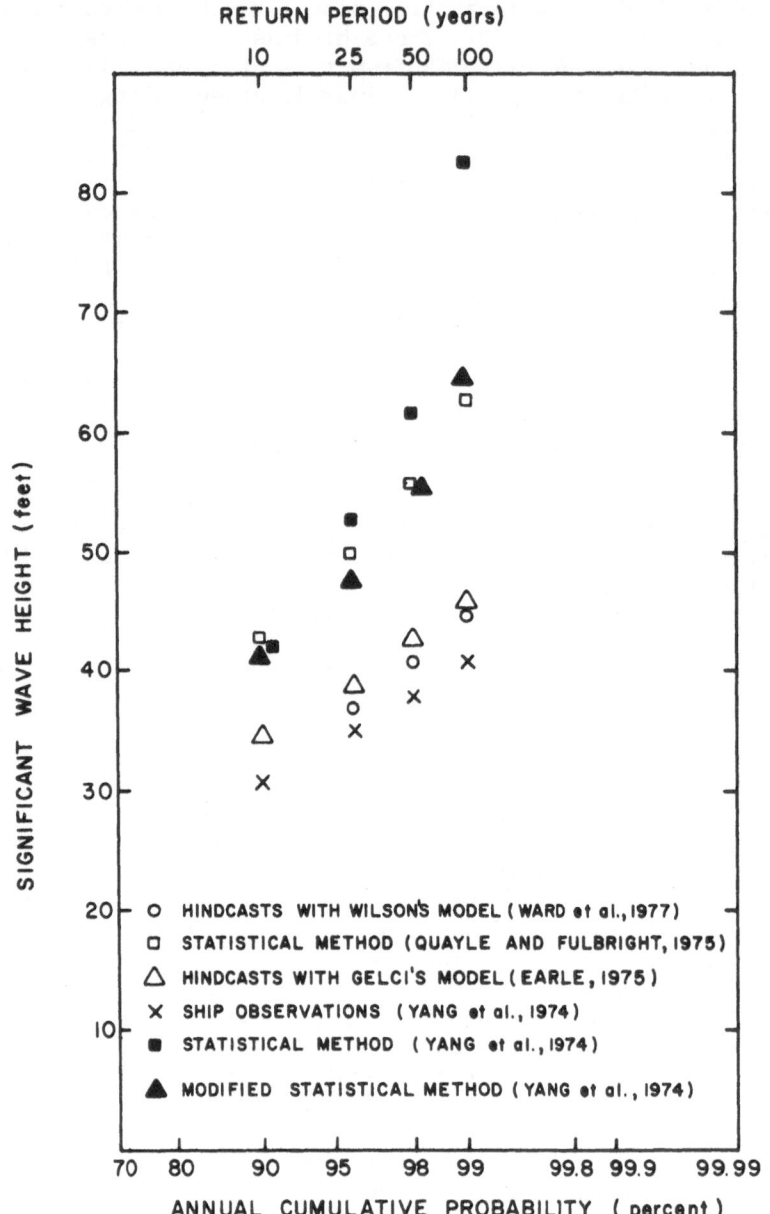

Fig. 6. Comparison of design wave conditions for deep water off Delaware. Wave heights based on hindcasts and observations are substantially lower than wave heights based on wind speed statistics. Wave hindcast results from two different models are consistent with each other.

Although the extrapolation from such a short series of measurements may have large errors, these errors cannot account for the discrepancy of 58 feet particularly since *Dattatri's* site is not affected by severe tropical or extratropical storms.

CONCLUSIONS

Although several wave models have been used to determine design wave conditions, the accuracy of these models requires further study. Because of the lack of measured long-term wave data particularly during severe storms and hurricanes, model calculations have not been compared to data for a wide variety of storm and hurricane conditions. Based on some comparisons between model results and wave measurements and between results from different models, the overall accuracy of the presently used hindcasting models is approximately 1-2 meters. The accuracy of *Wilson's* significant wave height model (e.g. *Wilson,* 1965, 1966) is comparable to accuracies of directional spectra models so that *Wilson's* model may be suitable for applications requiring only significant wave height and period information. Comparisons of design wave conditions show that the statistical method (*Thom,* 1973b) is generally not suitable for calculations of design wave conditions. In addition, there is a need to evaluate directional spectral models by comparison with measured long-term wave energy directional spectra. For the design of coastal and offshore structures, significant cost savings will result from better specification of the accuracy of present wave models and from future improvement of these models. With better accuracy specifications, confidence limits for final design conditions can be reduced. Moreover, the oceanographer or engineer can select the proper model for the project at hand. For example, models could be ranked in order of increasing computer requirements and, for some engineering projects, the use of the more advanced models may not be justified provided that the accuracy of the selected model is known and is sufficient. Many projects, however, will require the use of the best possible verified model.

REFERENCES

Baer, L. 1962. An experiment in numerical forecasting of deep water ocean waves, Lockheed Missile and Space Company Report LMSC-801296, San Diego, California

Barnett, T. P. 1968. On the generation, dissipation, and prediction of ocean wind waves, *J. Geophys. Res., 73(2):* 513-529.

Bea, R. G. 1974. Gulf of Mexico hurricane wave heights, Offshore Technology Conference, Houston, Texas, paper 2110.

Borgman, L. E. and D. T. Resio. 1977. Extremal prediction in wave climatology, Ports 77 Conference, Am. Soc. Civ. Engineers, Long Beach, California, Vol. 1, 394-412.

Bretschneider, C. L. 1952. Revised wave forecasting relationships. In: *Proceedings of 2nd Conference on Coastal Engineering*. Houston.

Bretschneider, C. L. 1963. Significant wave hindcasts for Station J North Atlantic storm, Technical Report National Engineering Science Company, SN-77-1.

Bretschneider, C. L. 1970. Revisions in wave forecasting, Look Laboratory Report, University of Hawaii.

Cardone, V. J. 1974. Ocean wave prediction: Two decades of progress and future prospects, Soc. of Naval Architects and Marine Engineers Seakeeping Symposium, New York.

Cardone, B. J., W. J. Pierson, and E. G. Ward. 1976. Hindcasting the directional spectra of hurricane generated waves. *J. of Petroleum Technology*, April, 385-394.

Cardone, V. J. and D. Ross. 1978. State-of-the-art wave prediction methods and data requirements. In: *Ocean Wave Climate*, M. D. Earle and A. Malahoff (Eds.), Plenum, New York (this volume)

Darbyshire, J. 1959. A further investigation of wind generated waves, *Deutsche Hydrographische Zeitschrift, 12(11)*. 1-13.

Darbyshire, J. 1961. Prediction of wave characteristics over the North Atlantic, *J. Inst. Navig., 14:* p. 399.

Dattatri, J, 1974. Discussion of extreme wave height distributions over oceans, *J. of Waterways Harbors and Coastal Engineering Division*, Am. Soc. of Civ. Engineers, WW4, 402-403.

Dexter, P. E. 1974. Tests on some programmed numerical wave forecast models, *J. of Phys. Oceanogr., (4(10)*, 635-644.

Earle, M. D. 1975. Oceanographic design conditions for Diamond Shoals Light Tower, U.S. Naval Oceanographic Office Technical Note 6110-3-75, Washington, D.C.

Earle, M. D. and D. A. Burns. 1975. Oceanographic design conditions for offshore light towers, U.S. Naval Oceanographic Office Technical Note 6110-8-75, Washington, D.C.

Enfield, D. B. 1973. Prediction of hazardous Columbia River Bar conditions, Final Report prepared for Techniques Development Laboratory (National Weather Service), Oregon State University.

Ewing, J. A. 1971. A numerical wave prediction method for the North Atlantic Ocean, *Deutsche Hydrographische Zeitschrift, 24(6)*, 241-261.

Feldhausen, P. H., S. K. Chakrabarti, and B. W. Wilson. 1973. Comparison of wave hindcasts at Weather Station J for North Atlantic Storm of December, 1959, *Deutsche Hydrographische Zeitschrift, 26(1)*, 10-16.

Gelci, R., H. Cazale, and J. Vassal. 1956. Utilisation des diagrammes de propagation à la prévision énergétique de la houle, *Bulletin de'information du Comité central d'océanographie et d'études des côtes,* number 8.

Gelci, R., H. Cazale, and J. Vassal. 1957. Prévision de la Houle. La méthode des densités spectroangulaires, *Bulletin d'information*

du Comité central d'océanographie et d'études des côtes,
number 9.

Gelci, R. and E. Devillaz. 1970. The numerical computation of
sea states (with discussions), LaHouille Blanche, 25(2), 117-136.

Hamilton, R. C. and E. G. Ward. 1974. Ocean data gathering program - quality and reduction of data, Offshore Technology Conference, Houston, Texas, paper 2108A.

Inoue, T. 1967. On the growth of the spectrum of a wind generated
sea according to a modified Miles-Phillips mechanism and its
application to wave forecasting, Ph.D. Thesis, New York University, New York.

Lazanoff, S. M. and N. Stevenson. 1975. An evaluation of a
hemispheric operational wave spectral model, Fleet Numerical
Weather Center Technical Note 75-3, Monterey, California.

Patterson, M. M. 1971. Hindcasting hurricane waves in the Gulf
of Mexico, Offshore Technology Conference, Houston, Texas,
paper 1345.

Patterson, M. M. 1972. Hindcasting hurricane waves in the Gulf
of Mexico. *J. for Petroleum Engineers,* August, 321-328.

Petrauskas, C. and P. M. Aagaard. 1971. Extrapolation of historical data for estimating design wave heights, *J. Soc. of
Petroleum Engineers, II,* 25-35.

Pierson, W. J., G. Neumann, and R. W. James. 1955. *Practical
Methods for Observing and Forecasting Ocean Waves by Means of
Wave Spectra and Statistics,* H.O. Publication 603, U.S. Naval
Oceanographic Office, Washington, D.C.

Pierson, W. J., L. J. Tick, and L. Baer. 1966. Computer based
procedures for preparing global wave forecasts and wind field
analyses capable of using wave data obtained by a spacecraft,
Sixth Naval Hydrodynamics Symposium, Office of Naval Research,
Washington, D.C.

Pore, N. A. and W. S. Richardson. 1969. Second Interim Report on
sea and swell forecasting, Weather Bureau Technical Memorandum
TDL-13, National Weather Service, Silver Spring, Maryland.

Quayle, R. G. and D. C. Fulbright. 1975. Extreme wind and wave
return periods for the U.S. coast, *Mariners Weather Log, 19(2),*
67-70.

Rabe, K. 1970. The French spectro-angular wave analysis/forecast-
ing program, Fleet Numerical Weather Center Program Note 6,
Monterey, California.

Salfi, R. E. and W. J. Pierson, 1977. A verification study of
hindcasted wave spectra based on wave data from weather ships J
and K and wind data obtained from December 1973 and January
1974 during Skylab. Final report to Spacecraft Oceanography
Project, National Environmental Satellite Service, National
Oceanic and Atmospheric Administration, Washington, D.C.

Sverdrup, H. U. and W. H. Munk. 1947. *Wind Sea and Swell:
Theory of Relations for Forecasting,* H.O. Publication 601,
U.S. Naval Oceanographic Office, Washington, D.C.

Thom, H. C. S. 1967. Toward a universal climatological extreme
    wind distribution, In: *Wind Effects on Buildings and Structures*,
    University of Toronto Press, 669-685.
Thom, H. C. S. 1971. Asymptotic extreme-value distributions of
    wave heights in the open ocean. *J. of Mar. Res., 29(1)*, 19-26.
Thom, H. C. S. 1973a. Distributions of extreme winds over oceans,
    J. of Waterways, Harbors and Coastal Engineering Div., Am. Soc.
    Civ. Engineers, WW1, 1-17.
Thom, H. C. S. 1973b. Extreme wave height distributions over
    oceans. *J. of Waterways, Harbors and Coastal Engineering Div.*,
    Am. Soc. Civil Engineers, WW3, 355-373.
U.S. Army Coastal Engineering Research Center, 1973. *Shore Protec-
    tion Manual, Vol. 1.*, U.S. Army Corps of Engineers, Fort Belvoir,
    Virginia.
Ward, E. G. 1974. Ocean data gathering program - An overview,
    Offshore Technology Conference, Houston, Texas, paper 2108B.
Ward, E. G., D. J. Evans, and J. A. Pompa. 1977. Extreme wave
    heights along the Atlantic Coast of the United States, Offshore
    Technology Conference, Houston, Texas, paper 2846.
Wilson, R. W. 1955. Graphical approach to the forecasting of
    waves in moving fetches. Beach Erosion Board Technical Memoran-
    dum No. 73, U.S. Army Corps of Engineers, Washington, D.C.
Wilson, B. W. 1961. Deep water wave generations by moving wind
    systems, *J. Waterways, Harbors and Coastal Engineering Div.*,
    Am. Soc. Civ. Engineers, WW2, 113-141.
Wilson, B. W. 1963. Deep water wave generations by moving wind
    systems (with discussions), Transactions, Am. Soc. Civ.
    Engineers, Vol. 128, part 4, 113-148.
Wilson, B. W. 1965. Numerical prediction of ocean waves in the
    North Atlantic for December 1959, *Deutsche Hydrographische
    Zeitschrift, 18(3)*, 114-130.
Wilson, B. W. 1966. Design sea and wind conditions for offshore
    structures, Proceedings of the Offshore Exploration Conference,
    666-708.
Wilson, B. W., S. K. Chakrabarti, and P. H. Feldhausen. 1972.
    Numerical prediction of storm waves and wave spectral densities,
    In: *Fifteenth International Conference on Great Lakes Research*,
    Madison, Wisconsin, abstract only.
Wilson, B. W., S. K. Chakrabarti, and P. H. Feldhausen. 1973.
    Hindcast of deep-water wave energy spectra for hurricane Audrey
    of 1957, Offshore Technology Conference, Houston, Texas,
    paper 1833
Yang, C. Y., M. A. Tayfun, and G. C. Hsiao. 1974. Extreme wave
    statistics for Delaware coastal water, In: *Proceedings of the
    International Symposium on Ocean Wave Measurements and Analysis*,
    Am. Soc. Civ. Engineers, Vol. 1, 352-361.

State-of-the-Art Wave Prediction Methods and Data Requirements

Vincent J. Cardone[1] and Duncan B. Ross[2]

[1]Oceanweather Inc. and [2]Atlantic Oceanographic and Meteorological Laboratories, National Oceanic and Atmospheric Administration

ABSTRACT

In the past decade, there has been a general shift in wave prediction methods from simple empirical techniques developed in the 1940's to the application of numerical models based upon the spectral energy balance equation. Such models simulate the physical processes governing the growth of surface waves which have been identified as a result of extensive theoretical and experimental study of wave generation. The importance of non-linear wave-wave interactions in wave generation has prompted the recent development of an alternate model context for wave pre-diction, which is based upon a parametric representation of the wave spectrum. This paper reviews the structure of state-of-the-art spectral and parametric methods and validation, and wave data requirements for further improvements in such methods.

INTRODUCTION

The serious study of ocean surface gravity wave prediction began over thirty years ago in response to the crucial need for wave forecasts to support the planning of amphibious operations during the Second World War. The imaginative wave prediction method developed by *Sverdrup and Munk* (1947) is the basis of a system, the significant wave method of wave prediction, that is still applied widely today. This method has been revised many times and is still the most widely used real-time wave forecasting method, largely because of its simplicity and efficiency. A contemporary significant wave model is reviewed by *Earle* (1978) in this volume.

The application of spectral concepts to the description of ocean surface waves by *Pierson and Marks* (1952) marked the beginning of a new era in ocean wave prediction. As early as 1953, spectral concepts had been incorporated into a practical wave forecasting method (*Pierson et al.*, 1955) and by 1957 digital computers had been applied to the generation of objective spectral wave predictions (*Gelci et al.*, 1957).

The past two decades have been marked by very active theoretical and experimental research into the basic processes governing ocean surface gravity wave generation by the wind, wave-wave and wave-current interaction, and wave energy dissipation through breaking and bottom interactions. Despite large remaining gaps in our knowledge of such processes, spectral ocean wave prediction models have been developed and applied to provide useful wave data for assessment of wave climate, design of ships and offshore structures, and for the preparation of marine forecasts. A review of the history of ocean wave prediction models is given by *Cardone* (1974).

The models that have been most widely implemented are variants of the model described by *Pierson et al.* (1966). Those models are based upon the numerical solution of the spectral energy balance equation in terms of a directional spectrum resolved in a matrix of finite bandwidth frequency and directional spectral components. The source function of the spectral energy balance equation contains terms that represent processes of wave generation, breaking, and dissipation. Two models (*Barnett*, 1968; *Ewing*, 1971) have included approximate representations of the wave-wave interactions.

An alternate model context for ocean wave prediction has recently been proposed as a result of the analysis of measurements of wave growth with fetch in the JONSWAP (Joint North Sea Wave Project) experiment, as described by *Hasselmann et al.* (1973). In particular, they concluded that the dominant process responsible for the development of the spectrum with fetch was the transfer of energy across the spectrum associated with nonlinear wave-wave interactions. The implication of this interpretation to the form of wave prediction models is profound, since an inclusion of a rigorous wave-wave interaction term in the source function is impossible with current computer speeds. Therefore, *Hasselmann et al.* (1976) proposed a simplified wave prediction model based upon a parametrical representation of the spectrum in which one parameter, the non-dimensional peak frequency, is used to characterize the stage of wave development.

This paper outlines the structure of contemporary discrete type spectral models and parametric models. Models recently developed and applied are reviewed briefly with an emphasis on model intercomparison and hindcast accuracy. Preliminary results of comparative spectral and parametric hindcasts of hurricane-generated sea states are presented. Finally, wave data requirements for the development and validation of ocean wave prediction models are discussed.

SPECTRAL ENERGY BALANCE

The framework of contemporary ocean wave prediction models is the spectral energy balance equation. A succinct discussion of the various forms of this equation is given by *Phillips* (1977). It is usually applied in its simplest form; that is, to surface gravity waves assumed to propagate through water of infinite depth that is otherwise at rest. In this form, it may be written

$$\frac{\partial E}{\partial t}(f,\theta,\vec{x},t) + \vec{C}_g(f,\theta)\cdot\nabla E(f,\theta,\vec{x},t) = S(f,\theta,\vec{x},t) \qquad (1)$$

where E is the energy density of the wave field described as a function of frequency, f, and direction of propagation, θ, position, $\vec{x}$, and time, t. $\vec{C}_g$ is the deep water group velocity. S, the source function, represents all physical processes that transfer energy to or from the spectrum.

The source function can be expressed as a sum of terms, each of which represents a class of physical processes. Schematically, it may be written (e.g., *Hasselmann et al.*, 1973)

$$S = S_{in}+S_{n\ell}+S_{ds} \qquad (2)$$

where S_{in} represents energy input from the atmosphere, $S_{n\ell}$ the transfer of energy across the spectrum associated with energy conserving nonlinear wave-wave interactions, and S_{ds} represents processes that serve to dissipate wave energy.

The nature of S_{in} has been a subject of active theoretical and experimental research since the appearance of the theories of *Phillips* (1957) and *Miles* (1957). *Phillips* explained the initial generation of gravity waves on an undisturbed sea surface through a resonant excitation by incoherent atmospheric turbulent pressure fluctuations convected by the mean wind. The mechanism leads to linear growth of wave energy in the early stages of wave development. *Miles* was the first to calculate, theoretically, the amplitude of the component of fluctuating atmospheric pressure in phase with wave slope, induced by a prescribed free surface wave in the air flow. The calculated energy transfer leads to exponential growth

of wave energy. More advanced theories for the rate of energy
transfer through this type of mechanism have been proposed by
Davis (1974) and *Gent and Taylor* (1976). The results of the
various theories, however, are not consistent and appear to be
quite sensitive to the individual assumptions made.

Numerous experimental studies of wave growth as a function of
fetch (*Snyder and Cox,* 1966; *Barnett and Wilkerson,* 1967; *Schule
et al.,* 1971; *Ross et al.,* 1971) have verified the linear-
exponential growth of spectral component energy with fetch predicted by the above mechanisms. The verification, however, has
been only qualitative. The observed linear and exponental growth
rates have usually differed from the theoretical predictions by
several orders of magnitude.

Direct measurement of the energy transfer between the wind and
surface waves has been attempted in wind tunnel-wave tank experiments and in the field (*Dobson,* 1971; *Elliot,* 1972; *Snyder,* 1974)
but results have been characterized by considerable scatter. However, the earlier experimental difficulties have apparently been
overcome in the recent experiment conducted in the Bight of Abaco,
Bahamas, preliminary results of which were reported by *Dobson et al.*
(1978). Their results tend to confirm earlier suspicions that S_{in}
alone cannot account for the rapid exponential growth of waves
observed in the field.

The nature of $S_{n\ell}$ has been studied theoretically by *Hasselmann*
(1962, 1963), and the results of the JONSWAP experiment were interpreted in terms of those theories (*Hasselmann et al.,* 1973). In
particular, they argued that most of the observed characteristics
of the development of the spectrum with fetch could be explained
by the nonlinear energy transfer. Those characteristics include:
the pronounced sharp peaks associated with spectra in the early
stages of wave development, a quasi-invariance of spectral shape
with fetch, the rapid growth of waves on the steep forward face of
the spectrum, and the tendency for growing spectral components to
overshoot their eventual equilibrium position.

An independent calculation of $S_{n\ell}$ for the mean observed JONSWAP
spectrum has been reported by *Fox* (1976), who used a method of calculation proposed by *Longuet-Higgins* (1975). The calculation showed
that $S_{n\ell}$ was of the same order as the net source function observed
in JONSWAP, but its distribution with frequency differed in an
important way from a numerical evaluation for the same spectrum as
performed by *Sell and Hasselmann* (1972). The new calculation suggests that the effect of non-linear interactions is to broaden
initially narrow peaks in the frequency spectrum and to distribute
energy to both lower and higher frequency waves. Direct wind input
of energy would therefore be principally responsible for the shape

and level of the spectral peak. *Webb* (1978) has presented an especially lucid discussion of how wave-wave interactions may simultaneously impart a stabilizing and diffusive influence on the spectrum.

Our knowledge of S_{ds} remains very limited and mainly qualitative. In a growing wind sea, energy is dissipated mainly by wave breaking in the high frequency part of the spectrum. Attempts to relate energy dissipation and whitecapping have been made by *Ross and Cardone* (1974) and *Hasselmann* (1974), while detailed analyses of the breaking process itself have been conducted by *Banner and Phillips* (1974). Other dissipative processes that have been proposed but not verified include attenuation of swell through interaction with short waves (*Phillips*, 1963) or scattering by turbulence (*Phillips,* 1959).

Fetch-limited wave growth studies, in particular JONSWAP, have provided reliable estimates of the net source function and revealed the importance of $S_{n\ell}$ to the source functions, at least for the conditions of JONSWAP (see Figure 1). *Hasselmann* (1974) has presented alternative qualitative assessments of the overall energy balance in a fetch-limited and fully developed spectra based upon assumed forms for S_{ds} and S_{in}. Quantitative specification of the energy balance must await precise determinations of at least one of S_{ds} and S_{in}. The results of *Dobson et al.* (1978) provide hope that the latter can be determined reliably and that more rigorous expressions for the source terms can be incorporated into practical wave prediction models.

Presently applied wave prediction models still contain source term specifications based upon empirical results of net wave growth. The next sections discuss representative contemporary discrete and parametric ocean wave prediction models.

DISCRETE OCEAN WAVE PREDICTION MODELS

Models of this type have been developed by *Gelci et al.* (1957), *Pierson et al.* (1966), *Barnett* (1968), and *Ewing* (1971). All are similar in that the numerical solution of Equation (1) involves the representation of the directional spectrum by a number of finite bandwidth spectral components and the successive simulation of wave propagation (the solution of the homogeneous part of Equation (1)), and calculation of local energy transfers over a series of time steps on a grid mesh overlaying the basin of interest.

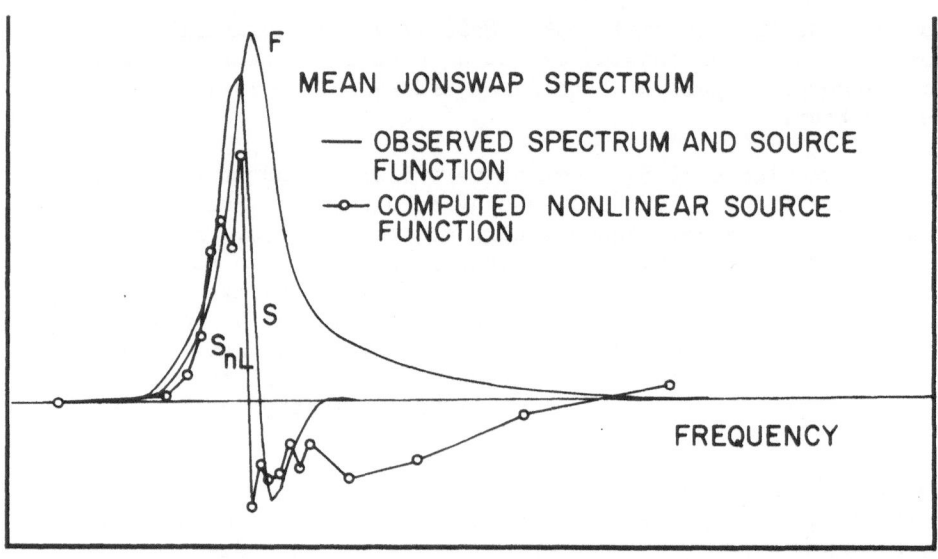

Fig. 1. Schematic diagram of the wave spectrum, F, net source function, S, and computed nonlinear transfer, $S_{n\ell}$, as a function of frequency for fetch-limited spectra according to JONSWAP. (From *Hasselmann*, 1974.) The importance of $S_{n\ell}$ to the source function S is clear.

Propagation

Various propagation schemes have been devised to represent the ideal of moving wave energy in each frequency-directional band with the group velocity appropriate to its frequency and along an arc of great circle appropriate to its direction. Methods have included simple first-order, upstream differencing (*Barnett*, 1968), fourth-order differencing (*Ewing*, 1971), discontinuous schemes which "jump" energy across a grid to approximate propagation (*Baer*, 1962), hybrid schemes which combine jump and differencing (*Pierson et al.*, 1966), and the explicit calculation of ray paths (characteristic curves) and wave packet positions (*Barnett*, 1969).

Recently, *Greenwood and Cardone* (1977) have reported a scheme that does not faithfully preserve monochromatic wave propagation. They argue that the error incident to a crude gradient scheme is of order $N^{\frac{1}{2}}$, where N is the number of time steps; the error inherent in neglecting bandwidth is of order N and is the dominant error for long propagation distances. They have proposed and tested successfully a scheme that employs downstream interpolation exclusively in a way that conserves energy rigorously, accounts for great circle propagation, and simulates lateral and longitudinal spreading associated with finite bandwidth spectral components. The method can be applied on almost any grid system but has not yet been implemented in an operational model.

Growth-Dissipation

All discrete-type spectral models include source function terms that operate to grow waves with a following wind and dissipate waves with an opposing wind. Most such source terms model non-linear effects implicitly, with the exception of the models of *Barnett* (1968) and *Ewing* (1971) which included in the source function a simplified parametric representation of $S_{n\ell}$.

The source function representation most widely used is a variant of the *Pierson et al.* (1966) method which treats spectral components that travel within $\pm 90°$ of the local wind according to the growth model

$$\frac{\partial E(f,\theta)}{\partial t} = [A(f,\theta,\vec{U}) + B(f,\theta,\vec{U}) \ E(f,\theta)] \times$$
$$\{1-[E(f,\theta)/E_\infty(f,\theta,\vec{U})]^2\} \tag{3}$$

where A represents the linear growth rate, B the exponential growth rate, E_∞ is the Pierson-Moskowitz (PM) spectrum for a fully developed sea at wind $\vec{U}$, which is coupled with the directional spreading function found in project SWOP (*Cote et al.*, 1960). Quite different growth rates result from the various algorithms for solving Equation (3) and from various choices of the forms of A and B.

It is usually stated that A represents the linear growth due to the resonance mechanism of *Phillips* (1957), and B represents the exponential growth rate resulting from the linear feedback theory of *Miles* (1957). In fact, the adopted magnitudes of A and B are empirically calibrated against field measurements of net observed wave growth. The A and B terms adopted in any application, therefore, are fitted constants and reflect in addition to the mechanism of *Phillips* and *Miles*, other mechanisms, in particular wave-wave interaction, albeit incorretly scaled, that contribute to the process of wave generation.

The Pierson-Moskowitz spectrum plays an important role in modeling the transition from a growing sea to a fully developed sea. Finally, spectral components traveling against the wind are attenuated at a rate that depends on wave frequency and the energy

of the local wind sea. Again, the dissipation algorithm is empirically based and probably accounts for the net effect of several
mechanisms, including wave-wave interactions.

Recent Model Applications

There have been a number of new applications of discrete
spectral wave prediction models in the past several years. The
model originally described by *Pierson et al.* (1966) has been
refined and successfully adapted on the U.S. Navy Fleet Numerical
Weather Central (FNWC) system for operational forecasting. Twice
daily, the model is used to prepare wave forecasts out to 72 hours
on a grid of points covering all Northern Hemisphere seas, with
input wind fields computed from sea level pressure forecasts provided by the numerical weather prediction model of FNWC. Before
each forecast run, observed wind data and analysis wind fields
prepared therefrom are used to update the wave spectral field via
a 12-hour hindcast run. A description and preliminary evaluation
of the Navy wave prediction system is given by *Lazanoff and
Stevenson* (1975).

The model has been applied to hurricane scale meteorological
systems and validated against data collected in an oil industry
sponsored measurement program known as the Ocean Data Gathering
Program (*Ward,* 1974). That version of the model is described by
Cardone et al. (1976a). The model has been applied in the hindcast
of all significant hurricanes known to have occurred in the Gulf
of Mexico. The hindcasts provided data used to estimate long-term
wave height statistics for the northern Gulf of Mexico (*Ward et al.,*
1978). It has also been tested in a real-time mode at the NOAA
Atlantic Oceanographic and Meteorological Laboratories to forecast
sea states in U.S. East Coast hurricane Belle, 1976, and hurricane
Anita, 1977, in the central Gulf of Mexico (*Cardone and Ross,* 1977).

The wave spectral prediction model originally described by
Isozaki and Uji (1973) has also been applied to tropical cyclone
wind fields by *Uji* (1975). The sensitivity of the wave solution
to storm forward speed of the hypothetical typhoon considered was
found to be large. In a later study (*Isozaki et al.,* 1976), the
hindcast model was applied to an actual typhoon with the hindcasts
compared to measurements of swell observed near the Japanese coast
some distance from the storm track. These models applied include
a spectral growth formulation similar to that of *Pierson et al.*
(1966).

A class of simpler spectral models formulated about the *Pierson
et al.* (1966) growth scheme has been proposed in one form or another
by *Baer et al.* (1970), *Hsu* (1973), and *Freeman* (1976). These models

attempt to avoid the computationally expensive solution of the spectral energy balance equation by finite differences for a large number of grid points when wave predictions at only a few sites are needed. For such applications, solutions are sought along ray paths projected outward from the sites. The ray model described by *Freeman* is documented only for stationary wind fields, but it apparently can be applied to nonsteady wind fields in a manner described by *Wilson* (1961) for a significant wave type model.

PARAMETRIC WAVE PREDICTION MODELS

The parametric method proposed by *Hasselmann et al.* (1976) is based upon the JONSWAP spectrum, which is of the form

$$E(f) = \alpha g^2 (2\pi)^{-4} f^{-5} \exp\left\{ -\frac{5}{4}\left(\frac{f_m}{f}\right)^4 + \ln\gamma \times \exp\left[-\frac{(f-f_m)^2}{2\sigma^2 f_m^2} \right] \right\} \quad (4)$$

where
$$\sigma = \begin{cases} \sigma_c & f \leq f_m \\ \\ \sigma_b & f \geq f_m \end{cases}$$

where f_m, the frequency of the spectral peak and α, the equilibrium range constant, are scale parameters. γ, the peak-enhancement factor, σ_a, the left peak width, and σ_b, the right peak width, are three shape parameters.

Evidence is presented that Equation (4) fits not only spectra obtained under stationary fetch-limited conditions but growing wind sea spectra in general. *Pierson* (1977) has criticized the procedures used to fit Equation (4) to measured spectra and suggests that sampling variability could bias the values of some of the fitting parameters. In reply, *Hasselmann et al.* (1977) claim that much of the scatter in the data observed in JONSWAP is geophysical rather than statistical in nature.

Justification for the parametric approach follows from the belief that nonlinear resonant wave-wave interactions control the evolution of the spectrum in a growing sea. In view of current and projected computer speeds, the rigorous numerical calculation of the nonlinear component of the source function is impossible in a discrete prediction model. The parametric model is predicated further on the assumption that the nonlinear interactions impart a stabilizing influence on the shape of the spectrum such that a quasi-universal form can be adopted.

The parametric model is formulated by the projection of the energy balance Equation (3), including the source terms, onto a set of prognostic equations in the parameters of the quasi-invariant spectral form (4). The procedure is shown for the five parameters of Equation (4) by *Hasselmann et al.* (1973). In *Hasselmann et al.* (1976) the directional wave distribution and spectral shape are prescribed, and the prognostic equations are derived only for the parameter pair $\nu = f_m U/g$ and α

$$\frac{1}{\nu}\left(\frac{\partial\nu}{\partial\tau} + P_{\nu\nu}\frac{\partial\nu}{\partial\eta}\right) + P_{\nu\alpha}\frac{1}{\alpha}\frac{\partial\alpha}{\partial\eta} = -N_\nu\alpha^2\nu + \frac{1}{U}\left(\frac{\partial U}{\partial\tau} + \frac{\partial U}{\partial\eta}\right),\tag{5}$$

$$\frac{1}{\alpha}\left(\frac{\partial\alpha}{\partial\tau} + P_{\alpha\alpha}\frac{\partial\alpha}{\partial\eta}\right) + P_{\alpha\nu}\frac{1}{\nu}\frac{\partial\nu}{\partial\eta} = I_{\nu^{7/3}} - N_\alpha\alpha^2\nu + \frac{0.2}{U}\left(\frac{\partial U}{\partial\eta}\right),\tag{6}$$

where

$$\begin{pmatrix} P_{\nu\nu} & P_{\nu\alpha} \\ & = \\ P_{\alpha\alpha} & P_{\alpha\nu} \end{pmatrix} \begin{pmatrix} 1 & -0.07 \\ \\ 0.47 & 0.2 \end{pmatrix},$$

$$N_\nu = 0.54, \quad N_\alpha = 5, \quad I = 5.1 \times 10^{-3},$$

and $\partial/\partial_\tau = (U/g)(\partial/\partial t)$, $\partial/\partial_\eta = (U/g)V_m \cdot \nabla$, with V_m parallel to the wind direction, $|V_m| = qg/(4\pi f_m)$, $q = 0.85$. The dimensionless gradient, ∂/∂_η, corresponds to the rate of advection of properties with the group velocity, V_m, of waves in the spectral peak. The correction factor, q, arises from averaging over the directional distribution of the spectrum, which was taken as frequency independent and proportional to the square cosine of the angle θ relative to the local wind direction for $|\theta| \leq \pi/2$, and zero in the upwind half-plane $\pi/2 \leq |\theta| \leq \pi$.

Special solutions to Equations (5) and (6) are given in *Hasselmann et al.* (1976) for simply prescribed wind fields. The full solution of Equations (5) and (6) requires numerical integration by finite difference techniques for a spatial grid on which time and space varying wind fields are prescribed. The spectrum can be recovered from ν and α through Equation (4) and the use of the mean JONSWAP values of γ, σ_a, and σ_b. A further simplification is made in *Hasselmann et al.* (1976), in which a quasi-equilibrium relation between α and ν is assumed, and only one prognostic equation in ν is required.

There are recognized to be situations when the parametric model context for ocean wave prediction is not appropriate. Examples are swell-dominant sea states, transitional sea states between wind sea and swell, and cases of rapidly fluctuating wind velocity (speed and/or direction) fields. *Hasselmann et al.* (1976) propose that for such cases a hybrid model will be required which combines the parametric model and the discrete method for swell prediction. A hybrid model has indeed been found to be necessary to hindcast accurately sea states in extratropical storms. That model was developed recently by *Weare and Worthington* (1978) as part of a study of the severe storm wave climate of the North Sea. Early in their study, a three-parameter (α, γ, ν) parametric scheme was found to be necessary, and swell was described in ten frequency and eight directional bands. The hybrid model included a set of rules that governed transfer of energy between sea and swell representations and vice-versa. The transition from growing seas to a fully-developed state was achieved in a manner analogous to the way the transition is modelled in Equation (3).

MODEL VALIDATION

Ocean wave prediction models are usually tested in two ways. First, they are used to simulate the evolution of the wave spectrum with fetch or duration for stationary, uniform wind fields. Such tests are used to reveal model sensitivity to alternate source function parameterizations or to compare model predictions with fetch-limited wave data.

A most revealing test of an ocean wave prediction model is its ability to specify the directional wave spectrum for realistic wind fields that vary in space and time. This is a very difficult test to carry out for several reasons. First, it is difficult to specify accurate wind fields over the ocean, and for specified wind fields, it is difficult to assess the errors in wave hindcasts caused solely by errors in the input winds. Second, measurements of the wave frequency spectrum are relatively scarce, and measurements of the directional spectrum are rare indeed. Finally, where wave measurements are available, the spectral analysis of the wave data provides only an estimate of the true spectrum. Predictions from hindcast models are of the expected value of the true spectrum, and any comparison of predicted and measured spectra must consider the effects of the sampling variability that characterizes the measured spectrum and the integral properties derived from the spectrum.

Intercomparison and Verification of Discrete Spectral Models

Several recent studies have intercompared different hindcast models. *Dexter* (1974) programmed three models: the growth-dissipation algorithm of *Pierson et al.* (1966) coupled to a grid propagation scheme devised by *Dexter*; the growth-dissipation algorithm of *Barnett* (1968) coupled to the same propagation scheme; the significant-wave hindcast model of *Wilson* (1961). Comparisons of predictions from different wave models against wave measurements in typically complex meteorological situations remain limited. *Feldhausen et al.* (1973) compared eleven hindcasts of the severe North Atlantic storm of December 1959. More recently, *Salfi and Pierson* (1977) compared hindcasts from the FNMC Pierson-Tick-Baer model during a storm period in the North Atlantic to wave measurements made by Ocean Weather Stations J and K. The above studies are discussed in the paper by *Earle* (1978) in this volume.

In a recent study, *Resio and Vincent* (1977) compared the predictions of the FNWC Pierson-Tick-Baer spectral growth algorithm and the algorithm of *Barnett* (1968). *Resio's* comparisons were made in a non-dimensional form to investigate how near the predictions are to dimensional similarity. Specifically, for constant wind input of 15 m/sec and 30 m/sec, *Resio* plotted the dimensionless wave height, $\bar{H} = gE^{\frac{1}{2}}/U^2$, against dimensionless fetch, $\bar{F} = gF/U^2$, and dimensionless duration, $\bar{T} = gT/U$, where U is wind speed, E is total variance, F is fetch, and T is duration. Although there is reasonably good evidence (*Mitsuyasu*, 1968; *Hasselmann et al.*, 1973) for a non-dimensional relation between $\bar{H}$ and $\bar{F}$, neither model tested reproduced that relation (Figure 2). *Barnett's* model, however, can be modified to approximately obey a $\bar{H}$-$\bar{F}$ law by introducing a more accurate propagation system and making the equilibrium-range constant dependent on the stage of wave development. *Resio* also showed that neither the FNWC model nor the modified *Barnett* model was consistent with a similarity relation between $\bar{H}$ and $\bar{T}$ (Figure 3); he suggests that a $\bar{H}$-$\bar{T}$ law does not hold in nature so that the JONSWAP formulation is inappropriate for duration-limited spectra. *Mitsuyasu and Rikiishi* (1978) have recently confirmed their earlier studies, which showed that duration-limited spectra are not self-similar and that exponential growth rates of duration-limited spectra are generally larger than those of fetch-limited spectra.

The version of the Pierson-Tick-Baer model described by *Cardone et al.* (1976a) was calibrated largely on the basis of wave measurements made in severe hurricane Camille, 1969, at six oil rigs in the Gulf of Mexico. Figure 4 compares the peak-hindcast and peak-measured frequency spectrum at the rig which experienced the severest sea states in Camille. Since the intended application of the model was the hindcasting of maximum individual wave height in hurricanes, a pertinent assessment of accuracy was obtained by comparing hindcast

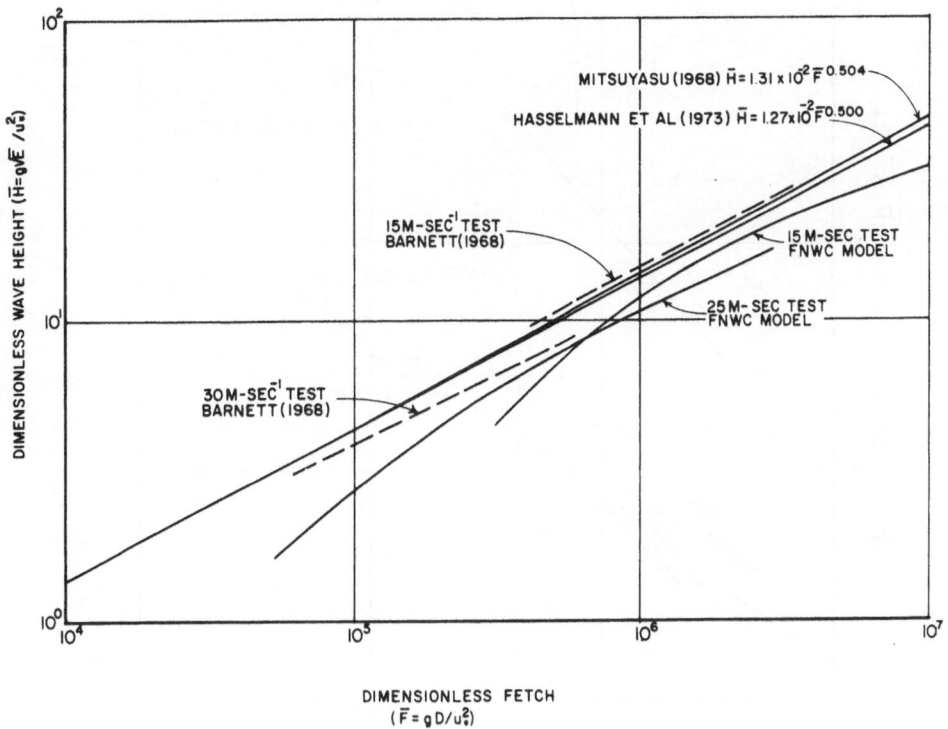

Fig. 2. Comparison of relationships between nondimensional wave height and nondimensional fetch from models to those derived empirically by *Mitsuyasu* (1968) and *Hasselmann et al.* (1973). (From *Resio and Vincent*, 1977.) Neither the *Barnett* model nor the FNWC model reproduce the empirical nondimensional curves.

and measured maximum heights at several locations in the Gulf of Mexico in storms Camille, 1969, and Felice, 1970, and comparing hindcast and estimated maximum heights in hurricanes Carla, 1961, Hilda, 1964, and Betty, 1965. In eighteen such comparisons, with measured heights ranging from 6 to 24 meters, the root-mean-square difference was 1.5 meters (Figure 5). The hindcast spectra associated with maximum wave heights agreed very well with the limited sample of measured spectra available.

The recent passage of Delia, 1973, over an offshore rig instrumented with wave staffs and electromagnetic current meters provided an opportunity to validate the directional characteristics of the hindcast spectra. *Forristall et al.* (1978), following the

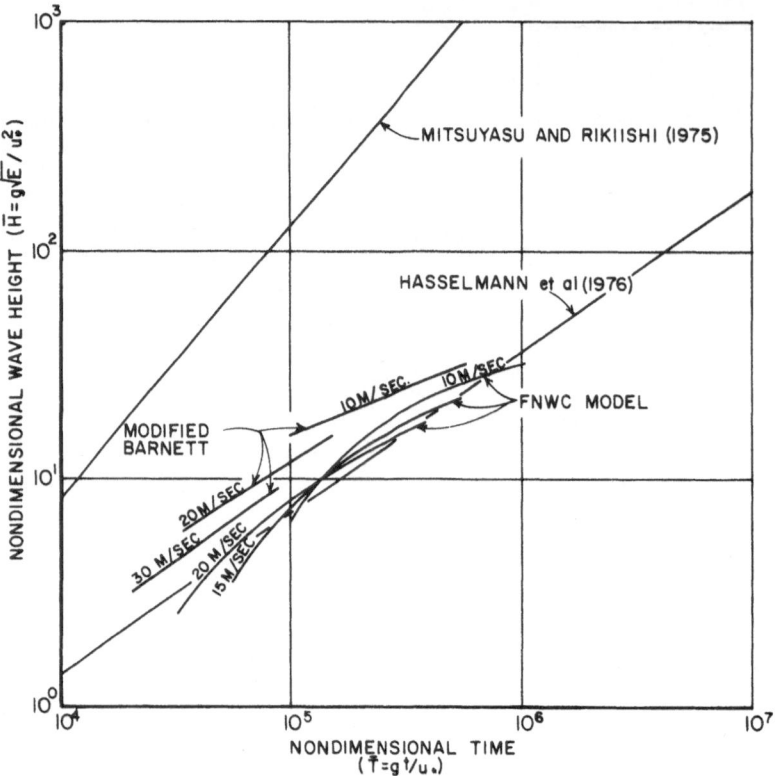

Fig. 3. Comparison of relationships between nondimensional wave height and nondimensional time from numerical models and laboratory and prototype scale observations. (From *Resio and Vincent,* 1977.) Neither the FNWC model nor a modified *Barnett* model is consistent with a similarity relation between nondimensional height and time.

method of *Bowden and White* (1966), calculated the directional spectrum from the surface elevation and orbital velocity measurements. The directional spreading function adopted was of the form proportional to

$$\cos^{2s}[(\theta-\theta_0)/2] \qquad (7)$$

where θ_0 is the mean direction of travel of the waves, s is a measure of the amount of spreading, and both parameters θ_0 and s are functions of frequency. The directional spectrum hindcast at a grid point near the measurement site at the peak of the storm is

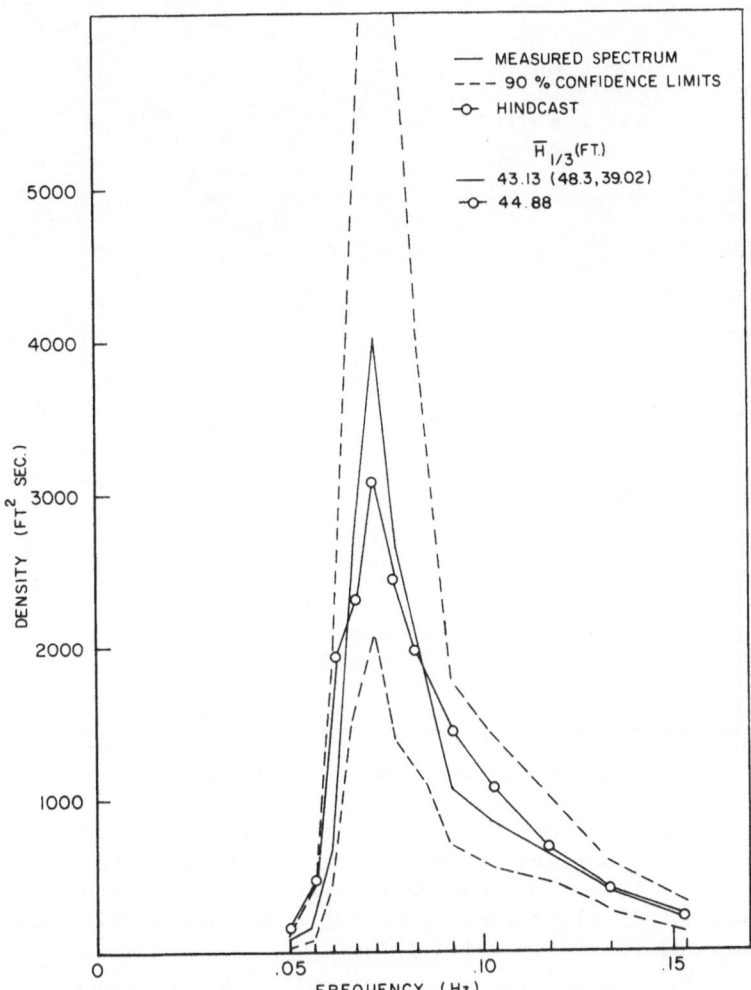

Fig. 4.   Comparison of peak measured (—) and discrete spectral model (-o-) hindcast spectrum in hurricane Camille, 1969 (From *Cardone et al.*, 1976).   The measured and hindcast significant wave heights $\overline{H}_{1/3}$ are within 2 feet of each other.   The significant wave heights corresponding to the 90% confidence limits of the measured data are indicated.

expressed in terms of θ_0 and s and compared to measured directional spectra in Figure 6.   Because of the distance between the model grid point and the tower, the measured spectrum at 1300 CDT (dotted line) should be compared to the hindcast spectrum at 1200 CDT.   The agreement is remarkably good considering the complexity of the sea state revealed in the measurements.

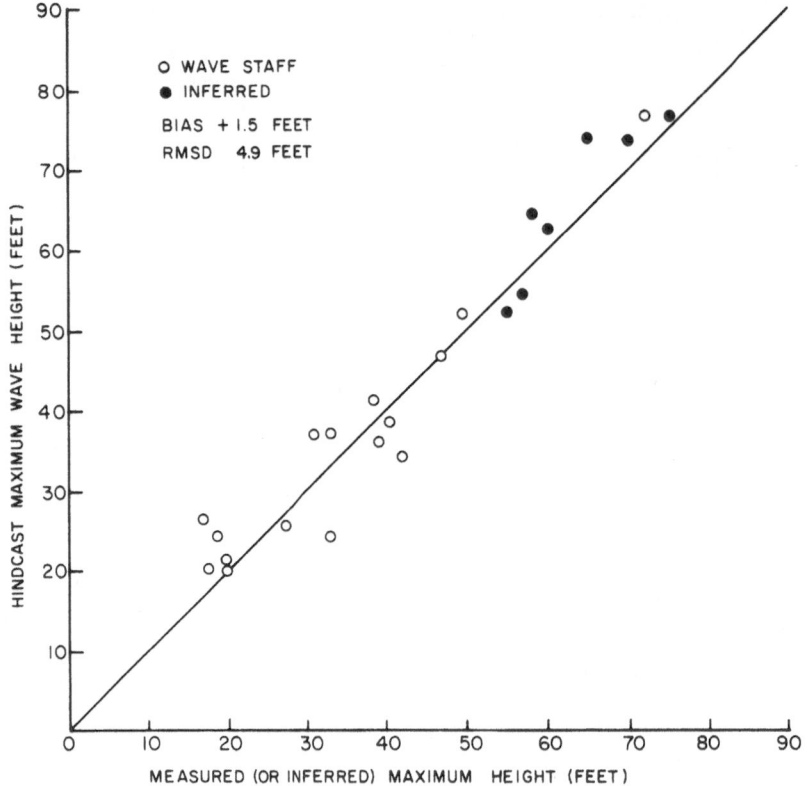

Fig. 5. Comparison of measured or inferred maximum wave heights in Gulf of Mexico hurricanes and hindcasts made with the model of *Cardone et al.*, 1976. Maximum wave heights were inferred by damage inspections of oil platforms. The root-mean-square difference (RMSD) is 4.9 feet and the bias is + 1.5 feet.

Intercomparison of Discrete and Parametric Spectral Models

Since it has only recently been proposed, the parametric approach has not been extensively tested. The model developed by *Weare and Worthington* (1978) was tested on the basis of hindcasts of 14 storms in the North Sea for which wave data was available from weather ships and oil production platforms. After some tuning of the parametric part of the model and revision of the sea-swell transfer algorithm, the model achieved root-mean-square differences of about 1 meter in significant height and insignificant bias.

Two studies are currently underway involving intercomparison of spectral and parametric models. In one, the wind fields used in the North Sea study will be used to drive a discrete spectral

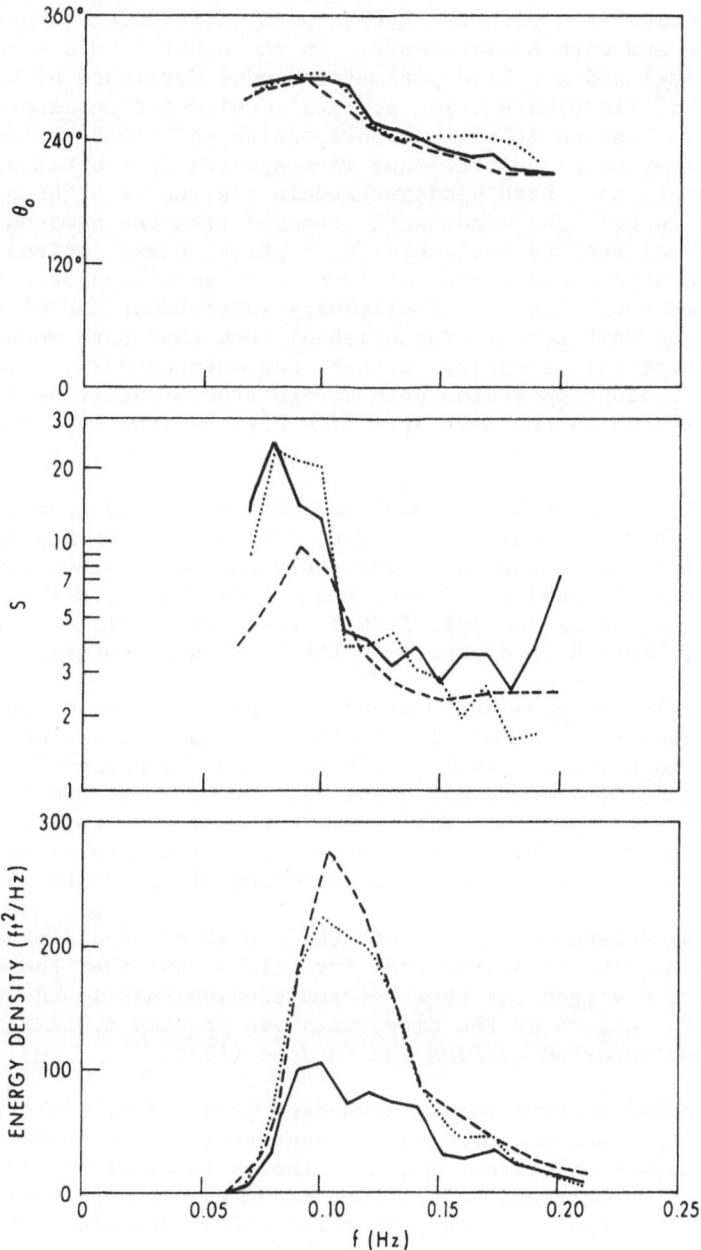

Fig. 6. Comparison of spectra measured at 1200 CDT (——) and 1300 CDT (••••) in hurricane Delia, 1973, and hindcast directional spectrum (– –) from *Forristall et al.* (1978). The measured results (••••) should be compared to the hindcast results (– –) due to the distance between the model grid point and the measurement location.

model for comparison with the hybrid-parametric model hindcasts noted above and with measurements. In the other, the discrete spectral model and a hybrid parametric model developed by *Gunther and Rosenthal* (1978) are being applied to hindcast sea states over the tropical Eastern Atlantic, where during GATE (GARP Atlantic Tropical Experiment), directional wave spectra were obtained at several locations. Both hindcast models are run with the same wind field input. The winds were prepared from the numerous meteorological reports available for a period characterized by complex sea states consisting of local wind-generated seas plus at least two swell trains. Preliminary intercomparison of hindcasts for the GATE series (unpublished) show that both methods provided comparable accuracy, though, not unexpectedly, grid domain limitations prevented both models from hindcasting swell which propagated to the GATE area from high latitudes of both hemispheres.

A simplified parametric model has been developed specifically for hurricane wave prediction by *Ross* (1976). The method involves the specification of the shape and scale parameters, non-dimensional peak frequency, ν, and non-dimensional total energy, $\hat{E}$ (= Eg/U^4, where E is the total energy), from a non-dimensional fetch parameter $\hat{R}$ (= gR/U^2, where R is distance to the hurricane center).

The simplified parametric model and the hurricane discrete spectral model (*Cardone et al.,* 1976b) were used to forecast wave conditions in hurricane Belle, 1976, a storm which passed over two NOAA Data Buoy Office weather buoys instrumented to record one-dimensional wave spectra. The two models were also applied to the hindcast of waves associated with severe historical storms in the Gulf of Mexico and along the Atlantic Coast of the United States.

Those hindcasts suggested that the two models could produce equivalent results in storms that move slowly but that the hindcast results diverged for those storms studied that moved very rapidly. The nature of the divergence was studied systematically in the study reported by *Ross and Cardone* (1978).

Such marked differences in hindcasts produced by different models for the same meteorological input reflect the quasi-empirical nature of current models. The validity of models should therefore be established in detailed hindcast studies involving wave data. The spectral and parametric models described above have been applied in studies of hurricanes Eloise, 1975, and Belle, 1976. In both storms, a wealth of meteorological data was available to describe the storm track and structure, while surface wind and wave measurements were available to verify the modelled wind fields and hindcast waves. The surface wind measurements and wave spectral measurements were made from NOAA Data Buoy Office buoys positioned in the Gulf of Mexico and off the U.S. East Coast.

While hurricanes represent a special class of meteorological
system, they may provide a rather general and critical test of wave
prediction methods. The range of wind speeds and wave numbers
excited by such storms exercise model parameterizations over a
broad range. Also, the wind field and wave generating regime are
far removed from the simple regimes (e.g., fetch-limited, fully-
developed) in which models are developed and tuned, a characteristic
noted recently by *Hasselmann* (1978) to be an important attribute for
model validation.

The Eloise hindcast is illustrated in Figures 7 through 11.
Figure 7 shows the observed storm track and the locations of buoys

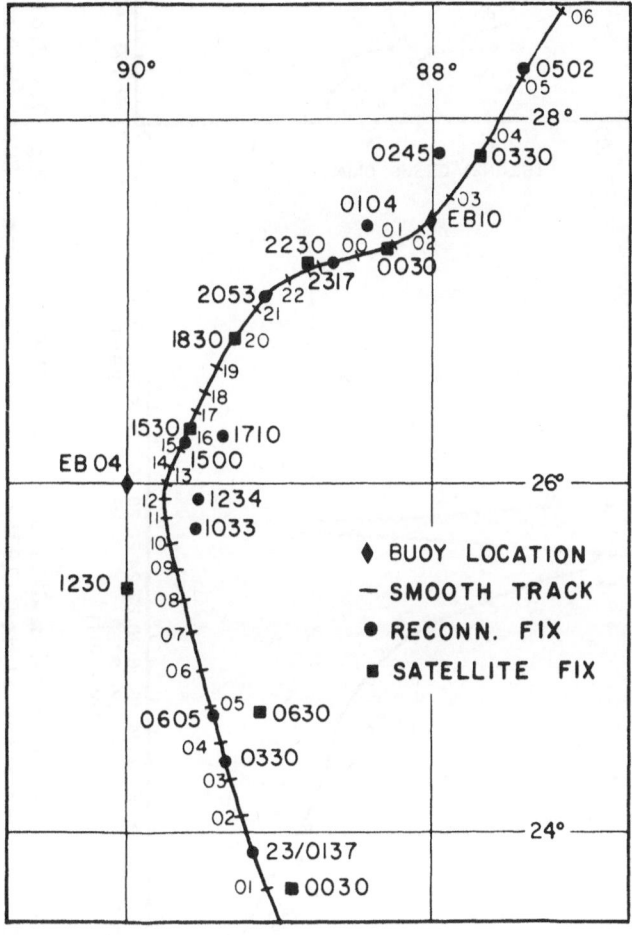

Fig. 7. Track of hurricane Eloise, 1975, in Gulf of Mexico
and locations of NDBO buoys EB04 and EB10. Aircraft reconnaissance
and satellite fixes are shown.

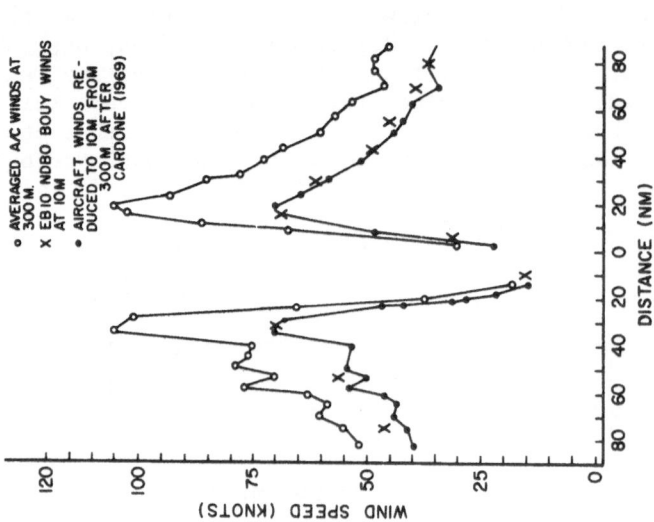

Fig. 9. Eloise EB10 wind speed comparison. NOAA aircraft winds reduced to the surface using planetary boundary layer theory compared to observations from NDBO buoy EB10 during hurricane Eloise.

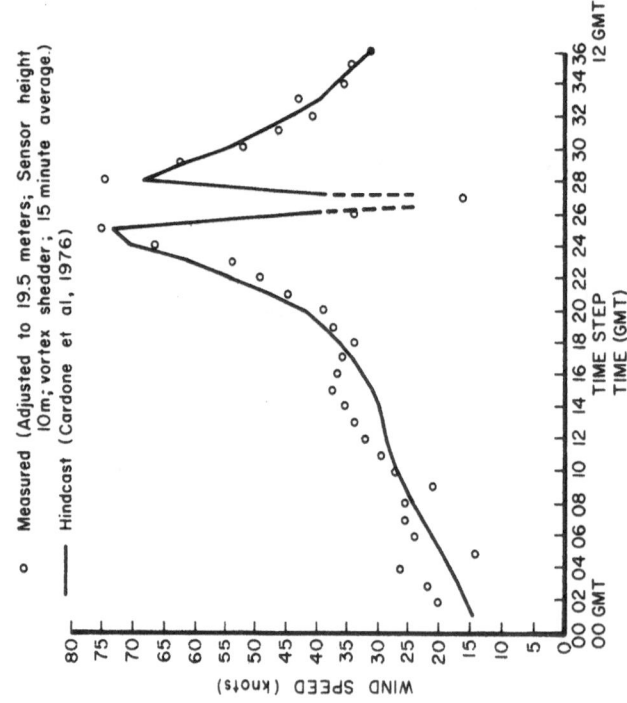

Fig. 8. Measured and modelled winds at NDBO buoy EB10 in hurricane Eloise, 1975.

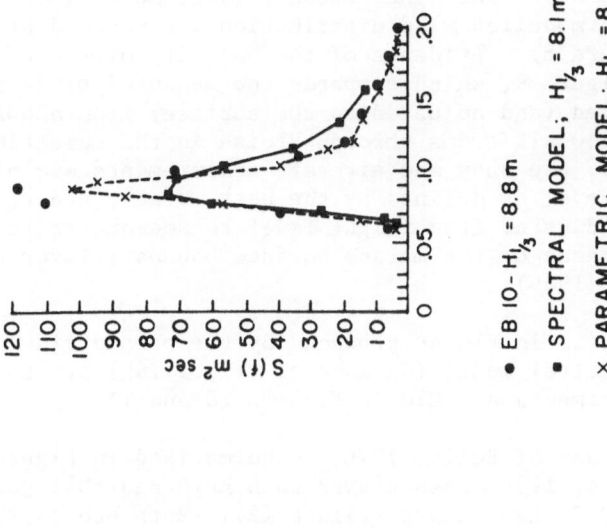

• EB 10- H₁/₃ = 8.8 m

■ SPECTRAL MODEL; $H_{1/3}$ = 8.1 m

× PARAMETRIC MODEL; $H_{1/3}$ = 8.2 m

Fig. 11. Comparison of peak measured spectrum (●) at NDBO buoy EB10 in hurricane Eloise to peak spectrum hindcast by special (■) and parametric (X) models.

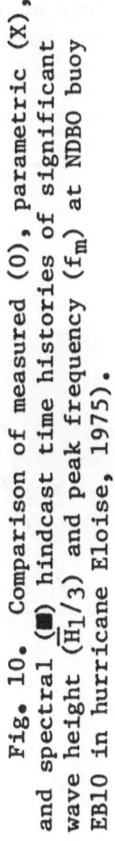

Fig. 10. Comparison of measured (0), parametric (X), and spectral (■) hindcast time histories of significant wave height (H₁/3) and peak frequency (fₘ) at NDBO buoy EB10 in hurricane Eloise, 1975).

EB04 and EB10. Information on the distribution of sea level
pressure about Eloise (and its time variation) and storm track was
used to initialize a hurricane wind boundary layer model (*Cardone
et al.*, 1976b). The modelled wind distribution was checked at the
buoy locations (Figure 8). Evidence of the validity of the buoy
winds is given in Figure 9, which compares the measured winds at
EB10 to winds measured (and adjusted to the surface) from a NOAA
aircraft flying at low altitudes through Eloise in the same time
period. In Figure 9, the buoy and aircraft measurements are plotted
with respect to the eye, as defined by the best track, and the
aircraft winds are adjusted from flight level to anemometer level
logarithmically by means of the marine surface boundary layer model
proposed by *Cardone* (1969).

The wave hindcasts in Eloise produced by the parametric model
(*Ross*, 1976) and spectral model (*Cardone et al.*, 1976b) are compared to wave measurements at EB10 in Figures 10 and 11.

The hindcast study of Belle, 1976, is summarized in Figures
12 through 16. Belle, 1976, passed over both EB15 and EB41 during
its journey up the U.S. East Coast (Figure 12). Both buoys provided
wind measurements for verification of the modelled wind field and

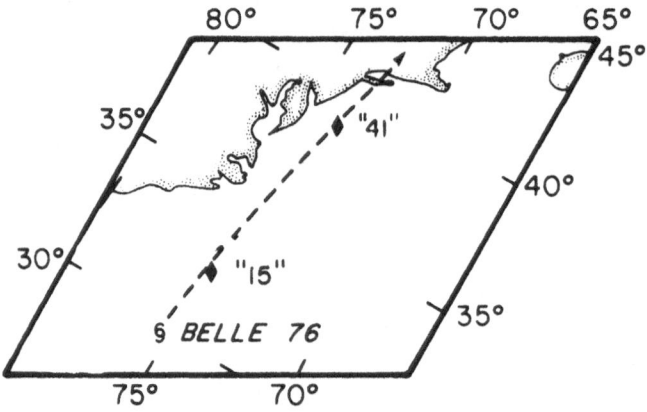

Fig. 12. Track of hurricane Belle, 1976, off U.S. East
Coast and locations of NDBO buoys EB15 and EB41.

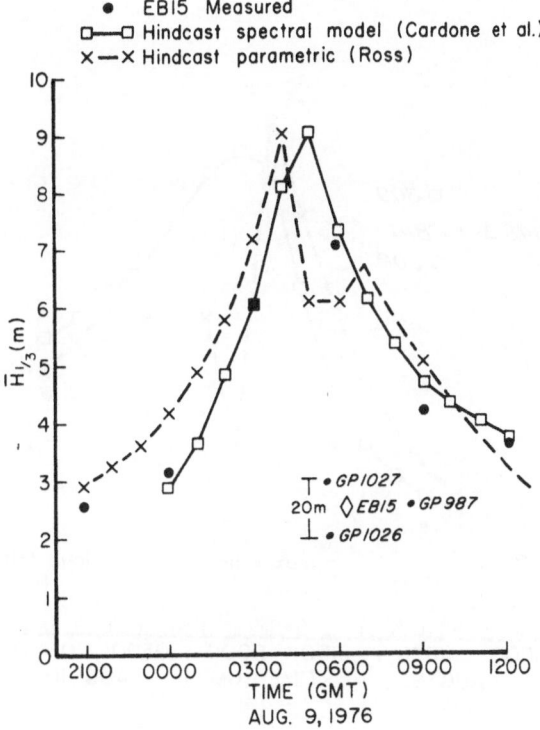

Fig. 13. Comparison of measured (•) parametric (X), and spectral (□) hindcasts of significant wave height at NDBO buoy EB15 in hurricane Belle, 1976. The models of *Cardone et al.*(1976b) and *Ross* (1976) were used for the hindcasts. Also shown are the spectral wave model grid points (separated by 20 miles) at which the spectral hindcast results were obtained.

wave measurements for verification of hindcast wave heights (Figures 13 and 14) and wave spectra (Figures 15 and 16) at the locations of the buoys.

In general, both models produced quite accurate wave hindcasts in both hurricanes, though the spectral model appeared to better represent the overall wave height distribution and spectral shapes. It should be noted that the spectral model used was calibrated on the basis of earlier storms, while the Eloise data set was used in the calibration of the parametric model.

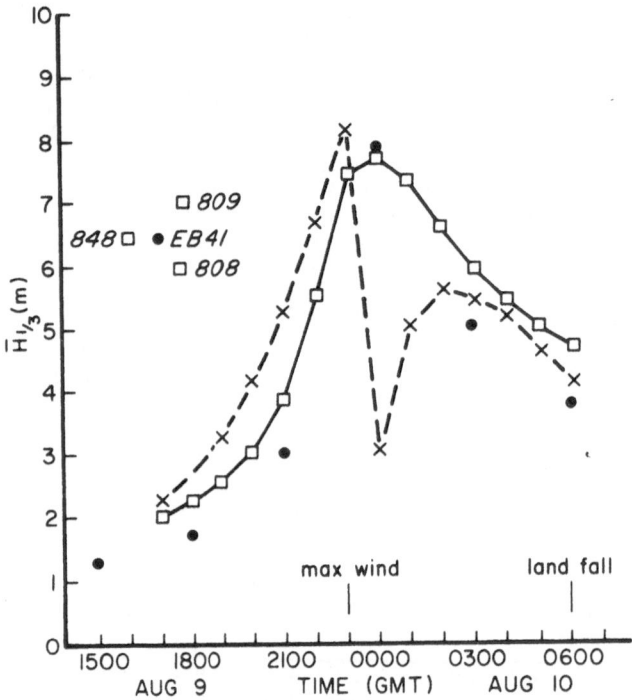

Fig. 14. Comparison of measured (•), parametric (X), and spectral (□) hindcast of significant wave height at NDBO buoy EB41 in hurricane Belle, 1976. The models of *Cardone et al.* (1976b) and *Ross* (1976) were used for the hindcasts. Also shown are the spectral wave model grid points at which the spectral hindcast results were obtained.

DATA REQUIREMENTS AND FUTURE DIRECTIONS

There has been a considerable expansion in the wave data base in the past several years. The new sources include operational programs such as the NOAA Data Buoy Office data buoys, with the potential to obtain tens of thousands of spectra yearly, experimental *in situ* programs such as the oceanographic program of the GARP Atlantic Tropical Experiment, the Ocean Data Gathering Program and other industry-sponsored measurement programs, and airborne and satellite remote sensing programs. While the quality and the nature of the data vary greatly, all such measurements can be utilized within the context of ocean wave prediction models.

Ocean wave models of the discrete type produce hindcast data in terms of directional spectra, from which can be derived frequency

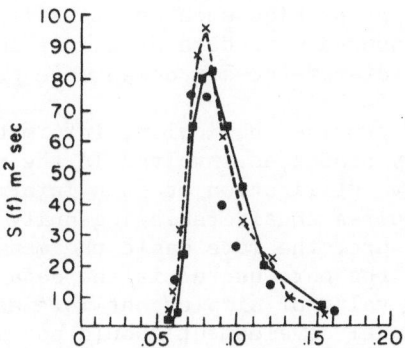

Fig. 15. Comparison of peak measured (●), spectral (■), and parametric (X) hindcast spectra at NDBO buoy EB15 in hurricane Belle, 1976.

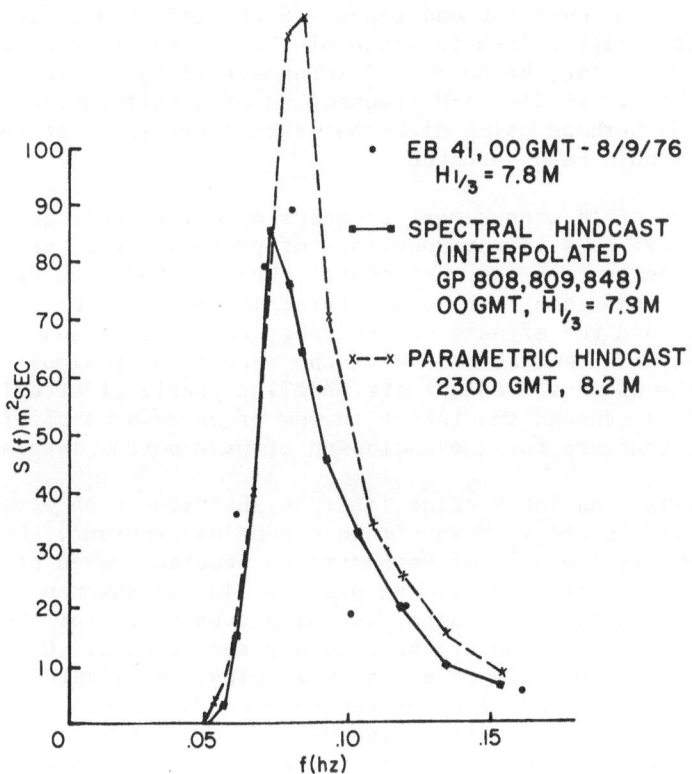

Fig. 16. Comparison of peak measured (●), spectral (■), and parametric (X) hindcast spectra at NEBO buoy EB41 in hurricane Belle, 1976.

spectra, and integral properties such as significant wave height
and period. Thus, measured wave data of almost any type have some
value in validating state-of-the-art ocean wave prediction models.

Wave measurements provide, basically, information that reflects
the net effect of many processes involved in the spectral energy
balance. The effective utilization of such information to improve
ocean wave models requires considerable ingenuity on the part of
the modeller. To be sure, the more basic the measurement (e.g.,
directional spectra), the more useful is the data for model development. However, the value of simpler but more extensive measurements to ocean wave model development should not be underestimated.
Intensive field programs involving sophisticated and delicate
instrumentation, deployed in a limited geographic area, provide
information usually over a small range of conditions. On the other
hand, the NOAA Data Buoy Office buoys in the Gulf of Alaska, Gulf
of Mexico, and Atlantic continental shelf can provide wave measurements in extratropical and tropical cyclones and other meteorological
scenarios that challenge hindcast models to exercise model
parameterizations over a broad range. Such tests can at least
identify critical failings in state-of-the-art models and perhaps
provide guidance for the design of intensive field programs. Basic
system capability studies and frequent *in situ* calibration of buoy
wave sensors, perhaps using airborne laser overflights at regular
intervals, should be undertaken.

Intensive field measurement programs appear to be required to
help fill large gaps in our knowledge of processes that are particularly important in the continental shelf region such as: wave
energy dissipation due to bottom interaction effects, wave-current
interaction, and the effects of atmospheric stability and wind
gustiness on wave generation. With the exception of simple refraction, the above mechanisms are modelled poorly if at all in
present models, though the recent review of *Shemdin et al.* (1978)
provides a structure for the inclusion of wave-bottom interactions.

There exists an interesting interplay between ocean wave
prediction models and microwave remote sensing programs. On the
one hand, spectral models of demonstrated accuracy, when properly
initialized meteorologically, can provide, in the absence of
in situ ocean surface wave data, useful ground truth for the interpretation and calibration of the remotely sensed data. On the
other hand, since *in situ* wave data will always be limited in
coverage, satellite and airborne remote sensors can provide
global scale data for comprehensive validation of existing models
and for an evaluation of the merits of alternate approaches.

To date, no single hindcast model has been tested extensively
enough over a wide enough range of conditions that it can be

applied generally to the determination of wave climate. Indeed, as suggested by *Hasselmann* (1978), different problems and applications may be best addressed by different numerical models. Thus, discrete spectral, parametric, and hybrid parametric methods will likely continue to be developed over the next several years and refined with more rigorous expressions of the individual physical processes represented in the source function. Objective model intercomparison and validation will be aided by extensive wave data sets which will become available as the data from the national and international programs cited above are processed, as industry data reach the public domain, and as global remotely-sensed wave data become available. The combination of such wave data and wave prediction model application should provide reliable global wave climate specification within the next decade.

REFERENCES

Baer, L. 1962. An experiment in numerical forecasting of deep water ocean waves. Lockheed Missile and Space Company Report LMSC-801296, San Diego, California.

Baer, L., L. C. Adamo, D. P. Hamm, R. F. Erickson, and J. Vermeulen. 1970. The LOCWAVE D-10130 numerical spectral prediction system. *Offshore Technology Conference*, Houston, Texas, paper 1246.

Banner, M. L. and O. M. Phillips. 1974. On the incipient breaking of small scale waves. *J. Fluid Mech*, *65*, 647-656.

Barnett, T. P. and J. C. Wilkerson. 1967. On the generation of wind waves as inferred from airborne measurements of fetch-limited spectra. *J. Mar. Res.*, *25*, 292-328.

Barnett, T. P. 1968. On the generation, dissipation and prediction of ocean wind waves. *J. Geophys. Res.*, *73*, 513-529.

Barnett, T. P. 1969. Wind waves in shallow water. Westinghouse Ocean Research Laboratory Report, San Diego, California.

Bowden, K. F. and R. A. White. 1966. Measurements of the orbital velocities of sea waves and their use in determining the directional spectrum. *Geophys. J. Roy. Astr. Soc.*, *12*, 33-54.

Cardone, V. J. 1969. Specification of the wind field distribution in the marine boundary layer for wave forecasting. Report TR-69-1, Geophys. Sci. Lab., New York University. Available from NTIS AD#702-490.

Cardone, V. J. 1974. Ocean wave predictions: Two decades of progress and future prospects. In: *Seakeeping 1953-1973/ Sponsored by Panel H-7 (Seakeeping Characteristics) at Webb Institute of Naval Architecture, Glen Cove, New York, October 28-29, 1973*, Soc. of Naval Arcitects and Marine Engineers, New York, 5-18.

Cardone, V. J., W. J. Pierson, and E. G. Ward. 1976a. Hindcasting the directional spectra of hurricane generated waves. *J. of Petrol. Technology*, *28*, 385-394.

Cardone, V. J., D. Ross, M. Ahrens, J. A. Greenwood, C. Greenwood, and R. E. Salfi. 1976b. Forecasting hurricane winds and waves - a pilot study. Final report to NOAA Sea-Air Interaction Laboratory and Shell Development Company. CUNY Institute of Marine and Atmospheric Sciences at the City College, New York.

Cardone, V. J. and D. B. Ross. 1977. An experiment in forecasting hurricane generated sea states. Proceedings of the 11th Technical Conference on Hurricanes and Tropical Meteorology, Am. Meteorological Soc., December 13-16, 1977, Miami, Florida, 688-695.

Cote, J., J. O. Davis, W. Marks, R. J. McGough, E. Mehr, W. J. Pierson, J. F. Ropek, G. Stephenson, and R. C. Vetter. 1960. The directional spectrum of a wind generated sea as determined from data obtained by the stereo wave observation project. New York University, College of Engineering, *Meteorological Papers 2: number 6*.

Davis, R. E. 1974. Perturbed turbulent flow, eddy viscosity and the generation of turbulent stresses. *J. Fluid Mech., 63,* 673-693.

Dexter, P. E. 1974. Tests of some programmed numerical wave forecast models. *J. Phys. Oceanogr., 4,* 635-644.

Dobson, F. W. 1971. Measurements of atmospheric pressure on wind generated sea waves. *J. Fluid Mech., 48,* 91-127.

Dobson, F. W., J. A. Elliot, R. B. Long, and R. L. Snyder. 1978. Bight of Abaco Pressure Experiment. To appear in Proceedings of the NATO Symposium on Turbulent Fluxes through the Sea Surface-Wave Dynamics and Prediction, Ile de Bendor, France, 12-16 September 1977, Plenum, New York.

Earle, M. D. 1978. Practical determinations of design wave conditions. In: *Ocean Wave Climate,* M. D. Earle and A. Malahoff (Eds.), Plenum, New York (this volume).

Elliot, J. A. 1972. Microscale pressure fluctuations near waves being generated by the wind. *J. Fluid Mech., 54,* 427-448.

Ewing, J. A. 1971. A numerical wave prediction method for the North Atlantic Ocean. *Deutsche Hydrographische Zeitschrift, 24,* 241-261.

Feldhausen, R. H., S. K. Chakrabarti, and B. W. Wilson. 1973. Comparison of wave hindcasts of weater station "J" for the North Atlantic storm of December 1959. *Deutsche Hydrographische Zeitschrift, 26,* 10-16.

Forristall, G. Z., E. G. Ward, V. J. Cardone, and L. E. Borgman. 1978. The directional spectra and kinematics of surface waves in Tropical Storm Delia. *J. Phys. Oceanogr., 8.*

Fox, M. J. H. 1976. On the non linear transfer of energy in the peak of a gravity wave spectrum II. *Proc. Roy. Soc. London Ser. A., 348,* 467-483.

Freeman, J. C. 1976. A state of art method for estimating extreme conditions in storms. Institute of Storm Research, Houston, Texas.

Gelci, R., H. Cazalé, and J. Vassal. 1957. Prévision de la houle. La Méthode des Densités Spectroangularies, Bulletin d'information du Comité central d'oceanographie et d' études des cotes, 9, 416.

Gent, P. R. and P. A. Taylor. 1976. A numerical model of air flow over waves. *J. Fluid Mech., 77*, 105-128.

Greenwood, J. A. and V. J. Cardone. 1977. Development of a global ocean wave propagation algorithm. Final report to U.S. Navy Fleet Numerical Weather Central, Monterey, California, Contract N-00338-76-C-3081.

Gunther, H. and W. Rosenthal. 1978. Parametrical numerical wave prediction tested in wind situations varying in space and time. To appear in Proceedings of the NATO Symposium on Turbulent Fluxes through the Sea Surface-Wave Dynamics and Prediction, Ile de Bendor, France, 12-16 September 1977, Plenum, New York.

Hasselmann, K. 1962. On the non-linear energy transfer in a gravity wave spectrum - Part 1. *J. Fluid Mech., 12*, 481-500.

Hasselmann, K. 1963. On the non-linear energy transfer in a gravity wave spectrum - Part 2. *J. Fluid Mech., 15*, 273-281; Part 3, *Ibid., 15*, 385-398.

Hasselmann, K., T. P. Barnett, E. Bouws, H. Carlson, D. E. Cartwright, K. Enke, J. A. Ewing, H. Gienapp, D. E. Hasselmann, P. Kruseman, A. Meerburg, P. Müller, D. J. Olbers, K. Richter, W. Sell, and H. Walden. 1973. Measurements of wind-wave growth and swell decay during the Joint North Sea Wave Project (JONSWAP). *Deutsche Hydrographische Zeitschrift, Supplement A8, 12.*

Hasselmann, K. 1974. On the spectral dissipation of ocean waves due to white capping. *Boundary Layer Meteorology 6*, 107-127.

Hasselman, K., D. B. Ross, P. Müller, and W. Sell. 1976. A parametric wave prediction model. *J. Phys. Oceanogr. 6*, 200-228.

Hasselman, K., D. B. Ross, P. Müller, and W. sell. 1977. Reply to "Comments on 'A Parametric Wave Prediction Model,'" by W. J. Pierson. *J. Phys. Oceanogr., 7.*

Hasselmann, K. 1978. Some comments on the physics and numerics of wave-prediction models. To appear in Proceedings of the NATO Symposium on Turbulent Fluxes through the Sea Surface-Wave Dynamics and Prediction, Ile de Bendor, France, 12-16 September 1977, Plenum, New York.

Hsu, F. 1973. Hindcast storm waves for compilation of wave statistics. *Soc. of Petroleum Engineers*, Paper SPE 4323.

Isozaki, I and T. Uji. 1973. Numerical prediction of ocean wind waves, *Papers in Meteorology and Oceanography, 24*, 207-231

Isozaki, I., T. Uji, T. Seto, and A. Kaietsu. 1976. Ocean waves on Sagami Bay caused by Typhoon 7410. *The Oceanographical Magazine, 27*, 1-23.

Lazanoff, S. M. and N. Stevenson. 1975. An evaluation of a hemispheric operational wave spectral model. Fleet Numerical Weather Central Technical Note 75-3, Monterey, California.

Longuet-Higgins, M. S. 1975. On the non-linear transfer of energy in the peak of a gravity wave spectrum. *Proc. Roy. Soc. London Ser. A, 347*, 311.

Miles, J. W. 1957. On the generation of surface waves by shear
     flows. *J. Fluid Mech.*, *3*, 185-204.

Mitsuyasu, H. 1968. On the growth of wind generated waves (I).
     Report Res. Inst. Appl. Mech. Kyushu Univ., *16*, 419-482.

Mitsuyasu, H. and K. Rikiishi. 1978. The growth of duration
     limited wind waves. *J. Fluid Mech.*, *85*, Part 4, 705-730.

Phillips, O. M. 1957. On the generation of waves by turbulent
     wind. *J. Fluid Mech.*, *2*, 417-445.

Phillips, O. M. 1959. The scattering of gravity waves by
     turbulence. *J. Fluid Mech.*, *5*, 177-192.

Phillips, O. M. 1963. On the attenuation of long gravity waves
     by short breaking waves. *J. Fluid Mech.*, *16*, 321-332.

Phillips, O. M. 1977. The Sea Surface. In: *Modelling and
     Prediction of the Upper Layers of the Ocean*, Pergamon Press,
     229-237.

Pierson, W. J. and W. Marks. 1952. The power spectrum analysis
     of ocean wave records, *Trans. Amer. Geophys Union*, *33*, 834-844.

Pierson, W. J., G. Neumann, and R. W. James. 1955. *Practical
     Methods for Observing and Forecasting Ocean Waves by Means
     of Wave Spectra and Statistics*. H. O. Publication 603, U.S.
     Naval Oceanographic Office, Washington, D.C.

Pierson, W. J., L. J. Tick, and L. Baer. 1966. Computer based
     procedures for preparing global wave forecasts and wind field
     analyses capable of using wave data obtained by a spacecraft.
     In: *Sixth Naval Hydrodynamics Symposium*, Office of Naval
     Research, Washington, D.C., 499-532.

Pierson, W. J. 1977. Comments on "A parametric wave prediction
     model," *J. Phys. Oceanogr.*, *7*, 127-134.

Resio, D. T. and C. L. Vincent. 1977. A numerical hindcast model
     for wave spectra on water bodies with irregular shoreline
     geometry, Part I: Test of non-dimensional growth rates. Misc.
     Paper H-77-9, Hydraulic Laboratory, U.S. Army Engineer Waterways
     Experiment Station, Vicksburg, Mississippi.

Ross, D. B., V. J. Cardone, and J. W. Conaway. 1971. Laser and
     microwave observations of sea surface conditions for fetch
     limited 17 to 25 m/s winds. *IEEE Trans. on Geoscience Elec-
     tronics*, *GE-8*, 326-336.

Ross, D. B. and V. J. Cardone. 1974. Observations of oceanic
     whitecaps and their relation to remote measurements of surface
     wind speed. *J. Geophys. Res.*, *79*, 444-452.

Ross, D. B. 1976. A simplified model for forecasting hurricane
     generated waves. (Abstract). Bull. Am. Meteorological Soc.,
     January, 113. Presented at conferences on Atmospheric and
     Oceanic Waves, Seattle, Washington, March 29-April 2.

Ross, D. B. and V. J. Cardone. 1978. A comparison of parametric
     and spectral hurricane wave prediction products. To appear in
     Proceedings of the NATO Symposium on Turbulent Fluxes through
     the Sea Surface-Wave Dynamics and Prediction, Ile de Bendor,
     France, 12-16 September 1977, Plenum, New York.

Salfi, R. E. and W. J. Pierson. 1977. A verification study of hindcasted wave spectra based on wave data from weather ships J and K and wind data obtained for December 1973 and January 1974 during Skylab. Final report to Spacecraft Oceanography Project, National Environmental Satellite Service, National Oceanic and Atmospheric Administration, Washington, D.C.

Schule, J. J., L. S. Simpson, and P. S. Deleonibus. 1971. A study of fetch-limited wave spectra with an airborne laser. *J. Geophys. Res., 76,* 4160-4171.

Sell, W. and K. Hasselmann. 1972. Computations of nonlinear energy transfer for JONSWAP and empirical wind-wave spectra. Report, Institute of Geophysics, University of Hamburg.

Shemdin, O., K. Hasselmann, S. V. Hsiao, and K. Herterich. 1978. Non-linear and linear bottom interaction effects in shallow water. To appear in Proceedings of the NATO Symposium on Turbulent Fluxes through the Sea Surface-Wave Dynamics and Prediction, Ile de Bendor, France, 12-16 September 1977, Plenum, New York.

Snyder, R. and C. S. Cox. 1966. A field study of the wind generation of ocean waves. *J. Mar. Res., 24,* 141-177.

Snyder, R. L. 1974. A field study of wave induced pressure fluctuations above surface gravity waves. *J. Mar. Res., 32,* 497-531.

Sverdrup, H. V. and W. H. Munk. 1947. *Wind, Sea and Swell: Theory of Relation for Forecasting.* H. O. Publication 601, U.S. Naval Oceanographic Office, Washington, D.C.

Uji, T. 1975. Numerical estimation of the sea wave in a typhoon area. *Papers in Meteorology and Geophysics, 26,* 199-217.

Ward, E. G. 1974. Ocean data gathering program - An overview. Offshore Technology Conference, Houston, Texas, paper 2108B.

Ward, E. G., L. E. Borgman, and V. J. Cardone. 1978. Statistics of hurricane waves in the Gulf of Mexico. Offshore Technology Conference, Houston, Texas, paper 3229.

Weare, J. and B. A. Worthington. 1978. A numerical model hindcast of severe wave conditions for the North Sea. To appear in Proceedings of the NATO Symposium on Turbulent Fluxes through the Sea Surface-Wave Dynamics and Prediction, Ile de Bendor, France, 12-16 September 1977, Plenum, New York.

Webb, D. J. 1978. The wave-wave interaction machine. To appear in Proceedings of the NATO Symposium on Turbulent Fluxes through the Sea Surface-Wave Dynamics and Prediction, Ile de Bendor, France 12-16 September 1977, Plenum, New York.

Wilson, B. W. 1961. Deepwater wave generation by moving wind systems. *J. Waterways, Harbors and Coastal Engineering Div.,* Am. Soc. Civ. Engineers, WW2, 113-141.

Wave Data Requirements for Ship Design and Operation

Susan Lee Bales and William E. Cummins

David W. Taylor Naval Ship Research

and Development Center

ABSTRACT

In order to accomplish its mission, any naval combatant must be able to withstand both the expected tactical loading and the occurring state of the natural environment. Overall ship as well as individual subsystem performance can be degraded (or enhanced) considerably by the state of the natural environment. This paper outlines the current practice for modelling the wave environment in naval ship design. Strengths and weaknesses of the Bretschneider wave spectrum are identified and ongoing work to improve both wave model and statistical data base quality is briefly described. A major part of this work is the development of a twenty year directional wave spectra climatology for the Northern Hemisphere.

INTRODUCTION

During the past few years, the ship design and engineering communities have paid an increasing amount of attention to the specification of the wave environment in which ships are expected to operate. The reasons for this are twofold. First of all, more wave data and wave models are being made available. Second, the need for using realistic seaway representations has been made apparent by the critical review of earlier procedures which in some cases biased the conclusions that guided actual ship design. In this regard, realistic seaways are those which permit the prediction of realistic ranges of ship responses.

A model of the seaway is considered to be an essential
element in any ship performance assessment, either analytic or
experimental. Thus, the critical decision to be made in embarking
on such a task is to decide which idealized wave spectral form
(or group of measured wave spectra) to use. Another essential
element is a knowledge of the probability distribution of the
important wave parameters for the operational area of concern.
A third essential element is the means or technology by which
to measure the ship's performance. As there is rarely time to
develop any of these requirements during the design of a single
ship, it is important that the necessary tools be developed a
priori and made ready for use by the designer on demand.

Work in all three of these areas has been actively going on
for the past several years as a result of the Navy's Seakeeping
Research and Development Program, first formally discussed in
the report of the *Seakeeping Workshop* (1975). The first two
areas, however, are of more relevance to this paper and are
briefly outlined. The status of recent achievements as well as
future goals are included in this discussion.

WAVE SPECTRA

The wave spectrum concept was first introduced to the ship
design community twenty-five years ago by *St. Denis and Pierson*
(1953). Since then, it has been most common to use idealized
wave spectra (as opposed to measured spectra) in analytic
studies of ship performance due to their inherent simplicity
and the resultant ease of calculation. Though several different
idealized forms have been used, *Bales* (1978) indicates that
currently the most widely used spectrum in Navy design work is
that developed by *Bretschneider* (1959). This spectrum can be
written in the form

$$S_\zeta(\omega) = A\,\omega^{-5}\,\exp\,[-B/\omega^4] \qquad ft^2\,sec$$

$$\text{where} \qquad A = 483.5\,(\tilde{\zeta}_w)_{\frac{1}{3}}{}^2/T_o{}^4 \qquad ft^2/sec^4 \qquad (1)$$

$$\text{and} \qquad B = 1944.5/T_o{}^4 \qquad 1/sec^4$$

This spectral form is defined here, for convenience, in terms of
a significant wave height, $(\tilde{\zeta}_w)_{\frac{1}{3}}$, and the modal or peak

period, T_0, and is a measure of the wave elevation at a point only.
Thus, no information as to the directional distribution of wave
origin is given, and it is usual to assume that all of the wave
energy comes from one direction -- a conservative assumption is
most but not all cases. The use of this spectral form has been
shown to produce generally realistic ranges of ship responses
(when compared with those predicted using measured spectra) by
Baitis, Bales, and Meyers (1974) and *Ochi and Bales* (1977) for
a variety of conventional displacement ships and a semi-submersible
platform. Further, though derived from measurements taken only
in the North Atlantic basin, recent results by *Ochi and Bales*
(1977) indicate the applicability of the form to other operating
areas in the world ocean.

    Though the use of the Bretschneider spectrum is widespread,
several items are of concern with regard to the shape of the
spectrum, for example its width or broadness and its "single-
peakedness." The exclusion of secondary peaks in the
Bretschneider formulation is of major concern. It has been well
illustrated by *Hoffman and Miles* (1976) that many measured open-
ocean spectra, for example for moderate to high sea conditions,
have at least two energy peaks corresponding to a local wind sea
and one or more swells of distant origin. For purposes of
conventional ship performance evaluations, this omission may be
important. For example, a secondary peak near some natural period
of some ship response mode may cause a much larger response than
for the primary peak. In roll, a large ship will detect and
respond to a long swell which may be virtually hidden to the
observer's eye by a local wind sea of much larger amplitude.
Another area where the use of single-peaked Bretschneider spectra
could potentially produce misleading results is in the case of
certain high performance craft which have multi-peaked response
amplitude operators (responses of vessel to wave of given height
and length). This usage could lead to poorly designed craft and
could bias results rather favorably for the advanced craft when
compared to a conventional displacement monohull during design
studies.

    Four courses of action are required to address this problem.
First, a comprehensive comparison of ship responses computed
using both Bretschneider and measured spectra should be conducted
for a variety of conventional and advanced ships. The responses
to the measured spectra could be used as a guide or base for
evaluating the validity of the responses to the Bretschneider
spectra for all sea severities. Work recently completed of
this nature for some conventional ships and a semi-submersible
platform by *Ochi and Bales* (1977) indicates that when a statistically derived range of Bretschneider spectra (varying modal

period) is used for each sea severity (significant wave height), agreement with results for the measured spectra is generally good, though calculated results may be somewhat low for wave heights of 15 feet or less. Figure 1, a typical result presented by *Ochi and Bales* (1977), shows probable extreme values of

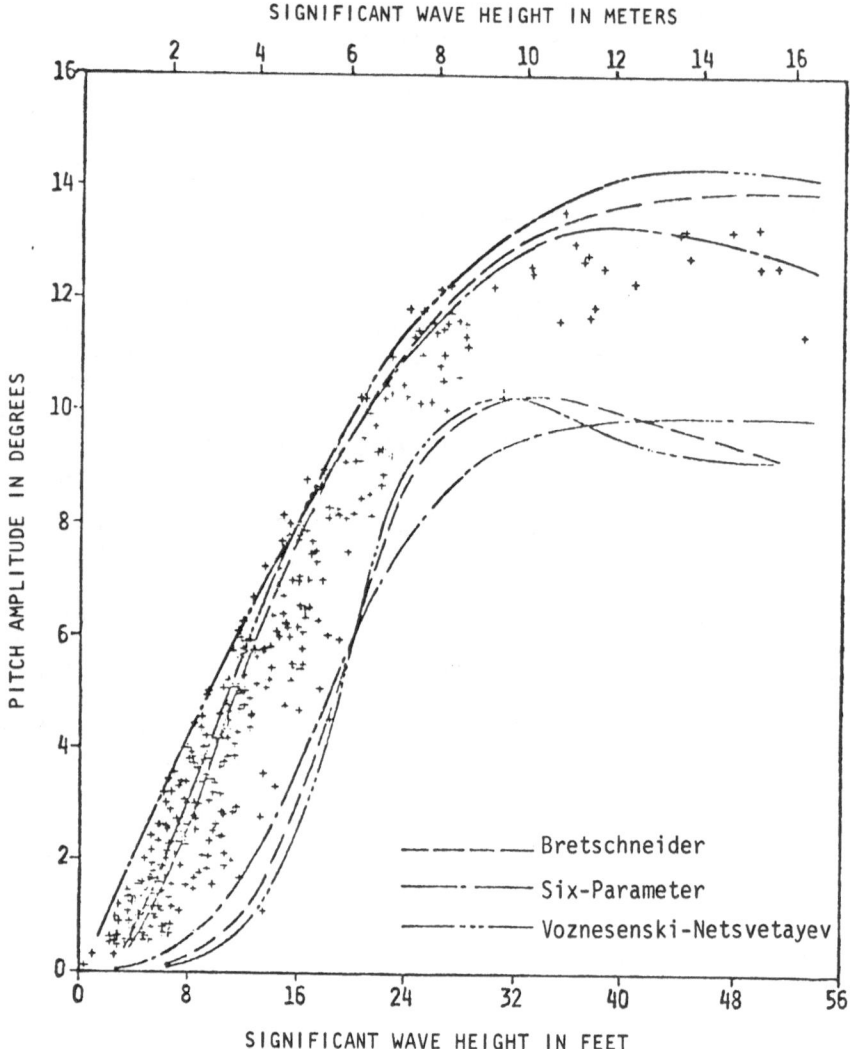

Fig. 1. MARINER probable extreme pitch amplitudes in head seas for measured spectra in North Atlantic and bounds of wave spectral formulations (from *Ochi and Bales*, 1977). The crosses represent predicted pitch motions based on data and the curves indicate 95% confidence limits for three spectral formulations.

the pitch motion of a conventional displacement ship using both
measured and idealized spectra. The plus marks on the figure
represent predicted pitch motions using the measured spectra
reported by *Hoffman and Miles* (1976) for Station India in the
North Atlantic. The curves indicating upper and lower bounds of
extreme pitch amplitudes were predicted using a confidence
coefficient of 0.95 with three idealized spectral types, for
example the Bretschneider, the six-parameter, which is discussed
subsequently, and the Voznesenski-Netsvetayev, a two-parameter
form which produces similar ship response predictions as the
Bretschneider spectral form.

     A second course of action that is required is the develop-
ment and validation of a spectral form which includes swell
(or secondary peaks) in its formulation. The six-parameter
spectrum developed by *Ochi and Hubble* (1976) provides one such
spectrum derived by the decomposition of the spectrum into
two parts, each expressed by a mathematical formula with the
three-parameters significant wave height, modal frequency,
and shape parameter. The six-parameter spectral family, derived
essentially by a statistical analysis of 800 spectra measured
in the North Atlantic, was also used in the ship response prediction analysis conducted by *Ochi and Bales* (1977). Though
Figure 1 shows no conclusive evidence to any advantage gained by
using the six-parameter spectrum, it is considered that the
inclusion of secondary wave peaks may become more necessary in
design problems relating to advanced naval craft as well as in
the area of providing real-time operational guidelines to ships
at sea.

     A third course of action that is needed is the development
of a knowledge of the likelihood or frequency of occurrence of
secondary wave peaks in nature. This item is addressed in the
next section of the paper along with the occurrence of all the
major wave parameters which have an influence on naval ships or
marine structures.

     Finally, the directional nature of wave spectra must be taken
into account. The use of uni-directional seas in performance predictions is known to sometimes bias design conclusions. For
example, as is pointed out by *Cox and Lloyd* (1977), the use of
uni-directional or long crested seas in the sizing of anti-roll
fins for conventional displacement ships, will result in un-
realistically large devices. A complicating factor is the
relatively wide directional distribution of wind seas and the
usually much narrower directional distribution of swells. At
present, very little data exist for studying this problem though
ongoing work, described in a subsequent section of this paper,
may relieve this situation.

WAVE STATISTICS

In order to attach any realism to the use of an idealized
wave spectrum such as the Bretschneider form described previously,
it is necessary to have some idea of the frequency of occurrence
of its important defining parameters such as the significant wave
height and the modal wave period. At present the best available
source of such data, from the viewpoint of naval operational
areas, is that found in *Hogben and Lumb* (1967) which is based on
estimates of sea conditions made from ships during the period
spanning 1953 to 1961. *Bates* (1978) describes the useage of these
data in ship design work and provides Figure 2 to identify the
number of observations included for each operational area.

Several advantages to the use of the *Hogben and Lumb* (1967)
wave statistics are that, with the exception of the North Pacific,
world-wide coverage of the more important Naval operational areas
is provided, and that the ranges of both height and period are
sufficiently defined to address resonant conditions for most
conventional ships. A disadvantage to the use of these data
is that due to a fair weather bias, for example ships try to
avoid stormy areas, the observations do not necessarily reflect
extreme occurrences. Further, the observations have usually been
made by relatively untrained observers, and also, as Figure 2
indicates, the observations definitely reflect trade routes of
British ships so that unweighted cumulation of statistics over
several areas may produce geographic biases. Regardless of
these possible difficulties, the *Hogben and Lumb* (1967) data
are currently the most widely used in the Navy's ship performance
evaluations described by *Bales* (1978).

Though the *Hogben and Lumb* (1967) data are broken into
distribution of height and period for each of twelve directions,
the reporting procedure does not permit the inclusion of
secondary wave trains (or directions) and it is thus impossible
to clearly separate sea and swell conditions. Additionally,
the preselected and fixed seasons, though useful in deriving
operational statistics, may camouflage the extreme values of
interest in some design problems.

A typical example of the usefulness of wave statistics is
given in Figure 3, adopted from *N. Bales* (1977). Observed wave
data, such as that reported by *Hogben and Lumb* (1967), for the
region and season of interest (in this case a coastal area of
the North Atlantic) are reduced to obtain contours of the joint
probability of occurrence of significant wave height, $(\tilde{\zeta}_w)_{1/3}$
and modal wave period, T_o. The contours shown are for percent
frequencies of occurrence of 1 percent and 0.1 percent. Of the

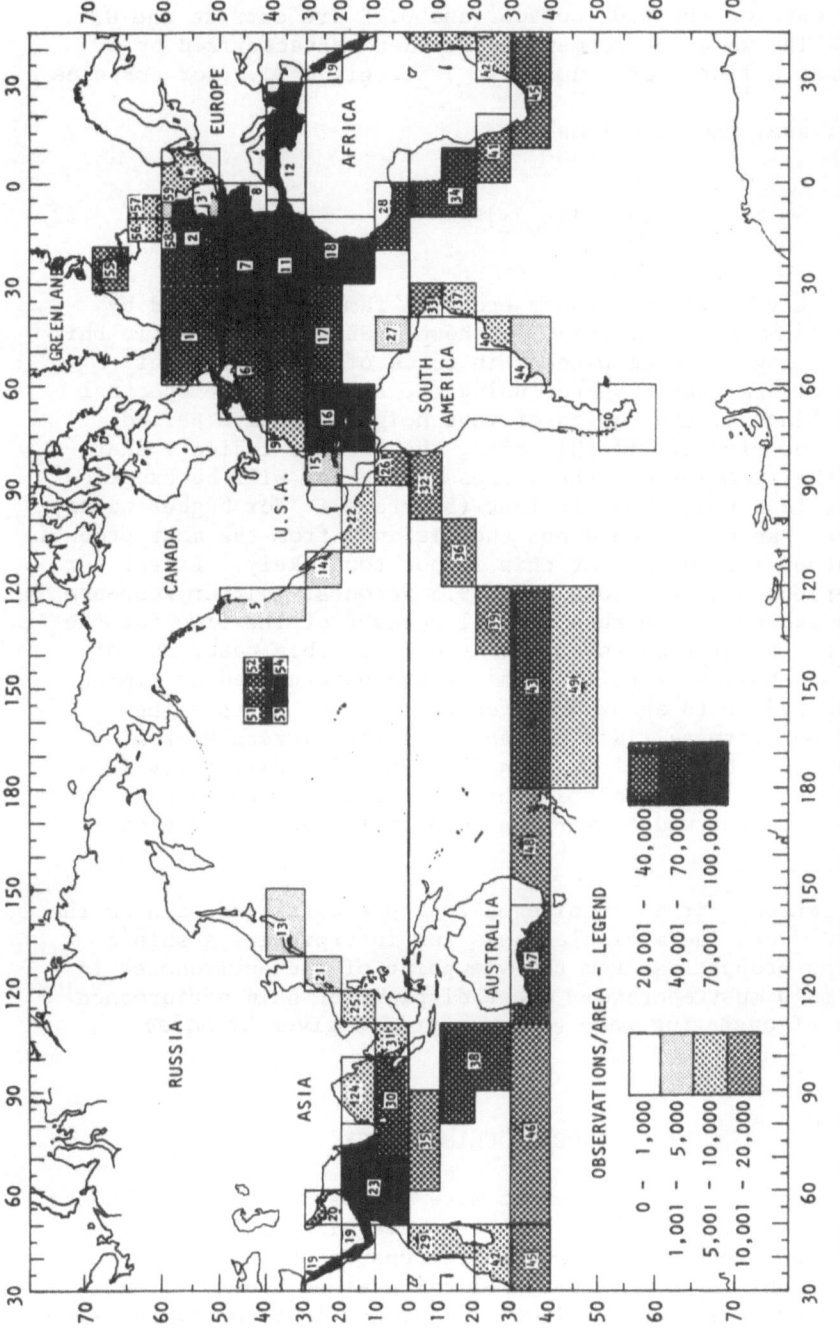

Fig. 2. Location and density of *Hogben and Lumb* (1967) observed wave conditions (from *Bales*, 1978). While the geographical coverage is not ideal, coverage of the more important Naval operational areas, with the exception of the North Pacific is provided.

observed significant wave height and modal wave period pairs,
1.0% are outside the 1.0% contour and 0.1% are outside the 0.1%
contour. The wave environment is further characterized by a
most probable line, for example $(\tilde{\zeta}_w)_{\frac{1}{3}}$ given T_o. For purposes

of comparison, the relationship

$$T_o = 2.76 \; [(\tilde{\zeta}_w)_{\frac{1}{3}}]^{1/2} \tag{2}$$

for the fully developed seaway spectral family, developed by
Pierson and Moskowitz (1964), is shown. Ship responses (in this
case, slamming and deck wetness in terms of probability of
occurrence) are then computed using the Bretschneider spectral
family defined by the ranges of wave height and wave period
shown in the figure. In this case, the results indicate that
neither the slamming nor the wetness criterion will be exceeded
for waves less than about 16 feet (5 meters). For higher waves,
the limits can be exceeded but the distance from the most probable
wave contour indicates that this is not too likely. Except for
modal periods between about 6 and 9.5 seconds, neither response
limit is exceeded more than about 1 percent of the time for the
ship operating at 6 knots in head seas. In this case, use of
the Pierson-Moskowitz relationship would have caused no limits
to be reached up to about 32.8 feet. That is, neither the
slamming nor wetness limit curves cross the Pierson-Moskowitz
curve for wave heights less than 32.8 ft. The difficulty in
applying such a fixed relationship in studies of performance
assessment in the widely varying potential wave environment is
evident.

In general, it is considered that presentations such as that
of Figure 3 provide a viable basis for interpreting a ship's
seakeeping properties from the viewpoint of the environment in
which a ship must operate. Other displays of ship performance
in terms of operating wave environment are given by *Bales*
(1978).

HINDCAST CLIMATOLOGY

The need for a date base of wave statistics tailored to the
needs of the naval engineer has been recognized and an ongoing
cooperative effort between several groups in the Navy is attempting
to address this need. Fleet Numerical Weather Central (FNWC) is
currently using its Spectral Ocean Wave Model (SOWM) to provide
a twenty year data set of directional wave spectra that are

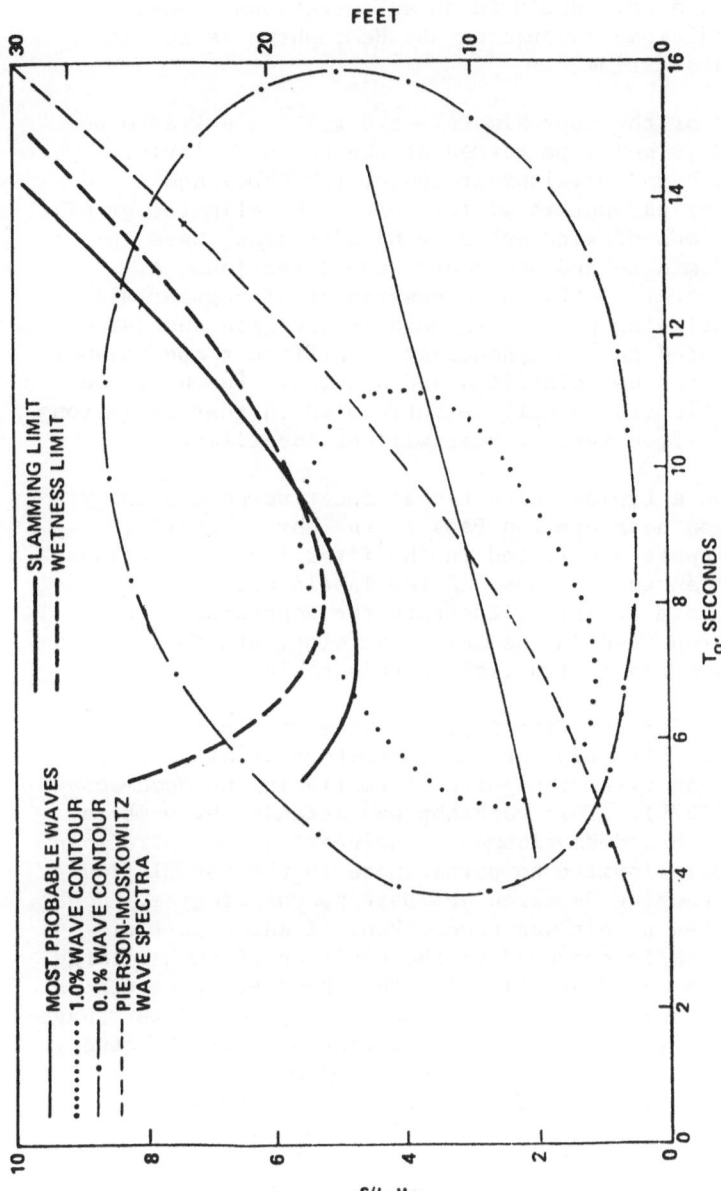

Fig. 3. Operational limits of medium endurance Coast Guard cutter at 6 knots in head seas for coastal North Atlantic wave conditions (from *Bales*, 1977). As described in the text, use of the Bretschneider spectrum indicates that neither the slamming nor the wetness limit will be exceeded for waves less than about 5 meters and use of the probable wave contours indicates that these limits will be seldom exceeded. Interpretation based on this type of analysis are considerably different than those based only on a Pierson–Moskowitz spectrum.

hindcast for over 2000 locations at from 40 to 80 nautical mile
intervals in the Northern Hemisphere at time steps of six hours.
This wave model, described by *Lazanoff and Stevenson* (1975),
is the result of the work of W. Pierson, V. Cardone, and others
and is currently the only one used on an operational basis to
forecast wave conditions throughout the Hemisphere as an aid
to ship routers and mariners.

The analysis of the approximately 5.8×10^7 spectra to be
generated by FNWC is being performed at the David W. Taylor
Naval Ship Research and Development Center (DTNSRDC) and a
climatology of wave parameters will result. The climatology will
contain distributions of wind speed, wind direction, wave height,
wave period, predominate and secondary wave directions, a
descriptor of spectral width, and a descriptor of angular width.
In addition, identifying parameters such as location and datetime will be included so that phenomena of different operational
areas and seasons can be identified and studied. Further, the
concept of a dynamic season will be introduced in that variations
in annual extremes from year to year will be identified.

Table 1 shows a typical directional spectrum generated by
SOWM for a location near Station Papa in the North Pacific.
Identifiers and inputs are listed in the first few rows followed
by the energy variances and sums. A few labels have been added
to the figure to more clearly illustrate the contents.   The
parameters to be reported in the new climatology are derived from
the directional spectra as typified in this table.

The potential for deriving improved wave statistics from a
sufficiently large collection of the parameters obtained from the
hindcast spectra was first recognized formally in the *Seakeeping
Workshop Report* (1975). This workshop was attended by members
of the ship research and development, engineering and operations
communities who were invited to participate in the formulation
of the Navy's Seakeeping Research and Development Program. Hence,
the effort described herein was undertaken. A major part of
the climatology work is consumed in the handling of such a large
data base. As such, a great deal of effort has been expended on
the development of suitably flexible data storage and data management software. However, the more interesting part of the study,
at least from the oceanographer's point of view, is in the
development of algorithms for selecting the parameters to be
included in the climatology. The algorithms must, by the sheer
volume of the data, be sufficiently cost effective to exercise,
and at the same time, produce data that are useful to the
designer. Some parameters, such as significant wave height, are
readily available from the data of Table 1. Others, such as

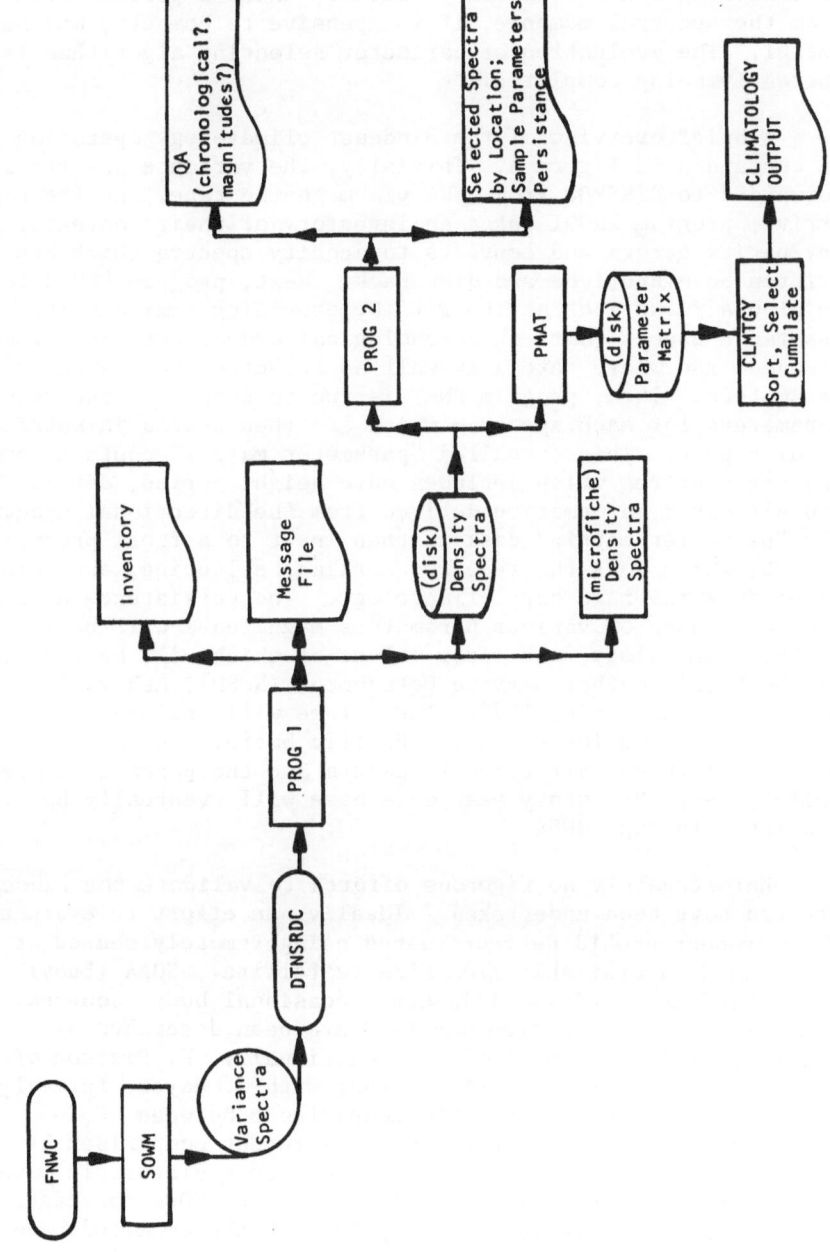

Table 1 - Typical directional variance spectrum generated by Fleet Numerical Weather Central Spectral Ocean Wave Model (SOWM) for a location Near Station Papa (50°N, 145°W).

wave period, are not so easily defined or else the possibilities
for definition are numerous. For example, the poor sampling
characteristics of the mode may tend to degrade the usefulness
of including a modal period parameter, while a period defined
from the spectral moments, if inexpensive to compute, may be more
useful. The evaluation of parameter selecting algorithms is
currently being completed.

A brief overview of the hindcast climatology operating system
is contained in Figure 4. Initially, the variance spectra are
forwarded to DTNSRDC from FNWC via magnetic tape. As the tapes
arrive, program PROG1 takes an inventory of their contents, notes
any parity errors and converts to density spectra which are then
written on microfiche and disk pack. Next, program PROG2 is run
to give a first look at the data by providing some quality
assurance checks (data in chronological order, certain values
exceeded and where, etc.) as well as selected spectra and sampled
parameters. Then, program PMAT is run to determine the required
parameters for each spectrum which are then stored in matrix form
on disk pack. This so-called "parameter matrix" contains one
row per spectrum which includes wave height period, direction,
and all other information derived from the directional spectra.
The "parameter matrix" disk is then input to a final program,
CLMTGY, which does the required sorting, selecting, and cumulating
to produce the hard copy climatology. The persistence as well as
the occurrence of various parameters magnitudes will be included
in the climatology, the first volume of which will be published
by the Naval Weather Service Detachment (NWSD), Asheville,
North Carolina during 1979. The volume will include data for a
five year period for the North Pacific basin. As many future
users of both the directional spectra and the parameters are
anticipated, the twenty year data base will eventually be
available through NWSD.

Unfortunately no rigorous efforts to validate the hindcast
spectra have been undertaken. Ideally, an effort to evaluate the
SOWM product should be coordinated using remotely sensed or
measured data available from NASA (satellite), NOAA (buoy), and
Navy, USCG and maritime industry (occasional buoy) sources.
However, a few validation samples have been described by
Lazanoff and Stevenson (1975). Additionally, W. Pierson of
City University of New York together with J. Hayes, formerly
of FNWC, have shown reasonable comparisons between GEOS-3
(satellite) and SOWM wave heights. A recent comparison by
DTNSRDC of the point spectrum derived from a disposable wave buoy
developed in the Netherlands, with a 48-hour SOWM forecast,
showed a very favorable agreement in both the amplitude and
frequency content of the two spectra. So, while only a few

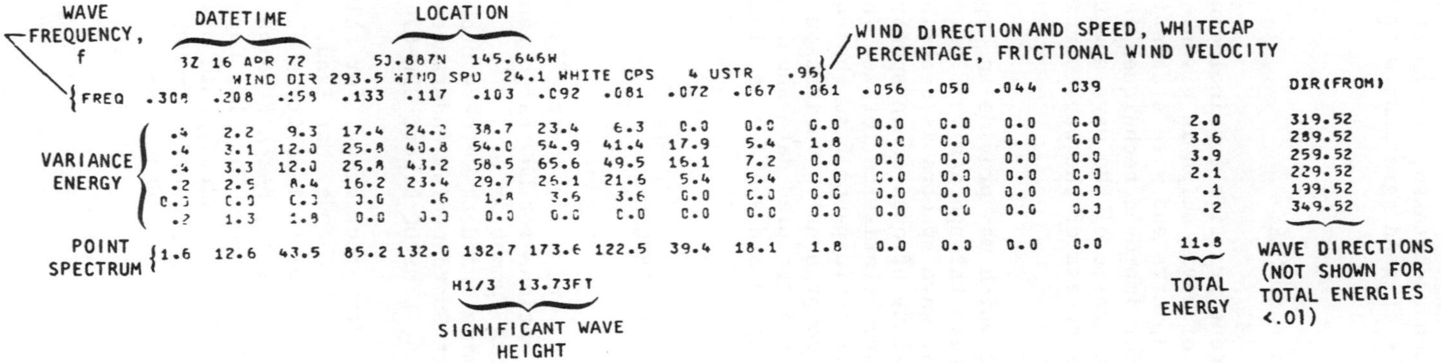

DATETIME 3Z 16 APR 72 **LOCATION** 50.887N 145.646W

WIND DIR 293.5   WIND SPD 24.1   4 USTR   WHITE CPS

WIND DIRECTION AND SPEED, WHITECAP PERCENTAGE, FRICTIONAL WIND VELOCITY

WAVE FREQUENCY, f FREQ	.308	.208	.159	.133	.117	.103	.092	.081	.072	.067	.061	.056	.050	.044	.039	DIR(FROM)
VARIANCE ENERGY	.4	2.2	9.3	17.4	24.2	38.7	23.4	6.3	0.0	0.0	0.0	0.0	0.0	0.0	0.0	2.0	319.52
	.4	3.1	12.0	25.8	40.8	54.0	54.9	41.4	17.9	5.4	1.8	0.0	0.0	0.0	0.0	3.6	289.52
	.4	3.3	12.0	25.8	43.2	58.5	65.6	49.5	16.1	7.2	0.0	0.0	0.0	0.0	0.0	3.9	259.52
	.2	2.5	8.4	16.2	23.4	29.7	25.1	21.6	5.4	5.4	0.0	0.0	0.0	0.0	0.0	2.1	229.52
	0.0	0.0	0.0	0.0	3.6	1.8	3.6	0.0	0.0	0.0	0.0	0.0	0.0	0.0	0.0	.1	199.52
	.2	1.3	1.8	0.0	0.0	0.0	0.0	0.0	0.0	0.0	0.0	0.0	0.0	0.0	0.0	.2	349.52
POINT SPECTRUM	1.6	12.6	43.5	85.2	132.0	132.7	173.6	122.5	39.4	18.1	1.8	0.0	0.0	0.0	0.0	11.8	TOTAL ENERGY

$H_{1/3}$ 13.73FT SIGNIFICANT WAVE HEIGHT

WAVE DIRECTIONS (NOT SHOWN FOR TOTAL ENERGIES <.01)

Fig. 4 – Overview of hindcast climatology operating system. The end result is a detailed wave climatology based on twenty years of directional wave spectra that are hindcast for the Northern Hemisphere at a six hour time step.

validation efforts have been undertaken, the spectra predicted by
SOWM generally appear to be in good agreement with measured
data.

SUMMARY

A brief outline of current practice for including the wave
environment in the design of naval ships is given. Weaknesses
and strengths of the wave spectra and wave statistics now used
are described. The need for improved techniques is discussed
with emphasis placed on the shape of the wave spectrum and the
reliability and comprehensiveness of the wave statistics that
are used.

An ongoing Navy project which may provide an improved capability in both of these areas through analysis of a twenty year
set of hindcast directional wave spectra is described. "Multi-
peakedness" and directionality of open-ocean spectra as well as
an improved data base of wave statistics for use in ship design
and engineering problems is anticipated from the early results
of this work. However, a comprehensive validation effort
utilizing remotely sensed and measured data should be initiated
to establish inherent strengths and weaknesses of the Spectral
Ocean Wave Model.

REFERENCES

Baitis, A. E., S. L. Bales, and W. G. Meyers, 1974. Design
    accelerations and ship motions for LNG cargo tanks, *Proceedings
    Tenth Naval Hydrodynamics Symposium.*
Bales, N. K., 1977. Slamming and deck wetness characteristics
    of a United States Coast Guard medium endurance cutter (WMEC)
    in longcrested head seas, DTNSRDC Report SPD-674-08, Bethesda,
    Maryland.
Bales, S. L., 1978. Sea environment manual for ship design,
    DTNSRDC Report SPD-720-01, Bethesda, Maryland.
Bretschneider, C. L., 1959. Wave variability and wave spectra
    for wind-generated gravity waves, Beach Erosion Board,
    Corps of Engineers, Technical Memo No. 118, Washington, D.C.
Cox, G. G. and A. R. Lloyd, 1977. Hydrodynamic design basis
    for Navy ship roll motion stabilization, *Trans. Soc. Naval
    Architects and Marine Engineers* (SNAME), 85.
Hogben, N. and F. E. Lumb, 1967. *Ocean Wave Statistics,* Her
    Majesty's Stationery Office, London.

Hoffman, D. and M. Miles, 1976. Analysis of a stratified sample of ocean wave records at station India, *Soc. Naval Architects and Marine Engineers* (SNAME) T and R Bulletin No. 1-35.

Lazanoff, S.M. and N. M. Stevenson, 1975. An evaluation of a hemispheric operational wave spectral model, Fleet Numerical Weather Central (FNWC), Technical Note 75-3, Monterey, California.

Ochi, M. K. and S. L. Bales, 1977. Effects of various spectral formulations in predicting responses of marine vehicles and ocean structures, Offshore Technology Conference, Houston, Texas, paper 2743.

Ochi, M. K. and E. N. Hubble, 1976. On six-parameter wave spectra, *Proceedings of the 15th Conference on Coastal Engineering.*

Pierson, W. J. and L. Moskowitz, 1964. A proposed spectral form for fully developed wind seas based on the similarity theory of S. A. Kitaigorodsky, *J. Geophys. Res.*, 69, 24.

Report of the Seakeeping Workshop at the U.S. Naval Academy, 1975. Seakeeping in the ship design process, NAVSEC and DTNSRDC report, Bethesda, Maryland.

St. Denis, M. and W. J. Pierson, 1953. On the motion of ships in confused seas, *Trans. Soc. Naval Architects and Marine Engineers*, 61.

Extreme Waves and Loadings on Floating Vessels

Daniel Hoffman

Hoffman Maritime Consultants Inc.

9 Glen Head Rd., Glen Head, N.Y. 11545

ABSTRACT

The application of extreme wave data in the design and operation of floating vessels is discussed and the concept is defined. Sources of extreme wave data and its possible application as input to vessel performance assessment are evaluated and it is suggested that the design and operation of floating vessels should be based on extreme loads and responses rather than just wave heights. Several techniques available are cited and illustrations of typical representations for such data are shown. A technique referred to as an equivalent wave approach is discussed as a combination of the extreme wave and load approaches.

INTRODUCTION

The tendency in the offshore industry sector and, to a lesser extent in shipping circles, to define wave data requirements in terms of extreme waves rather than extreme loads has led to some rather specific wave data formats and different emphasis in data collection and analysis programs. While the offshore industry selected as a design criteria the 100-year wave height with an associated wave period (e.g., *Department of Energy (U.K.)*, 1974, and *Det Norske Veritas*, 1975), ship loading has customarily been expressed in terms of long-term response predictions based on wave spectral definitions and wave height exceedence statistics. The need to define the wave associated with an extreme load is not identified in most cases. Various statistical models are being used to extrapolate from observed, measured, or hindcast wave height data and to predict long-term responses.

In order to evaluate correctly the relative merit of wave data formats a better definition of the problem is needed. With regard to fixed structures, the extreme wave which has the highest height expected in 100 years, or the lifetime of the structure, is of interest. With regard to ships, however, the extreme wave conditions, though very often quoted in relation to survivability, are not the ideal format to describe the environment since it is more likely that a certain load level will be exceeded at a somewhat less severe condition which may prevail for a longer time. Many phenomena leading to shipping or offshore disasters cannot be analytically simulated by the use of extreme waves only. Such phenomena as the probability of shipping water or slamming require much more extensive wave data than merely extreme wave data. Similarly, damage can occur at wave heights well below the extreme wave condition. In order to understand this approach to the analysis of extreme conditions relevant to floating structures and ships, a more detailed discussion is given in the following sections.

EXTREME WAVE DATA

The available extreme wave data is extracted from three major sources--observation, measurements, and hindcasting. It is less than likely that extreme conditions can be accurately reported by visual observation, though it has been shown that correlation of observed wave height and wind speed or measured wave height, and exceedence tables based on observations can be of important value in the mid to upper range of the heights, i.e., excluding very small and extreme waves. Hence, it is possible to extrapolate that data which is considered reliable to yield extreme conditions by applying a constant multiplier for the short term and a more elaborate statistical model for the longer return period. Measurements of wave data in general and extreme values in particular are often affected by the characteristics of the device used and one should carefully examine limitations imposed by the instrumentation.

A list of maximum wave heights recorded by different sources is given in Table 1, computed from data presented by the *International Ship Structure Committee* (1973). The data is plotted in Figure 1 as maximum wave height versus wave period, defined as the zero crossing period of the maximum wave (*Sellars*, 1976). Curves of constant wave steepness (defined as the ratio of wave height, H, to wave length, λ) are included in this figure. The curve labeled $H/\lambda = .143$ represents a limit associated with the breaking of a traveling wave and the curve labeled .22 represents a limit associated with the breaking of a standing wave. These limits are usually assumed for regular waves, but have been shown to be applicable to irregular waves too (*Sellars*, 1976).

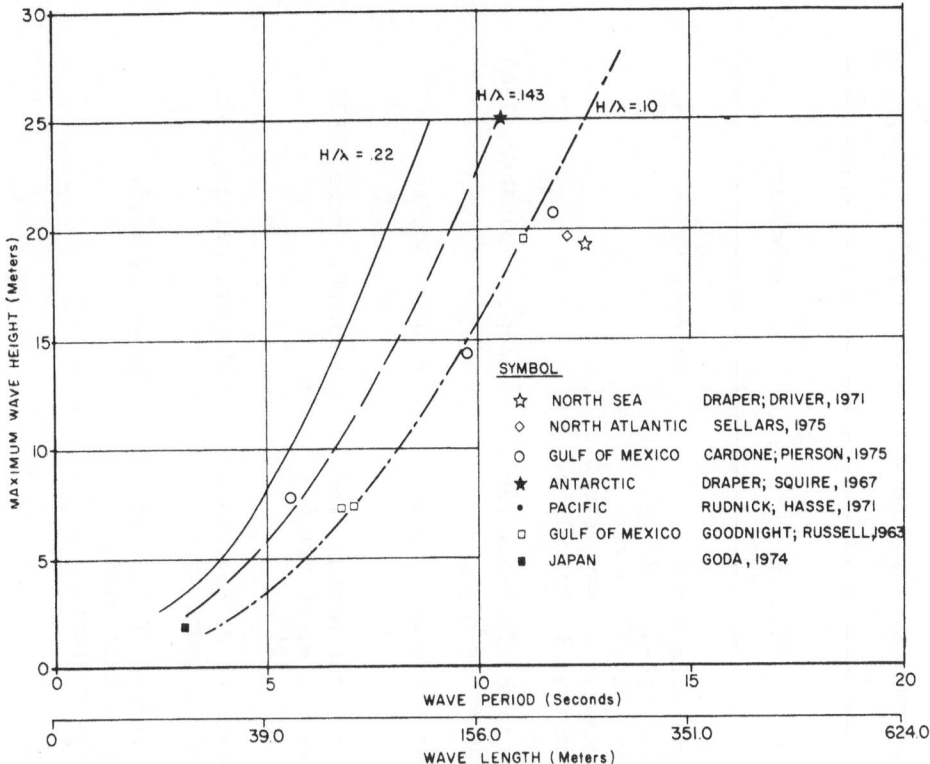

Fig. 1. Range of maximum wave proportions for measured wave data. The limiting curve labeled H/λ = 0.143 applies to breaking progressive waves and the limiting curve labeled H/λ = 0.22 applies to breaking standing waves. The data are better fit by the curve labeled H/λ = 0.10.

Wave spectra for high wave conditions are also given in *Sellars* (1976) and shown in Figure 2. The results shown are for hurricanes Camille and Eloise in the Gulf of Mexico, a storm at Station "India" in the North Atlantic, two North Pacific spectra and two spectra whose peak frequencies are greater than .16 hz. A theoretical equilibrium spectrum for an upper bound on the wave spectrum (*Phillips*, 1966) is also shown $S_\xi(\omega) = \beta\ g^2\ \omega^{-5} \approx 1.127\omega^{-5}$ where β is a constant with a value of about $1.2.10^{-2}$ and g is the acceleration due to gravity. When the wave heights are in meters, $\beta\ g^2$ is approximately 1.127. The maximum wave height in a specific sea area or along a route and its corresponding period is necessary in order to determine the upper limit of wave height and period as well as their probability of occurrence. Such values are used along with the probability of occurrence of other pairs of height and period, not necessarily extremes, to determine the response level expected to be exceeded once over the specified return period.

TABLE 1. Characteristics of extreme waves for several ocean areas.

Location	Height ft.	Period sec.	Steepness	Comment	References
Gulf of Mexico	71.5 (21.8m)	12.0	0.097	Hurricane Camille wind speed = 120 mph (109 knots). Waves over top of wave staff Occurred 17 Aug. 1969.	Cardone et al., 1975
North Atlantic	67.0 (20.4m)	--	--	Occurred 12 Sept. 1961, Tucker gauge.	Draper; Whitaker, 1965
Antarctic	81.7 (24.9m)	11.0	0.131	Data from stereophoto	Draper; Squire, 1967
Pacific	80.0 (24.4m)	16.2	0.060	Data from motion picture film, occurred 1 Dec. 1969.	Rudnick; Hasse, 1971
North Sea	61.0 (18.6m)	13.3	0.067	Occurred 23 Nov. 1969, Tucker gauge.	Draper; Driver, 1971
North Atlantic Station India	65.0 (19.8m)	12.5	0.081	Occurred 16 Feb. 1962 in 44 knot wind, Tucker gauge.	Sellars, 1975.

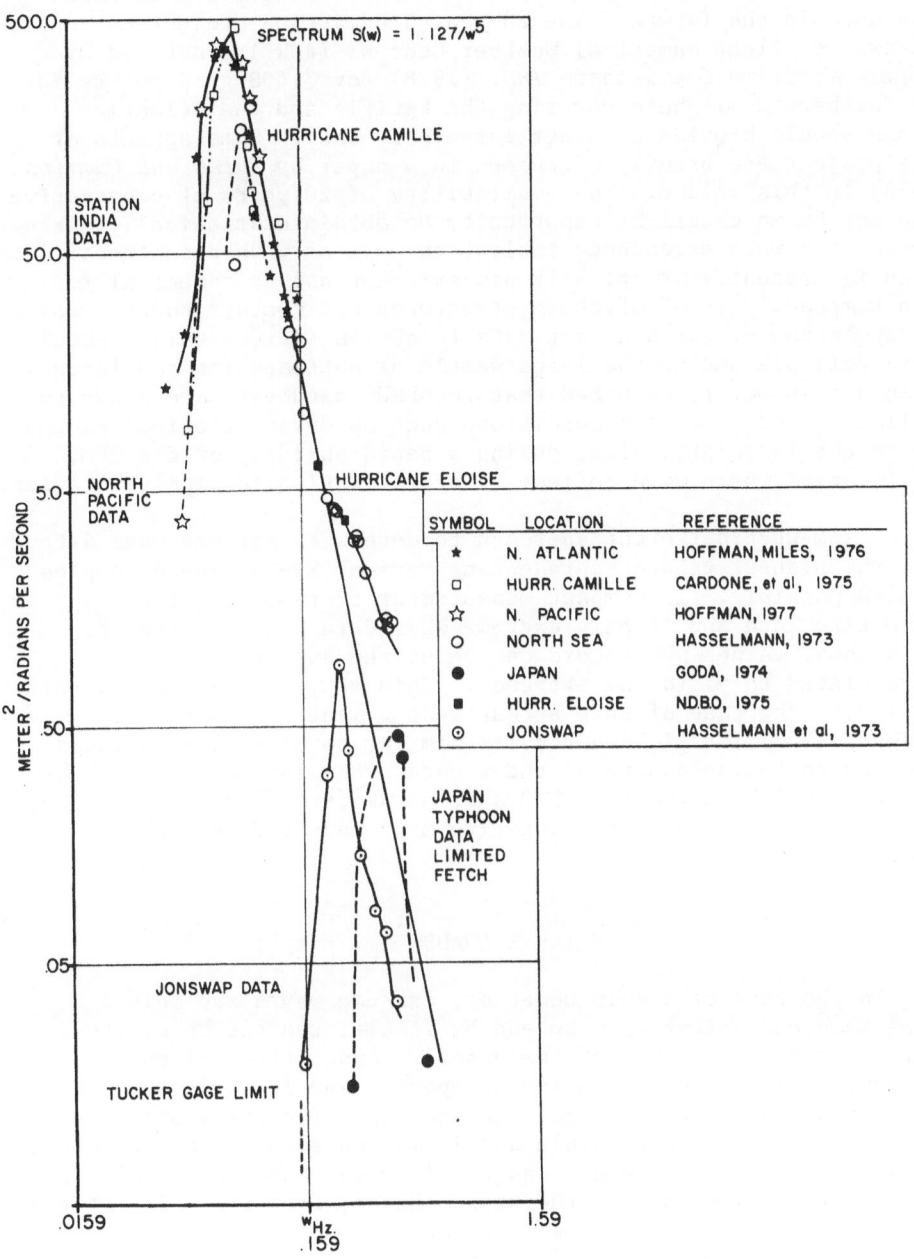

Fig. 2. Comparison of wave spectra for high wave conditions.
The extreme wave spectra in the upper part of the figure are typical
of spectra used to determine extreme waves and vessel responses.

Hindcast data should constitute a major source for extreme
wave data in the future. The 20-year hindcasting project currently
underway at Fleet Numerical Weather Central (FNWC) conducted by
Hoffman Maritime Consultants Inc. (1978) for 3,400 grid points in
the Northern Hemisphere covering the Pacific and the Atlantic
oceans should provide a rather large data base. Some apsects of
this project are briefly discussed in a paper by *Bales and Cummins*
(1978) in this volume. The availability of 20 years of consecutive
data may be an excellent opportunity to obtain statistically adequate
samples for wave exceedence tables, as well as wave persistence data
which is presently practically non-existent and is essential for
down time analysis of offshore structures and weather bound vessels.
Extrapolation of the hindcast data to obtain extreme values should
prove reliable due to the large sample of data and its consistency.
It should, however, be noted that the FNWC hindcasts are known to
be limited under certain conditions such as during tropical storms
due to the large grid size, during a rapid build-up of a storm,
and in areas where wave reflection or refraction is likely to exist.

A somewhat different approach to determine extreme wave data
from the highest values representing typical 20-30 minute samples
is also possible. If compact measurement systems cannot store
the entire data and if only extreme wave data are of interest,
the highest values per record can be stored and these can be
extrapolated to yield the extremes. This approach was successfully
applied in the case of ship stress measurements (*Lundgren and
Hoffman,* 1967) and yielded results similar to those extrapolated
from the root-mean-square of the record. Similar application can
also be applied to the Tucker-Draper parameters (*Draper,* 1966)
which have been used to characterize many measured wave data
records.

EXTREME LOADS

In the case of a ship underway, extreme waves are only a
means to an end rather than an end by itself; the end being the
extreme load or response of the vessel. Under these circum-
stances, it is easier to measure responses and loads in the open
environment rather than waves. In the case of a stationary vessel,
wave measurements are feasible and it may be advantageous to monitor
waves as well as responses so as to allow the application of the
data to other vessels or structures. Hence, from the point of view
of the ship designer, extreme wave data may not necessarily be the
critical parameter to use. Similarly, for an offshore vessel
operator the wave height which is likely to induce the limiting
operational response is of consequence rather than the maximum
wave.

Different vessels react differently to waves and even the same vessel exhibits different characteristics at different heading angles depending on the wave characteristics, depth of immersion, and weight distribution. It is therefore unreasonable to associate a structure or vessel with a specific extreme wave since, depending on the design, different ships and floating structures would respond to waves in different manners. A typical example has been recently demonstrated (*Hoffman and Zielinski*, 1977) where a double articulated single anchor leg mooring (SALM), consisting of a disc-type base, a pipe and a spar buoy, exhibits no significant motions or loads under seas of 20 meters and above significant wave height. The survivability of one vessel or another is associated with different heights of waves and periods and, in many cases, these values are well below what are generally considered extreme waves.

The prediction of extreme load values has been discussed in detail in several previously published reports (*Hoffman and Lewis*, 1969; *Hoffman, Williamson, and Lewis*, 1972; *Hoffman, van Hooff, and Lewis*, 1972). *Hoffman and Lewis* (1969) have shown that such loads can be obtained based on a large measured data bank covering response to low as well as high waves. In the cases where the responses to extreme waves are missing due to the limited measurement period or other reasons, the data can be extrapolated in a reliable way and the extreme load can be estimated. Similarly, equivalent analytical techniques have been developed where a family of spectra covering the entire range of wave heights likely to occur, and represented in a form to take account of the variability of spectral shapes in each group of wave heights, can be used to simulate short-term trends and their scatter. Hence, long-term predictions can be made using wave exceedence data and a statistical model which was developed from full-scale measurements. The full-scale measurements and the analytical model yield good correlations for linear or nearly linear responses. An alternative approach using the maximum value, rather than the rms ratio recorded during a 20-30 minute interval and representing a four hour steady state condition is given by *Lundgren and Hoffman* (1967). A good comparison of predicted extremes, the actual recorded and the calculated data is shown in Figure 3, illustrating the adequacy of the method used to extrapolate the measured and calculated values to predict extreme loads. A more extensive study was later carried on and summarized (*Hoffman, van Hooff, and Lewis*, 1972) showing consistent trends using either rms stress or extreme values to represent each short-term record.

Other long-term extrapolation techniques available are those using the Weibull distribution (*Nordenstrom*, 1972) which has been used for wave height extrapolation. Both the Weibull distribution and the statistical model discussed above, which is based on a normal distribution of the Rayleigh distributed short-term values,

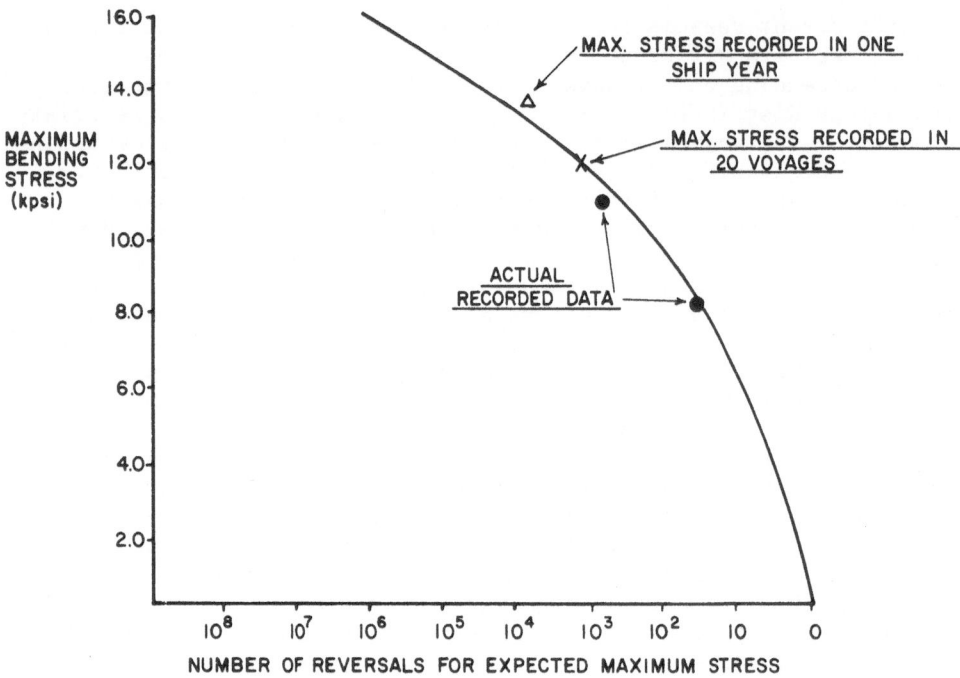

Fig. 3. Calculated and recorded trends of long-term stress illustrating good agreement. The abscissa is the number of stress reversals.

give extremely close results for most cases (*Stephan and Hoffman,* 1978). In both cases, the statistical distribution is used to extrapolate short-term responses such as recorded over a steady state period to long-term prediction. In the case of the Weibull distribution, the variables are usually determined graphically, while in the latter case the mean rms values and the standard deviation about it are used as input to determine the value expected to be exceeded once during the specified return period.

EXAMPLES OF EXTREME LOAD PRESENTATION

Long-term predictions are usually expressed in terms of the highest level expected to be exceeded once in a specific return period such as a year, ten years, or the lifetime of a vessel or structure. The highest expected value is usually determined by weighing the long-term values at the various wave heights according to their probability of occurrence. Table 2 illustrates a typical example of long-term predictions for ship roll responses for several wave heights as well as the highest expected value at 24

TABLE 2. Typical long-term prediction output for roll response.
The first part of this table provides the highest
expected roll (degrees) as a function of significant
wave height $H_{1/3}$ and the second part gives the highest
expected roll (degrees) during two bi-weekly periods
each month. An example of this table's use is given
in the text.

600' Barge   Speed = 6.0000 knots
Roll (Degrees)   Short Crested Seas

Group	$H_{1/3}$ Meters	High Occurrence In		
		24 Hrs.	48 Hrs.	168 Hrs.
1	0.7	2.0	2.1	2.4
2	1.1	2.7	2.9	3.1
3	1.4	3.7	3.9	4.3
4	1.7	4.6	4.8	5.2
5	2.0	6.6	7.0	7.7
6	2.2	7.1	7.5	8.1
7	2.6	7.5	7.8	8.4
8	2.9	9.1	9.6	10.5
9	3.2	8.9	9.3	10.0
10	3.5	12.3	13.0	14.2
11	4.3	14.3	15.0	16.2
12	5.5	18.6	19.5	21.3
13	7.2	23.4	29.5	26.3
14	8.8	27.0	28.2	30.3
Month				
Jan 1		18.3	19.3	20.9
Jan 2		17.8	18.7	20.4
Feb 1		17.3	18.3	20.0
Feb 2		17.9	18.8	20.5
Mar 1		17.1	18.1	19.8
Mar 2		16.1	17.1	18.9
Apr 1		15.4	16.5	18.3
Apr 2		15.1	16.1	18.0
May 1		10.7	11.3	12.3
May 2		10.7	11.3	12.4
Jun 1		9.8	10.5	11.7
Jun 2		7.8	8.5	9.6
Jul 1		9.4	10.0	11.2
Jul 2		9.2	9.9	11.2
Aug 1		8.9	9.6	10.7
Aug 2		9.6	10.3	11.5
Sep 1		11.1	11.8	12.9
Sep 2		12.9	13.9	15.8
Oct 1		15.4	16.4	18.3
Oct 2		15.5	16.6	18.4
Nov 1		16.6	17.7	19.4
Nov 2		16.6	17.7	19.4
Dec 1		16.9	17.8	19.6
Dec 2		17.4	18.4	20.1

bi-weekly periods over the year. From Table 2, the highest
expected roll in 24 hours during the first two weeks in September
(Sept.) is 11.1 degrees, which is equivalent to a significant
wave height of 3.4 meters persisting continuously for 24 hours.
This in turn can be expressed in terms of the maximum expected
wave height in 24 hours, e.g., 7.6 meters--assuming steady state
conditions over the period.

The long-term predictions can also be used to yield a
histogram showing the number of occurrences expected for different
response levels. A typical example is given in Table 3 using vertical acceleration response for the purpose of illustration.
Assuming 10^8 reversals of acceleration in the ship lifetime, the
probability of exceeding .25g is 4.8×10^{-5} which can be interpreted that this level is expected to be exceeded about 5,000 times
in the lifetime of the ship. This form of presentation puts the
extreme values in a useful form convenient for other types of
analysis such as fatigue loads.

Another format for representing extremes is shown in Table 4.
The case shown is based on calculations performed for a barge in
short-crested seas at a forward speed of 6 knots. The values
shown are the number of expected occurrences of exceeding a eight
degree roll once in 48 hours under the conditions of wave height
and relative heading angle between the ship and the primary wave
system. The results shown in Table 4 are a simplified operability
chart giving the sensitivity of the vessel to changes in heading
and wave height.

Operability is best measured in terms of down time and in
order to predict such values wave persistence is required rather
than wave exceedence. Any continuous measurement program should
yield information sufficient to create such data. Similarly,
hindcasting can also be used. Persistence charts are somewhat
similar to exceedence tables. Each case is given for a specific
duration, for example 8 hours a day, or a week, and illustrates
the probability of the wave height persisting below a certain
level for the specified duration. The use of such data leads to
the type of information summzrized in Table 5. For each of the
roll response levels indicated, the number of 48-hour work periods
expected for two work periods during each month of the year is
given as well as the total operability over the year based on the
specific criteria set, i.e. 4°, 6° roll, etc. This form of extreme
values utilization is indeed more meaningful than referring to the
maximum wave height. Though it is fairly common among mariners
to express the limits of operations in terms of the wave height
rather than the response level, it is primarily due to the lack
of quantified response measurements on board. The observed wave
height is therefore often used as criteria in spite of its known
limitations. Recent introduction of on-board monitoring and pre-

TABLE 3. Example of long-term acceleration prediction.
During the ship's lifetime, 10^8 acceleration
reversals occur. An example of this table's use
is given in the text.

Acceleration ft/sec^2	Probability of Exceedence	Expected Number of Exceedences in Ship's Lifetime	Histogram (Number of occurrences)
0.0	$1.0 \cdot 10^0$	$1.0 \cdot 10^8$	0.0
0.9	$3.2 \cdot 10^{-1}$	$3.2 \cdot 10^7$	$6.8 \cdot 10^7$
1.8	$1.0 \cdot 10^{-2}$	$1.0 \cdot 10^6$	$2.2 \cdot 10^7$
2.7	$3.5 \cdot 10^{-2}$	$3.5 \cdot 10^6$	$6.7 \cdot 10^6$
3.6	$1.3 \cdot 10^{-2}$	$1.3 \cdot 10^6$	$2.3 \cdot 10^6$
4.4	$4.4 \cdot 10^{-3}$	$4.4 \cdot 10^5$	$8.1 \cdot 10^5$
5.3	$1.5 \cdot 10^{-3}$	$1.5 \cdot 10^5$	$2.9 \cdot 10^5$
6.2	$4.9 \cdot 10^{-4}$	$4.9 \cdot 10^4$	$1.0 \cdot 10^5$
7.1	$1.5 \cdot 10^{-9}$	$1.5 \cdot 10^4$	$3.3 \cdot 10^4$
8.0	$4.8 \cdot 10^{-5}$	$4.8 \cdot 10^3$	$1.1 \cdot 10^4$
8.9	$1.4 \cdot 10^{-5}$	$1.4 \cdot 10^3$	$3.3 \cdot 10^3$
9.8	$4.2 \cdot 10^{-6}$	$4.2 \cdot 10^2$	$1.0 \cdot 10^3$
10.7	$1.2 \cdot 10^{-6}$	$1.2 \cdot 10^2$	$3.0 \cdot 10^2$
11.5	$3.6 \cdot 10^{-7}$	$3.6 \cdot 10^1$	$8.8 \cdot 10^1$
12.4	$1.0 \cdot 10^{-7}$	$1.0 \cdot 10^1$	$2.5 \cdot 10^1$
13.3	$2.9 \cdot 10^{-7}$	$2.9 \cdot 10^1$	$7.3 \cdot 10^0$
14.2	$8.4 \cdot 10^{-9}$	$8.4 \cdot 10^{-1}$	$2.1 \cdot 10^0$
15.1	$2.4 \cdot 10^{-9}$	$2.4 \cdot 10^{-1}$	$6.0 \cdot 10^{-1}$
16.0	$6.7 \cdot 10^{-10}$	$6.7 \cdot 10^{-2}$	$1.7 \cdot 10^{-1}$
16.9	$1.9 \cdot 10^{-10}$	$1.9 \cdot 10^{-2}$	$4.8 \cdot 10^{-2}$
17.8	$5.2 \cdot 10^{-11}$	$5.2 \cdot 10^{-3}$	$1.4 \cdot 10^{-2}$
18.6	$1.4 \cdot 10^{-11}$	$1.4 \cdot 10^{-3}$	$3.8 \cdot 10^{-3}$
19.5	$3.8 \cdot 10^{-12}$	$3.8 \cdot 10^{-4}$	$1.0 \cdot 10^{-3}$
20.4	$1.0 \cdot 10^{-12}$	$1.0 \cdot 10^{-4}$	$2.8 \cdot 10^{-4}$
21.3	$2.7 \cdot 10^{-13}$	$2.7 \cdot 10^{-5}$	$7.5 \cdot 10^{-5}$
22.2	$7.0 \cdot 10^{-14}$	$7.0 \cdot 10^{-6}$	$2.0 \cdot 10^{-5}$
23.1	$1.8 \cdot 10^{-14}$	$1.8 \cdot 10^{-6}$	$5.2 \cdot 10^{-6}$
24.0	$4.6 \cdot 10^{-15}$	$4.6 \cdot 10^{-7}$	$1.3 \cdot 10^{-6}$
24.9	$1.2 \cdot 10^{-15}$	$1.2 \cdot 10^{-7}$	$3.4 \cdot 10^{-7}$
25.7	$2.8 \cdot 10^{-16}$	$2.8 \cdot 10^{-7}$	$8.6 \cdot 10^{-8}$
26.5	$6.9 \cdot 10^{-17}$	$6.9 \cdot 10^{-9}$	$2.1 \cdot 10^{-8}$

dicting systems as an aid to operation (*Hoffman*, 1978) illustrated
the advantages of measured and predicted responses as criteria
rather than the estimates which must be used in the absence of
such tools.

TABLE 4.   Long-term operability chart.   The number of times
           that the roll angle exceeds 8° during 48 hours is
           shown as a function of wave height and heading angle.

600' Barge   Short Crested Seas
Roll (Degrees)   Speed = 6.0000 knots

Wave Ht. Ft.	Heading Angle (Degrees)												
	0	15	30	45	60	75	90	105	120	135	150	165	180
3.0	0	0	0	0	0	0	0	0	0	0	0	0	0
4.0	0	0	0	0	0	0	0	0	0	0	0	0	0
5.0	0	0	0	0	0	0	0	0	0	0	0	0	0
6.0	0	0	0	0	0	0	0	0	0	0	0	0	0
7.0	0	0	0	0	0	0	0	0	0	0	0	0	0
8.0	0	0	0	0	0	1	1	1	0	0	0	0	0
9.0	0	0	0	0	0	2	3	2	0	0	0	0	0
10.0	0	0	0	1	11	30	38	26	9	1	0	0	0
11.0	0	0	0	0	9	27	37	25	8	0	0	0	0
12.0	0	0	2	42	183	371	463	388	209	61	7	0	0
16.0	0	0	29	352	1139	1933	2226	1855	1050	322	34	0	0
21.0	0	9	237	1236	2763	3998	4416	3892	2637	1205	291	35	9
27.0	56	296	2082	5829	9609	9999	9999	9999	9727	5984	2259	429	132
34.0	199	838	4244	9547	9999	9999	9999	9999	9999	9999	5711	2065	1032

EQUIVALENT WAVE HEIGHT CONCEPT

In contrast to the extreme wave height criteria such as the
one associated with a 100-year wave or a typical winter storm, the
concept of equivalent wave height is based on extreme load calculations and was originally developed in order to relate conventional
static bending moment calculations for ship hulls and dynamic
response calculations or full-scale measurements.   The purpose was
to provide the designer with a quantity he is familiar with.
Hence, the equivalent wave, H_e, was defined as the height of a
trochoidal wave of length equal to that of the ship, which by conventional static bending moment calculations gives a bending moment
equal to one half of the peak to trough values calculated theoretically or measured experimentally.   The peak to trough bending moment

TABLE 5. Example of down time chart for northern North Sea.
The number of 48 hour work periods as a function of
limiting roll responses are given for two bi-weekly
periods each month and are based on 1 exceedence in
48 hours

Month	Roll Response (Degrees)					
	4.00	6.00	8.00*	10.00	12.00	14.00
Jan 1	0.0	0.6	2.3	3.3	3.5	4.2
Jan 2	0.0	0.6	2.3	3.4	3.5	4.3
Feb 1	0.0	0.7	2.4	3.4	3.5	4.3
Feb 2	0.0	0.8	2.5	3.6	3.7	4.5
Mar 1	0.1	1.0	2.7	3.7	3.9	4.6
Mar 2	0.4	1.3	3.0	4.0	4.2	4.9
Apr 1	0.9	1.7	3.4	4.5	4.6	5.4
Apr 2	1.5	2.3	4.0	5.0	5.2	6.0
May 1	2.1	2.9	4.6	5.7	5.8	6.5
May 2	2.8	3.7	5.4	6.4	6.5	7.0
Jun 1	3.5	4.4	6.0	7.1	7.2	7.5
Jun 2	4.1	4.9	6.6	7.5	7.5	7.5
Jul 1	4.1	4.9	6.6	7.5	7.5	7.5
Jul 2	3.7	4.6	6.3	7.3	7.4	7.5
Aug 1	3.3	4.2	5.9	6.9	7.1	7.4
Aug 2	2.9	3.7	5.4	6.5	6.6	7.1
Sep 1	2.3	3.1	4.8	5.9	6.0	6.6
Sep 2	1.7	2.5	4.2	5.3	5.4	6.1
Oct 1	1.0	1.8	3.5	4.6	4.7	5.5
Oct 2	0.6	1.4	3.1	4.2	4.3	5.1
Nov 1	0.3	1.1	2.8	3.9	4.0	4.8
Nov 2	0.1	0.9	2.6	3.7	3.8	4.6
Dec 1	0.0	0.8	2.5	3.5	3.6	4.4
Dec 2	0.0	0.6	2.3	3.4	3.5	4.3
TOTAL	35.2	54.8	95.3	120.1	123.2	137.6

*Typical operational limits. 95.3 is equivalent roughly to 190
days a year

value is the long-term response expected or the highest value
measured over a specific duration. A more detailed explanation
of this method is given in *Hoffman and Lewis* (1969).

A slightly different approach to the definition of an
equivalent wave height along the same line of thought is given
by *Hoffman and Zeilinski* (1974). The equivalent wave height in
this case is defined as the ratio of the predicted long-term
response value at 10^{-8} probability level (i.e., the lifetime of
the vessel) divided by the maximum acceleration due to unit wave
height as obtained from the transfer function of the response.
The probability level of 10^{-8} is representative of a typical ship
lifetime of 20-25 years, taking into consideration the mean period
of response and the number of hours spent on the open seaway.

Since the transfer function is given as a function of frequency,
heading, speed, draft, and other parameters, the problem can be
narrowed by considering specific operational conditions or can remain generalized by searching and interpolating among all the values
available for the various conditions. Hence, the equivalent wave
height can be determined for different headings and speed combinations as shown in Table 6. It is apparent that for each condition

TABLE 6. Example of instantaneous roll and pitch angles at
maximum vertical accelerations computed by the
equivalent wave height approach.

600,000-ton VLCC Design

Loading Condi-tion	Speed (kts)	Heading (°)	Max Accel. a_o (ft/sec^2/ft)	Resonance Freq. (rads/sec)	Equiv. Wave Ht (ft)	Roll Angle (°)	Pitch Angle (°)
Loaded	0	120	.334	.475	31.5	0.84	3.22
	18	120	.496	.425	34.0	1.00	4.28
		150	.494	.489	34.0	.33	5.00
Ballast	0	60	.500	.425	29.3	38.03	2.80
		90	.640	.433	22.9	49.18	0.06
		120	.414	.450	35.5	30.77	3.44
		150	.218	.400	51.9	13.50	6.44
	21	90	.630	.435	30.3	65.86	0.29
		120	.510	.450	37.4	7.60	4.55

a different equivalent wave height is obtained. The purpose of
the analysis in *Hoffman and Zeilinski* (1974) was not to determine
the equivalent sinusoidal wave height but to determine the phasing
between the extreme motions. If the total acceleration (gravitational and dynamic) must be determined, the static component of
the acceleration due to the instantaneous angle of roll must be
known, hence the angle of roll expected to occur at the instant
of maximum vertical or lateral acceleration must be determined.
Since maximum acceleration of a sine wave $\sin(\omega_e t + \varepsilon)$ occurs at
the instant of $(\omega_e t + \varepsilon) = 90°$, or $270°$, where ε is the acceleration phase angle, $\omega_e t$ can be calculated and the corresponding
instantaneous roll, pitch or any other response for which the
transfer function and phase angles are given can be determined.
In the case where the maximum acceleration does not differ greatly
between one heading and another, more than one equivalent wave
height may be obtained. From these, the sensitivity of the results
to slight variations in heading and the reliability of this approximate approach can be determined.

The form of presentation shown in Table 6 can be directly
compared with some of the classification societies' rules which
specify amplitudes and periods of motions as limiting criteria
using extreme wave height concepts where each motion component
is specified independently with no relation to each other.

It is the author's experience that the type of data presented
in Table 2 is the ideal format for extreme load applications, and
that neither the equivalent wave height nor the 100-year wave is
the answer to the designer and it is the actual extreme responses
that are of consequence. The use of the equivalent wave height as
an intermediate tool in response calculation is certainly beneficial though special care should be exercised due to the approximate nature of the method.

CONCLUDING REMARKS

The concept of extreme wave height, though not directly
applicable in all cases to floating vessels design, naturally
has its merits in the application of envrionmental conditions to
design and analysis of shipping and other types of accidents.
Extreme loads seem to be the preferable criteria for the designer
because these relate directly to the behavior of the vessel rather
than generalized environmental conditions at a location. An
intermediate utilization of the two approaches is the use of the
equivalent wave height concept which combines the best of the two
approaches though not necessarily in a mathematically justified
rationale.

In general, shipping accidents are associated with extreme conditions; however, the definition of the range of such environmental conditions must be clearly stated for each vessel. Accidents are very often controlled, at least to a certain extent, by human judgment and the severity of the environment must be evaluated in relation to it. In contrast, survivability assumes conditions under which minimum control can be exercised and could be somewhat closely tied with the extreme wave concept in the absence of a more precise definition.

REFERENCES

Bales, S. L. and W. E. Cummins. 1978. Wave data requirements for ship design and operation, In: *Ocean Wave Climate*, M. D. Earle and A. Malahoff (Eds.), Plenum, New York, (this volume).

Cardone, V. J., W. S. Pierson, and E. G. Ward. 1975. Hindcasting the directional spectrum of hurricane generated waves, Offshore Technology Conference Houston, Texas, paper OTC 2332.

Department of Energy, (U.K.). 1974. Guidance on the design and construction of offshore installation, HM Stationery Office, London, U.K.

Det Norske Veritas. 1975. Rules for the construction and inspection of mobile offshore units, Oslo, Norway.

Draper, L. 1966. The analysis and presentation of wave data -- a plea for uniformity, *Proc. of the 10th International Conference of Coastal Engineering*, Tokyo.

Draper, L. and J. S. Driver. 1971. Waves at ocean weather ship station Juliett, *Deutsche Hydrographische Zeitschrift*.

Draper, L. and J. S. Driver. 1971. Winter waves in the northern north sea at 57° 30'N 3.00'E recorded by M. V. FAMITA, National Institute of Oceanography. internal report A48.

Draper, L. and E. M. Squire. 1967. Waves at ocean weather ship station India; Reply to discussion, Transactions of Royal Institution of Naval Architects, 109:1.

Draper, L. and M. A. B. Whitaker. 1965. Waves at ocean weather station Juliett, *Deutsche Hydrographische Zeitschrift*.

Goda, Y. 1974. Estimation of wave statistics for spectral information, International Symposium of Ocean Wave Measurement and Analysis, New Orleans, Louisiana, Am. Soc. Civil Engineers.

Goodnight, R. D. and R. L. Russell. 1963. Investigations of the statistics of wave heights, *J. of Waterways, Harbors and Coastal Engineer Div.*, Am. Soc. Civ. Engineers, Vol. 89.

Hasselmann, K., T. P. Barnett, E. Bouws, H. Carlson, D. E. Cartwright, K. Enke, J. A. Ewing, H. Gienapp, D. E. Hasselmann, P. Kruseman, A. Meerburg, P. Müller, D. J. Olbers, K. Richter, W. Sell, and H. Walden. 1973. Measurements of wind-wave growth and swell decay during the Joint North Sea Wave Project (JONSWAP), *Deutsche Hydrographische Zeitschrift Suppl. A8:#12*.

Hoffman, D. 1977. Analysis of wave spectra at station "Papa," National Maritime Research Center, Kings Point, N.Y., Report NMRC-KP-173.

Hoffman, D. 1978. Heavy lift monitoring and prediction in the construction of North Sea platforms, Offshore Technology Conference, Houston, Texas, OTC paper 3147.

Hoffman Maritime Consultants Inc. 1978. Twenty years hindcast of directional wave spectra in the northern hemisphere, project no. 7891.

Hoffman, D. and E. V. Lewis. 1969. Analysis and interpretation of full-scale data on midship bending stresses of dry cargo ships, Ship Structure Committee Report 196.

Hoffman, D. R. van Hooff, and E. V. Lewis. 1972. Evaluation of methods for extrapolation of ship bending stresses data, Ship Structure Committee Report 234, AD 753224.

Hoffman, D. and M. Miles. 1976. Analysis of a stratified sample of ocean wave records at station "India," Soc. of Naval Architects and Marine Engineers (SNAME), T & R Bulletin No. 1-35.

Hoffman, D. A. Williamson, and E. V. Lewis. 1972. Correlation of model to full-scale results in predicting wave bending moment trends, Ship Structure Committee Report 233, AD753223.

Hoffman, D. and T. E. Zielinski. 1974. Load analysis of atomic powered vessels, Hoffman Maritime Consultants Inc. Report 7423.

Hoffman, D. and T. E. Zielinski. 1977. Summary of response characteristics of a single anchor leg mooring (SALM) under tow, Hoffman Maritime Consultants Inc. Report 7770.

International Ship Structure Committee. 1973. Report of Committee 1 - Environmental Conditions, Hamburg, Germany.

Lundgren, J. and D. Hoffman. 1967. Analysis of extreme value data to predict long-term ship stress probability; Discussion of Committee 1 Report, 3rd ISSC, Oslo.

NDBO (NOAA Data Buoy Office). 1975. Buoy observation during hurricane Eloise: Data report, National Oceanographic Data Center, National Oceanic and Atmospheric Administration, Washington, D.C.

Nordenstrom, N. 1972. Methods for predicting long-term distributions of wave loads and probability of failure of ships; Det Norske Veritas Report No. 17-3-5, Oslo, Norway.

Phillips, O. M. 1966. The dynamics of the upper ocean, Cambridge University Press, London, U.K., 109-119.

Rudnick, P. and Hasse, R. 1971. Extreme Pacific waves, December 1969, *J. Geophys. Res.*, *76:3*, 742-744.

Sellars, F. 1975. Maximum heights on ocean waves, *J. Geophys. Res.*, *80:3*, 398-404.

Sellars, F. 1976. Maximum wave conditions for design; prepared for Soc. of Naval Architects and Marine Engineers (SNAME), Hydrodynamics Committee.

Stephan, B. H. and Hoffman, D. 1978. Mathematical properties of several long-term distributions and some of their applications; paper under preparation at Webb Institute of Naval Architecture.

II

WAVE MEASUREMENTS

Ocean Surface Features Observed by HF Coastal Ground-Wave Radars: A Progress Review

Donald E. Barrick[1] and Belinda J. Lipa[2]

[1]Wave Propagation Laboratory, National Oceanic and Atmospheric Administration and [2]SRI International

ABSTRACT

High Frequency (HF) coastal radar systems measure and employ the Doppler spectrum of the sea-echo signal to extract relevant ocean surface parameters. The dominant first-order spectral peaks provide mean surface currents; at medium frequencies (MF) they have been used to yield directional information about the wave height spectrum. The second-order spectral continuum is related to the wave height directional spectrum through a nonlinear integral equation. This integral equation has been inverted for narrow-beam coastal radars having large antennas to demonstrate the validity of the theoretical model upon which inversion techniques are based.

The present paper reviews progress which has been made to date in the determination of ocean current and wave conditions from HF radar sea echo. The results have associated statistical variance due to finite sampling of a random distribution of wave heights. A 200-sample average allows the amplitude spectrum and the directional distribution of wind-driven waves to be determined within confidence limits of 10%. The corresponding accuracy of ocean current determination is about 20 cm/s.

INTRODUCTION

Sea echo at frequencies below VHF* has been observed by radar since World War II. *Crombie* (1955) first discovered the interaction mechanism of radio waves with the sea surface by spectrally processing experimental records. The prominent spectral peaks he observed

at symmetric Doppler shifts from the transmitted frequency originated
from ocean waves half the radar wave length, moving toward and away
from the radar. This was ascertained because water waves of this
length have a unique phase velocity (given by the lowest-order
gravity-wave dispersion relation), which in turn produces the precise Doppler shifts he observed. This simple diffraction-grating
explanation of MF/HF* sea scatter has been termed "first-order Bragg
scatter" within the radio-oceanographic community. Subsequent
theoretical analyses of deterministic (*Wait*, 1966) and statistical
nature (*Barrick*, 1972a) have confirmed the validity of this explanation and established the quantitative relationships between wave
height and the echo strength.

Subsequent theoretical investigations (*Hasselmann*, 1971;
Barrick, 1971a; *Barrick*, 1972b) indicated that in addition to the
dominant first-order spectral peaks, a "second-order" spectral continuum should also be present. Radar observations confirmed the
existence of this continuum (*Tyler et al.*, 1972; *Barrick*, 1972b;
Barrick et al., 1974). With exact expressions available relating
the first- and second-order sea-echo Doppler spectrum to the wave
height directional spectrum, it has become obvious over the last
decade that MF/HF radars of various types can be used to remotely
measure sea state by spectrally processing the received signal, and
its subsequent appropriate inversion (*Lipa*, 1977, 1978). A summary
of these various MF/HF radar techniques is given in *Barrick* (1972b),
with a more recent review available in *Barrick* (1978). The use and
potential of ionospheric HF over-the-horizon radars and synthetic-
aperture MF/HF radars are discussed in subsequent papers in this
book by *Maresca* (1978) and *Teague* (1978).

In addition to wave motions, underlying currents will impose
observable Doppler shifts on the first-order sea-echo signals.
Crombie (1972) first made and interpreted such observations,
followed by experiments by *Stewart and Joy* (1974) and *Barrick et al.*
(1974). Subsequently, a prototype of a transportable operational
coastal HF radar system has been built and demonstrated which maps
near-surface ocean currents (*Barrick and Evans*, 1976; *Barrick et al.*,
1977).

The purpose of this paper is to review progress which has been
made to date in the measurement and interpretation of radar sea
echo, and to indicate the direction of future research. The following section reviews conventional HF radar systems, gives the mathematical expressions for their echoes and discusses how different

---

* MF refers to the radio band with frequencies, f, between 300 kHz
and 3 MHz, having radio wave heights, λ, between 1000 m and 100 m;
HF, where 3 MHz $<$ f $<$ 30 MHz, 100 m $>$ λ $>$ 10 m; VHF, where 30 MHz
$<$ f $<$ 300 MHz; 10 m $>$ λ $>$ 1 m.

frequency bands of the echo spectrum are used to derive the desired
oceanographic quantities. The inversion of the continuum frequency
spectrum obtained from large, narrow-beam HF coastal radars to give
parameters of the directional ocean wave spectrum is described.
Next, a small transportable HF radar system which obtains reliable
real-time surface current maps in coastal waters is described.
Finally, the need for a similar system which could provide coastal
wave height directional spectra on a routine basis is discussed.
The theoretical development of such a system is described in the
following paper (*Barrick and Lipa*, 1978).

CONVENTIONAL HF SURFACE-WAVE RADARS

A conventional radar, microwave or HF, is a backscatter system
(i.e., the transmitter and receiver are colocated), and the station
is not in motion (so that any echo-signal Doppler shift is due to a
moving target). A coastal radar installation looks out over the
ocean at an angle of incidence very near grazing (with the exception of ionospheric over-the-horizon radars, which are discussed
by *Maresca* (1978, this volume). In this grazing configuration, the
propagation mode is known as "surface wave" or "ground wave."
Because of the high conductivity of sea water at HF, vertical polar-
ization must be employed. Radar distances in excess of 100 km are
achieved at MF/HF, even with installations right at sea level, due
to diffraction beyond the horizon. *Sommerfeld* (1909) first analyzed
this mode of propagation above a flat earth, while *Norton* (1941)
presented results for a spherical earth. *Barrick* (1971b) reviews
these results and extends the analyses to include the effect of
roughness on the spherical sea.

The conventional radar system isolates a target's position in
polar coordinates by (a) relating the range or distance from the
target to the elapsed time between pulse transmission and echo
reception, and (b) determining the azimuth angle to the target by
forming and rotating a narrow beam with the antenna system. Thus
a "radar resolution cell" which defines a patch of sea surface seen
by the radar at a given time should be as shown in Figure 1. Such
a (nearly) rectangular cell has radial width $c\tau/2$, where τ is the
effective radar pulse width and c is the free-space radio propaga-
tion speed ($\sim 3 \times 10^8$ m/s). Its transverse width is $R\Delta\phi = \frac{cT}{2}\Delta\phi$,
where T is the elapsed time between pulse transmission and reception,
and $\Delta\phi$ is the effective beamwidth of the anetnna system. A rough
relationship between antenna length, L (projected in the desired
direction, ϕ), beam width $\Delta\phi$, and wave length λ, is $\Delta\phi \simeq \lambda/L$ (radians).

The common description of the target scattering strength is
the "radar cross section" (e.g., *Skolnik*, 1962, or *Ruck et al.*,
1970). For the sea echo where we desire a received power spectral

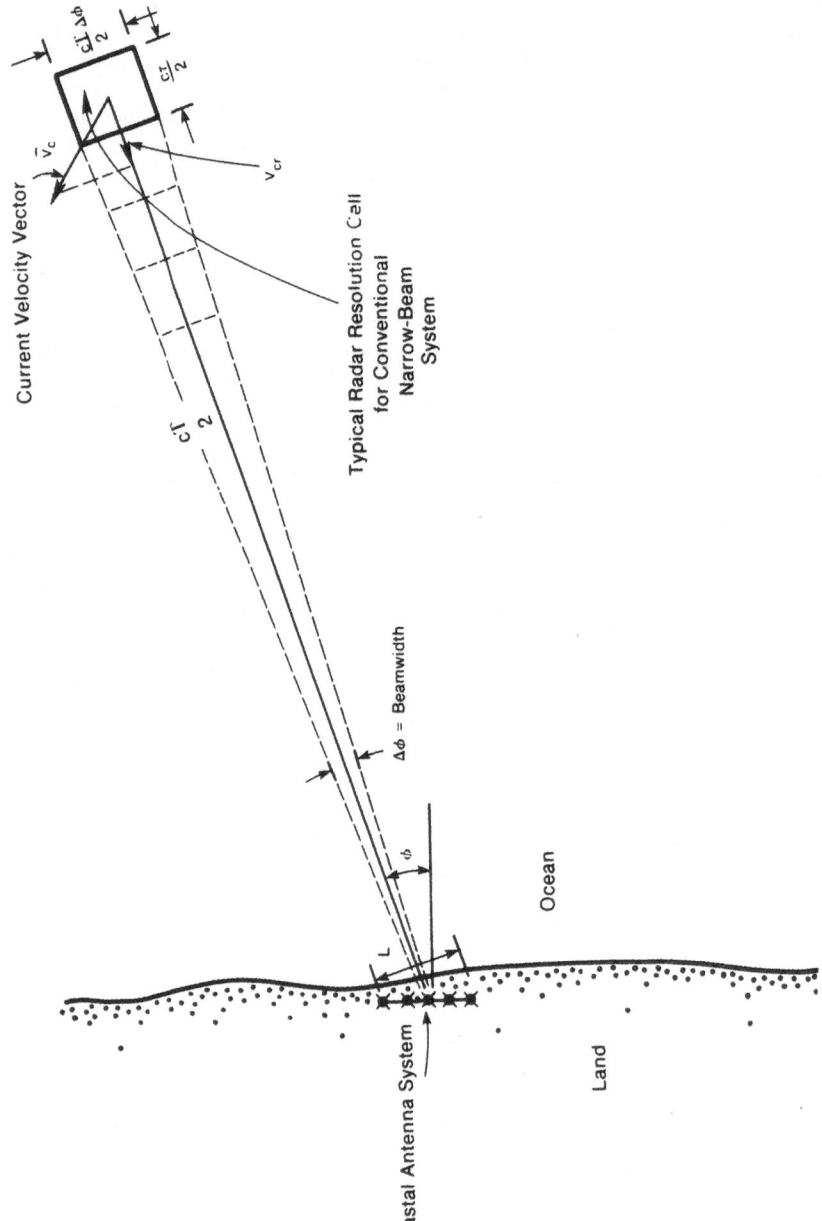

Fig. 1. Diagram showing geometry of typical narrow-beam coastal backscatter radar. The system can observe current velocity component v_{cr} radial to radar via Doppler shift.

density, the appropriate quantity is the (average) radar back-scattering cross section per rad/s bandwidth, defined as $\sigma(\omega,\phi)\Delta A$ where ΔA is the area within the radar resolution cell and ω = the frequency in rad/s. This is related to the average radar cross section per unit area of the sea, σ_o, as $\sigma_o = \frac{1}{2} \int_{-\infty}^{\infty} \sigma(\omega)d\omega$ (e.g., *Barrick*, 1972a and 1972b).

To first order, $\sigma_{(1)}(\omega,\phi)$ is given in terms of $S(\overline{\kappa})$, the spatial wave height spectrum of the sea as (*Barrick*, 1972a; *Barrick and Weber*, 1977)

$$\sigma_{(1)}(\omega,\phi) = 2^6\pi k_o^4[S(\overline{\kappa}_r)\delta(\omega_o - \sqrt{2gk_o}) + S(-\overline{\kappa}_r)\delta(\omega_o + \sqrt{2gk_o})], \quad (1)$$

where $\overline{\kappa}_r$ is the Bragg spatial wave vector pointing along the line of sight from the patch toward the radar (which we take for convenience as the x-axis), i.e.,

$$\overline{\kappa}_r = (2k_o,0) \quad\quad\quad\quad (2)$$

and k_o is the radio wave number $2\pi/\lambda$, with ω_o being the radian carrier frequency ($\omega_o = 2\pi f = k_o c$). We understand the spectrum $S(\overline{\kappa})$ to be a function of angle, ϕ, with respect to some convenient direction, such as North or, as shown in Figure 1, with respect to the perpendicular to the coastline. The quantity $\delta(x)$ is the Dirac-delta function of argument x. The above equation expresses quantitatively the simple Bragg-scatter relationship observed by *Crombie* (1955): The first-order sea echo shows up as two spectral "spikes" symmetrically spaced at shifts $\pm \sqrt{2gk_o}$ about the carrier ω_o, whose strengths are proportional to the wave height spectrum evaluated at the Bragg wave number $2k_o$ in direction ϕ. The position of these spikes are given by the deep-water dispersion equation, and is referred to as the Bragg frequency, $\omega_B \equiv \sqrt{2gk_o}$, where g is the acceleration due to gravity.

If the waves within the scattering patch are being transported by a current with velocity $\overrightarrow{v}_c$, having a "radial" component v_{cr} pointing toward the radar, the two spikes will be shifted by an amount $\omega_B\Delta \equiv 2v_{cr}/\lambda$ in the positive directions. (This means the arguments of the delta functions in (1) will be $\omega_o - \omega_B - \omega_B\Delta$ and $\omega_o + \omega_B - \omega_B\Delta$, respectively.) Figure 2 is an example of a measured sea-echo Doppler spectrum at 13.4 MHz, normalized to the Bragg

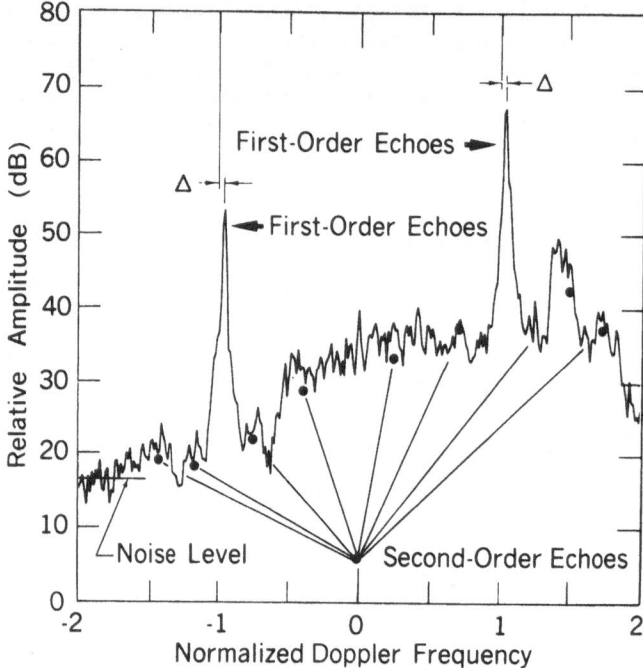

Fig. 2. HF sea-echo Doppler spectrum at 13.4 MHz measured from San Clemente Island. The frequency shift Δ is due to a current with a radial component speed of 22.5 cm/s.

frequency $\omega_B/2\pi$ = .374 Hz. This record was made with a large, permanent HF surface-wave radar facility looking westward from San Clemente Island in the Pacific. Both the dominant first-order spikes and their small current-induced shift are clearly in evidence. The radial component of current in this case corresponding to Δ was 22.5 cm/s, and this has been confirmed from simultaneous drifter measurements made by *Stewart and Joy* (1974).

From Equation (1) it can be seen that energy in the first-order echo is directly proportional to the ocean wave height directional spectrum at the wave vector of the Bragg-scattering wave train. A permanent coastal radar installation based on this first-order relationship is not practical, however, for the follow-ing reasons. (a) It would be necessary to vary the radar frequency over a wide region in the MF band in order to sample the wave height spectrum at the important longer wave lengths. Besides being very difficult to obtain accurate values of system gains and path losses at all these frequencies, the MF region is heavily used throughout the world for communications purposes, implying an intolerable interference problem. (b) It would be necessary to scan a narrow beam over azimuth, ϕ, to measure the directional nature of the wave height spectrum. This would require a huge antenna array tens of kilometers in length.

Because of these practical difficulties in using the first-order lines, attention has turned to the interpretation of the continuum portion of the radar spectrum. Although the required mathematics are complicated, the continuum is caused by scatter from all of the ocean wave trains and contains far more information on ocean surface properties.

Hasselmann (1971) suggested that the second-order sidebands around the first-order line should be proportional to the ocean wave amplitude spectrum. Although this idea is now known to be an over-simplification, it did indicate the importance of a rigorous understanding of the second-order structure. *Hasselmann* also suggested a convenient normalization in which the continuum is divided by the first-order energy. This removes unknown path losses and system gains from the signal, allowing absolute values of ocean wave parameters to be readily obtained.

The exact relationship between the second-order radar cross section and the directional ocean wave spectrum was given by *Barrick* (1972b). For backscatter of vertically polarized radar of wave number β, the second-order radar cross section at a Doppler shift ω from the carrier frequency is given by

$$\sigma_{(2)}{}^{(\omega,\phi)} =$$

$$2^6 \pi k_o^4 \sum_{m_1,m_2=\pm 1} \int\!\!\int_{-\infty}^{\infty} |\Gamma|^2 \delta(\omega - m_1\sqrt{gk_1} - m_2\sqrt{gk_2}) S(m_1\tilde{k}_1) S(m_2\tilde{k}_2) dpdq \quad (3)$$

where the two scattering ocean wave vectors are: $\tilde{k}_1 = \frac{1}{2}\bar{\kappa}_r + (p,q)$ and $\tilde{k}_2 = \frac{1}{2}\bar{\kappa}_r - (p,q)$ and k_1, k_2 are the corresponding wave numbers. The delta function enforces the second-order Bragg resonance condition. The quantity Γ is a coupling coefficient which is determined theoretically (*Weber and Barrick*, 1977) and given in *Barrick* (1972b) and which includes the effects of both hydrodynamic and electromagnetic nonlinearities carried to second order in a perturbation theory. This coupling coefficient is a function of $\tilde{k}_1$, $\tilde{k}_2$, and ω. The second-order regions of the measured spectrum are also shown in Figure 2.

INVERSION OF THE SECOND-ORDER ECHO

Successful inversion of the integral equation, Equation (3), could in principle yield the complete ocean wave spectrum from a single radar observation. However, because of the formidable nature of the integral, progress in inversion has been delayed until recently.

Nondirectional Information

The first step was taken by *Barrick* (1977a,b). In the approximation he used, the coupling coefficient is removed from the integral and expressed as a weighting function. Further, it is assumed that the directional dependence of the ocean wave spectra that appears under the integral sign in Equation (3) may be ignored. Integration is carried out over one variable using the delta function, and *Barrick* shows that the remaining integral is proportional to the nondirectional wave height spectrum, deriving the following closed-form approximation for the temporal ocean wave spectrum,

$$S_t(\omega-\omega_B) \approx \sigma_{norm}(\omega) \qquad (4)$$

where S_t is the temporal wave height spectrum at a frequency equal to the Doppler shift from the first-order Bragg frequency ω_B, and σ_{norm} is the second-order radar spectrum normalized by the first-order energy, divided by the factor $|\overline{\Gamma^2}|/2$; the latter is a "weighting function" obtained by integrating or averaging the coupling coefficient over the integration variable as a function of Doppler frequency, ω. The relation (4) is similar to that proposed by *Hasselmann* but accuracy is increased by the inclusion of the weighting function. The mean square wave amplitude is then obtained by integrating the energy of the normalized second-order sidebands over frequency. *Barrick* (1977b) has tested values of wave height obtained from measured radar spectra using this formula against estimates from a Waverider buoy. Results of this comparison for nine-sample radar averages are given in Figure 3 which shows that agreement is satisfactory for sufficiently high seas and/or radar frequencies. The standard deviation of the measured data points is $\sim23\%$, which is consistent with the $1/\sqrt{N} \simeq 33\%$ predicted standard deviation from finite-sample averaging theory.

Integral Inversion

Ultimately, we require an inversion technique that will give the complete ocean wave spectrum as a function of direction and

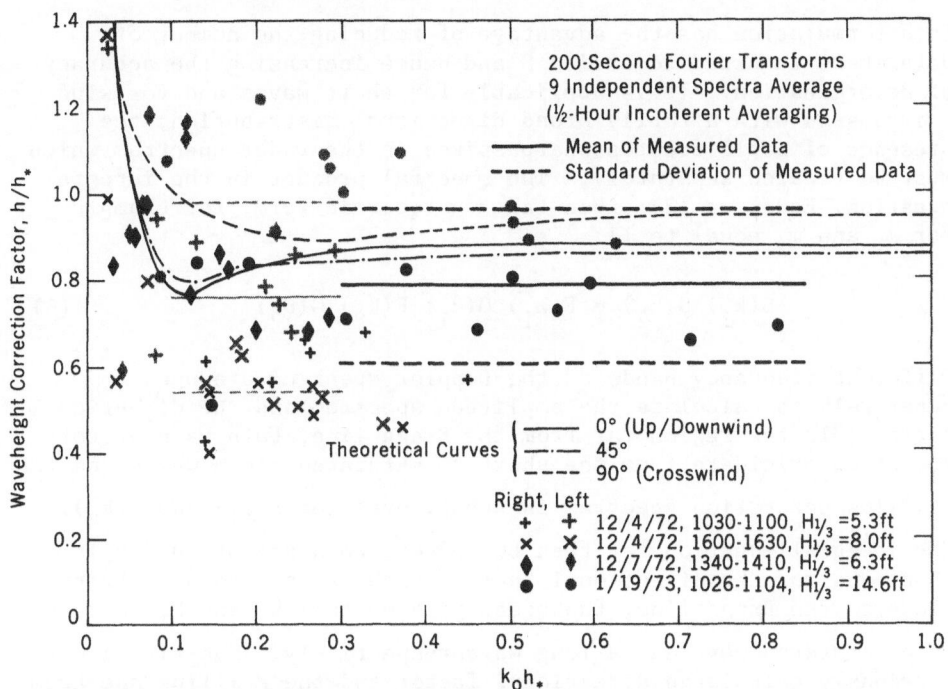

Fig. 3. Theoretical and experimental results of the use of the wave height inversion model; h is radar-deduced rms waveheight, and h_* is the wave height measured by a Waverider Buoy. k_o is the radar wave number. (After *Barrick*, 1977b)

wave length. As a step in this direction, *Lipa* (1977) has considered a model in which the complete directional ocean wave
spectrum wave vector k is expressed as the product of independent
factors, an amplitude spectrum dependent on wave number k and a
directional factor dependent on direction θ:

$$S(\tilde{k}) = F(k) \, G(\theta) \, . \tag{5}$$

This formulation has the advantage of reducing the number of
parameters required to define S and hence increasing the accuracy
of determination. It is applicable for short waves and for wind
driven seas with a farily broad directional distribution; the
presence of swell creates sharp spikes in the radar spectrum which
must be treated separately. The spectral product in the integral
equation, Equation (3), then takes a quartic form, for example
for m_1 and m_2 equal to $+1$:

$$S(\tilde{k}_1) \, S(\tilde{k}_2) = F(k_1) \, G(\theta_1) \, F(k_2) \, G(\theta_2) \, . \tag{6}$$

Different frequency bands of the Doppler spectrum are used
separately to calculate the amplitude spectrum and the directional
factor. In the region far from the Bragg line, both wave vectors
$\tilde{k}_1$ and $\tilde{k}_2$ originate from the shorter, saturated ocean waves and the
Phillips saturation spectrum is substituted for $F(k_1)$ and $F(k_2)$.

The integral equation may then be reduced to a set of quadratic
equations for the directional factor which are solved by iteration.
Close to the Bragg line, the ocean wave vectors $\tilde{k}_2$ and $\tilde{k}_1$ correspond
to a saturated wave and a long wave, respectively. Substituting the
previously calculated directional factor and the Phillips spectrum
for $F(k_2)$ linearizes the integral equation which may then be solved
for the low-frequency amplitude spectrum $F(k_1)$.

Testing the Inversion Method with Simulated Data

The inversion method has been tested by application to
theoretical power spectra corresponding to model wave height spectra.
Real signals were simulated by allowing the sea-echo spectrum to be
a random variable and then adding a finite number of independently
randomized spectra to obtain a sample average. Parameters of the
wave height spectrum derived by inversion were then compared with
the ideal values. Testing the method in this fashion is necessary,
as in practice one must always deal with a finite-sized sample
average.

Theoretical power spectra were obtained from Equation (3) by numerical integration, using a Pierson-Moskowitz model for the amplitude spectrum

$$F(k) \simeq 0.01 \exp[-0.74 \ (k_c/k)^2]/k^4 \tag{7}$$

and a cardioid normalized over angle for the directional factor

$$G(\theta) = \cos^s \left(\frac{\theta - \theta_w}{2} \right) \Big/ \int_0^{2\pi} \cos^s \left[\frac{\theta}{2} \right] d\theta \ . \tag{8}$$

The constant s defines the spread of the waves about the wind direction, θ is direction relative to the radar beam and θ_w is the wind direction. Randomness was included in the simulated signal by assuming the sea surface amplitude to be a Gaussian random variable. With a sample average of 200 independent Doppler spectra, the directional ocean wave spectrum was reproduced to within 10% and the wind direction to within 5°. Figures 4 and 5 show examples of recovered ocean wave parameters, plotted with the input theoretical functions. The inverted points are plotted with one standard deviation confidence limits computed from the signal variance corresponding to a Gaussian sea surface. For ocean wave spectra which are peaked in direction, the second order structure is extremely sensitive to wind direction; Figure 6 gives the standard deviation in wind direction estimated from a 200-sample average for s = 4.

From theoretical tests such as this, it was concluded that significant information on the ocean surface can be obtained by inversion of the second-order structure. Use of inversion as a practical tool depends on the validity of approximating the radar echo by first- and second-order terms. This is at present under investigation both theoretically by estimation of third-order terms and experimentally by inverting radar echoes obtained in conjunction with tilt buoy measurements. We now describe one such comparison (*Lipa*, 1978).

Test Application to Measured Data

The radar data, provided by the Stanford Center for Radar Astronomy were measured on the California coast using a coherent pulse-Doppler radar (*Teague et al.*, 1977). The narrow-beam system was operated with a transmitter frequency of 22 MHz and a half-power beam width of 10°. The power spectrum was computed with a

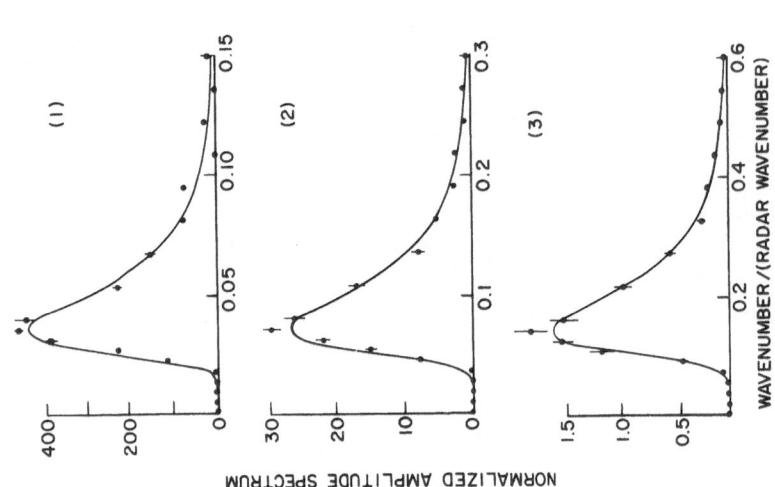

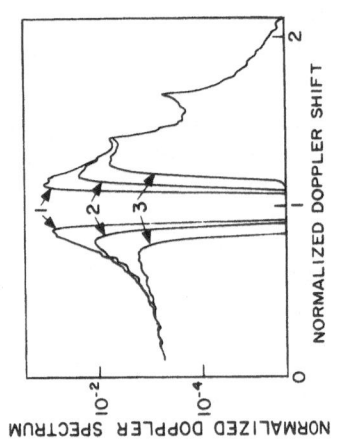

Fig. 4. Integral inversion for the amplitude spectrum at different wind speeds. Radar spectra on the left were calculated for a Pierson-Moskowitz model for the ocean wave spectrum: with a $\cos^4(\theta/2)$ directional distribution where the wind radar angle is 60°, and the wind speed for the three cases is given by (1) 31 knots (2) 22 knots (3) 16 knots. The model amplitude spectra (on the right) are the continuous lines and points derived by inverting the radar spectra and are plotted with their standard deviations. (After *Lipa*, 1977.)

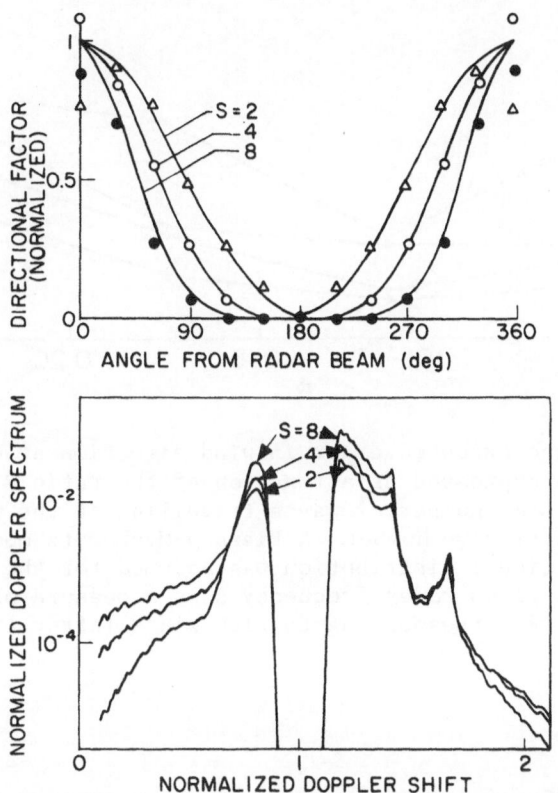

Fig. 5. Inversion for the directional factor. The theoretical radar spectrum are computed for a Pierson-Moskowitz amplitude spectrum and a $\cos^s(\theta/2)$ directional factor for values of s equal to 2, 4, 8. The radar frequency was 30 MHz, wind speed 22 knots, and wind/radar angle 0°. Model directional factors are normalized at the origin and are drawn as continuous lines together with the points derived by inverting the radar spectrum (after *Lipa*, 1977).

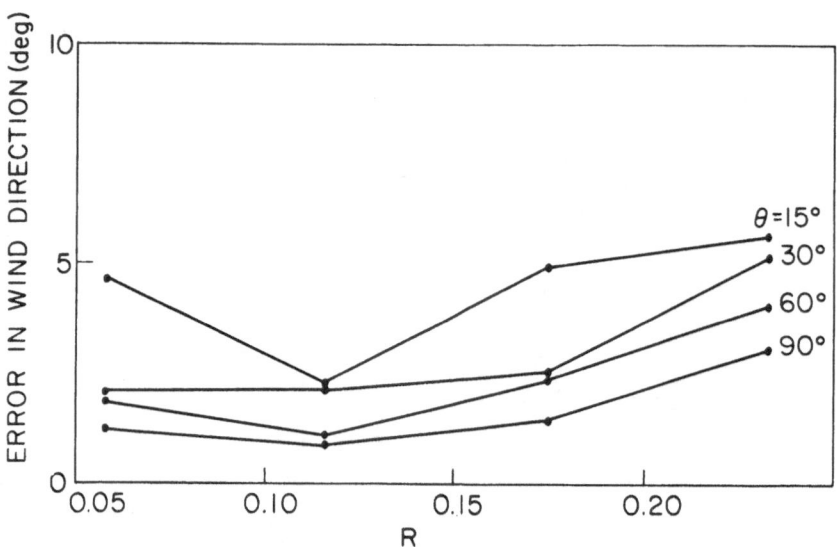

Fig. 6. The expected error in wind direction at different
wind directions expressed as a function of the ratio R = k_c/β
where k_c is the wave number of waves travelling at the wind speed
and β is the radar wave number. A Pierson-Moskowitz model with a
$\cos^4(\theta/2)$ directional distribution was assumed for the ocean wave
spectrum. At a given radar frequency R is a measure of wind speed,
small values of R corresponding to high winds (after *Lipa*, 1977).

frequency resolution of 0.01 Hz and 48 independent spectra were
averaged to give the spectrum used for inversion, which is shown
in Figure 7.

The tilt buoy measures the first six Fourier coefficients
$N_{pq}(k)$ of the directional ocean wave spectrum $S(k,\theta)$ where

$$N_{pq}(k) = \int_0^{2\pi} \cos^p(\theta)\sin^q(\theta)\ S(k,\theta)d\theta$$

(9)

$$p,q = 0,\ 1,\ 2\ |p + q| \leq 2\ .$$

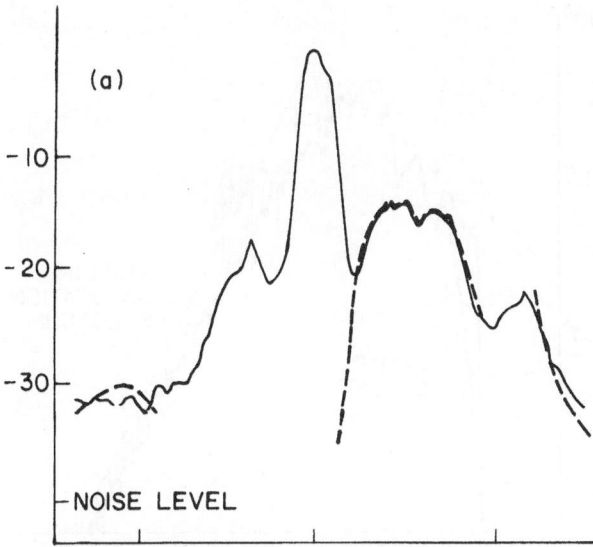

Fig. 7.   A measured radar spectrum for positive Doppler shifts.
The transmitter frequency is 22 MHz and the frequency resolution is
0.01 Hz.   The continuum is normalized by the first-order spectral
energy and inverted to give a model of the directional ocean wave
spectrum shown in Figures 8 and 9.   The dashed line is the second-
order radar spectrum that corresponds exactly to the derived model
ocean wave spectrum for the frequencies used for inversion (after
Lipa, 1978).

Of these, five are independent as $N_{00}(k) = N_{02}(k) + N_{20}(k)$. $N_{00}(k)$
is just the amplitude spectrum of the ocean waves, which may be
compared with that derived by inversion of the radar spectrum.
Results of the comparison are shown in Figure 8.   The directional
factor derived from the radar data is shown in Figure 9 with a
wave direction opposite the radar beam, approximately from the
northwest.   The next five Fourier coefficients were obtained by
integrating the directional factor; they are compared with the
buoy results in Table 1, which also contains comparisons for wind
direction and rms wave height.   The radar and tilt buoy results are
consistent within the standard error for the 48 sample average
(<20% for the ocean wave spectrum, 10° for the mean wave direction).

Finally, the derived model ocean wave spectrum was checked for
consistency with the radar data.   This test was performed by computing that portion of the radar spectrum corresponding to the
derived model ocean wave spectrum; the resulting radar spectrum is

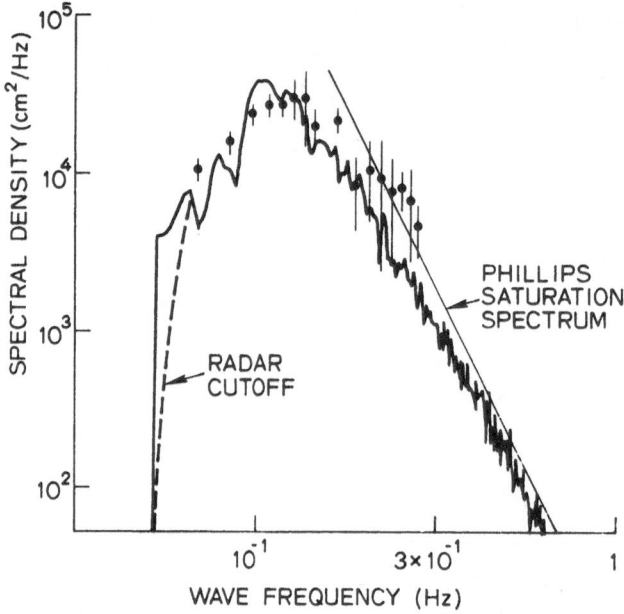

Fig. 8. Comparison of the temporal spectral obtained by inversion of the radar spectrum shown in Figure 7 with that measured by a tilt buoy (continuous line). Inverted points are plotted with one standard deviation error bars. (After *Lipa*, 1978.)

shown by the dotted lines in Figure 7 and agreement with the data is considered reasonable. This test case illustrates the fact that important ocean wave parameters may be readily derived from the sea echo from a narrow-beam HF radar.

THE NOAA CURRENT-MAPPING RADAR SYSTEM

NOAA'a Wave Propagation Laboratory undertook a project two and one-half years ago aimed at exploiting the Doppler shift added to the echo by underlying currents. The developmental portion of the project had the following specific objectives of interest here: (a) The construction of a small, low-powered, transportable, rugged solid-state radar system; (b) The inclusion and use of a minicomputer system both to control the entire operation of the radar and also to process the data at the site, producing current maps in near-real time; (c) The development of a compact antenna system easily erectable within an hour, which circumvents the large permanent structures discussed previously. This section discusses engineering details of this system and the antennas. Technical details related to the measurement and interpretation of current

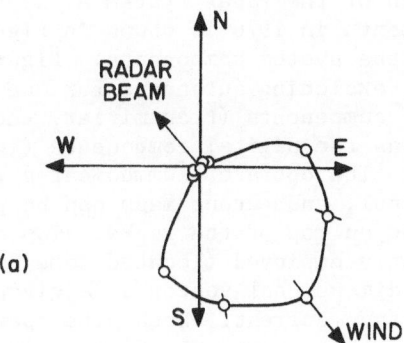

(a)

Fig. 9.   The directional factor obtained by inversion of the radar spectrum.   The wind direction is computed to be down the radar beam.   Moments of the directional factor are compared with those measured by the tilt buoy in Table 1.   (After *Lipa*, 1978.)

TABLE 1.   Comparison of radar and buoy observations.
(After *Lipa*, 1978.)

	Radar	Buoy
RMS wave amplitude:	0.61 ± 0.1 m	0.54 m
Wind direction:	$312° \pm 5°$	310°
Moment Ratios:		
N_{10}/N_{00}	-0.56 ± 0.1	-0.59
N_{01}/N_{00}	0.51 ± 0.09	0.49
N_{20}/N_{00}	0.51 ± 0.09	0.61
N_{02}/N_{00}	0.49 ± 0.08	0.58
N_{11}/N_{00}	-0.12 ± 0.03	-0.19

data with this system have been published elsewhere *Barrick and
Evans,* 1976; *Barrick et al.,* 1977).

An artist sketch of the radar system as it operated in Florida
for current measurements in 1976 is shown in Figure 10, put together
from photographs of the system components. Figure 11 is a photo of
the actual hardware, excluding antennas, for one site; the right
rack contains the RF components (transmitter, receiver, etc.) and
the left rack contains the digital components (including the PDP
11/34 minicomputer). The operator communicates with the system via
the typewriter terminal, and output maps can be produced by the pen
plotter, both resting on top of the racks. Two complete sites of
such gear are presently employed (located some 30-40 km apart) to
deduce a total, two-dimensional vector at a given point on the sea
representing the surface current. Each site operates completely
independently of the other, however. Figures 10 and 11 illustrate
that the total hardware for an HF radar system can be made compact
and transportable.

The major departure of this system from conventional HF radars
is in the use of a novel antenna concept, eliminating the need for
the large, permanent, expensive array farms employed in the past.

Fig. 10. Sketch of NOAA current-mapping radar system site as
operated in Florida.

Fig. 11. Photograph of complete radar hardware. Including
digital and RF components.

The transmitting antenna is simple in concept, merely flooding the
sea in front of the radar with a nearly uniform angular density of
energy. The receiving system, however, consists of a group of
three or four independent "citizens-band" whips. The spacing
between these whips and their orientation on the beach are measured
and input to the computer. The complex voltages from each antenna
are sampled on a time-multiplex basis each millisecond, so that for
all practical purposes the antenna voltages are recorded simultaneously relative to sea-surface motions. Then we employ a closed-
form mathematical expression (available in *Barrick et al.*, 1977) to
extract the angle of arrival of the sea-echo signals at each Doppler
frequency output from the digital FFT (fast Fourier transform).
This receiving system is not an array which forms a beam in the conventional sense; if it were operated as a phased array, its beam
width at 25 MHz would be > 90°, far too large to resolve currents

of a fine grid scale. Rather, it is in concept similar to a radio
direction finder. Using the algorithm, for example, one can precisely determine the angles of arrival of two signals at the same
Doppler frequency. With additive noise present along with the
desired signals, the angular accuracy decreases. Our simulations
and experiments, however, show that even for mean signal-to-noise
ratios as low as 10 dB, rms angular extraction errors are better
than ± 2.

     To extract and map currents, this system operates on the first-
order portion of the sea-echo Doppler spectrum only. In the absence
of current, our previous discussions show that this echo is a very
narrow spectral peak, approximating a Dirac-delta function. Before
angular extraction, however, the signal at a given time delay corresponds to the sea echo from within a semi-circular ring of surface
area. With a current such as the South-North Gulf Stream flowing
through this ring, the added Doppler shifts due to the radial component of current will tend to "smear" this first-order narrow spike
into a broad spectral band. Figure 12 is an example showing the
stronger, positive first-order sideband measured by this system
looking eastward into the Gulf Stream from Ft. Lauderdale, and the
expected position of the first-order echo. Hence the Doppler shift
of each point output from the FFT retained after thresholding can be
directly related to the radial current velocity; a scale showing
this relationship is superposed above the spectrum. Then the angle
of arrival of each one of these radial current velocities is determined from the complex voltages originating from the receiving
antennas. Thus by using the first-order echo we have derived a
data array in polar coordinates giving the radial current velocity
pattern as a function of range (time delay) and azimuth angle. Such
data is combined with an analogous array from the other radar site
to produce current maps such as that shown in Figure 13 at a preset
pattern of grid points.

     The point is emphasized that surface-current maps have been
successfully constructed with quite fine areal resolution, although
"conventional wisdom" and a brute-force approach would have called
for antenna system exceeding 350 m in length. The discovery of the
novel, simple antenna concept to do this particular job was made
possible by understanding the physics of the sea echo to be used
for the analysis, and the nature of the typical current patterns
likely to be encountered in coastal waters.

                              CONCLUSION

     Theoretical models have been derived relating the first- and
second-order HF sea-echo Doppler spectrum to the wave height directional spectrum. The underlying physical interaction mechanism at

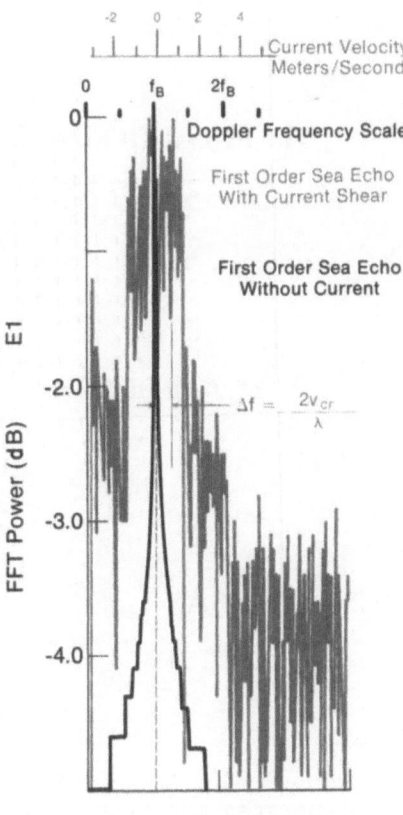

Fig. 12. Plot of FFT spectral power output. The black
spectrum is the idealized test sea-echo spectrum in the absence
of a current. The gray spectrum is the measured sea echo at 37.5
km range from Ft. Lauderdale, Fla., as modified by Gulf Stream
current field.

the sea surface can be described as Bragg scatter: to first order,
the radar wave interacts with a single-ocean wavetrain whose wave
length is half that of the radar, while to second order, a continuous
summation of two-ocean wave train sets interact with the radar wave.
Narrow-beam radar measurements which are made with coastal HF
installations having large antenna systems confirm that these models
correctly explain the sea echo. Inversion techniques employing the
nonlinear integral equation for second-order scatter have successfully demonstrated that sea directional spectral information can be
extracted from the echo Doppler.

A recent NOAA developmental program has shown that transportable, low-power HF radars are feasible for coastal ocean-surface

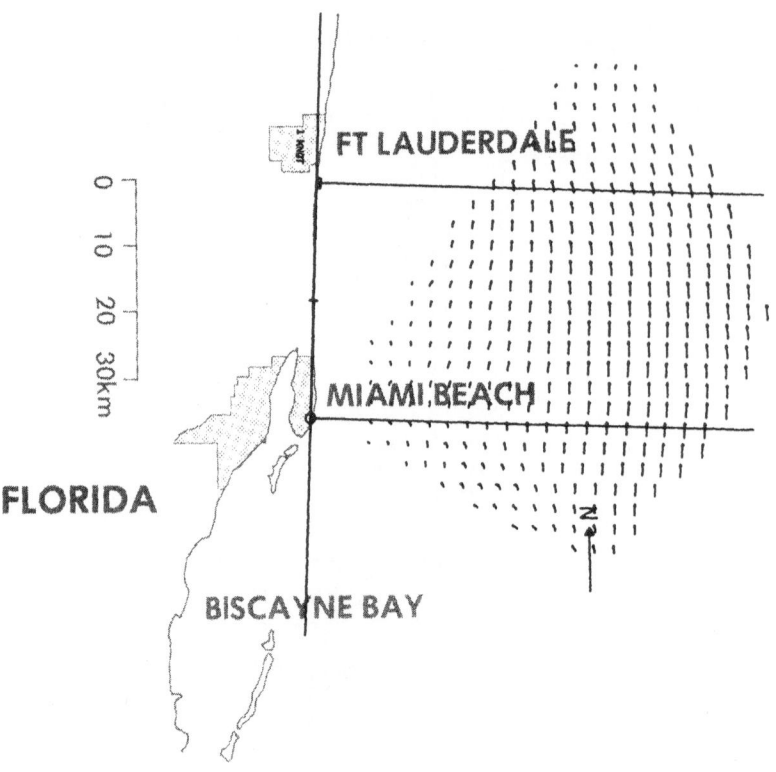

Fig. 13.   Computer-generated map of the radar-deduced Gulf Stream surface current on 20 October 1976.

measurements. In particular, two such minicomputer-controlled units have produced surface-current vector maps in Florida and Alaska in near-real time from the first-order portion of the sea echo. The antenna system for this radar, in contrast to the earlier huge conventional installations, is small and easily erectable on the beach in less than an hour.

The convenience and portability of the current measurement system is in strong contrast to systems which at present operate to measure the directional ocean wave spectrum. Even in the upper HF region, antenna sizes required to form a narrow beam are sufficiently great that permanent, expensive coastal installations would be needed. An array antenna to form and steer a beam of 2° width at 25 MHz, for example, would be 350 meters in length. While novel traveling-wave antennas such as a single vertical half-rhombic held aloft by a helium-filled balloon such as that used by Stanford (*Teague et al.*, 1977) are less gigantic, they nonetheless still preclude the possibility of simple, routine measurements requiring

a minimum of staff.   Furthermore, such a system has a wider beam
(10° - 20°), which cannot easily be steered in angle; this implies
that reasonable angular accuracy in obtaining the directional
spectrum via integral inversion is difficult at its important lower
end.   In the following paper *(Barrick and Lipa,* 1978), we propose
and test a novel, small antenna scheme which uses the same radar
hardware as that used for mapping currents.   This system is expected
to yield the wave height directional spectrum from inversion of the
second-order Doppler spectrum, and would then eliminate the requirement of large coastal installations.

ACKNOWLEDGEMENT

    Part of this work was performed under Office of Naval
Research contract N0014-77C-0356 while Dr. Lipa was at Stanford
University.

REFERENCES

Barrick, D. E.   1971a.   Dependence of second-order sidebands in HF
    sea echo upon sea state, 1971 IEEE G-AP International Symposium
    Digest, Sept. 21-24, Los Angeles, Calif., 194-197.

Barrick, D. E.   1971b.   Theory of HF/VHF propagation across the
    rough sea, Parts I and II, *Radio Science,* 6: 517-533.

Barrick, D. E.   1972a.   First-order theory and analysis of MF/HF/VHF
    scatter from the sea, *IEEE Trans. on Antennas and Propagation,*
    AP-20: 2-10.

Barrick, D. E.   1972b.   Remote sensing of sea state by radar, In:
    Remote Sensing of the Troposphere, V. E. Derr, ed., Chapter 12,
    U.S. Government Printing Office.

Barrick, D. E., J. M. Headrick, R. W. Bogle, and D. D. Crombie.
    1974.   Sea backscatter at HF: interpretation and utilization
    of the echo, *Proceedings of the IEEE, 62:* 673-680.

Barrick, D. E. and M. W. Evans.   1976.   Implementation of coastal
    current-mapping HF radar system.   Progress report No. 1, NOAA
    Tech. Report ERL 373-WPL 47.

Barrick, E. W., M. W. Evans, and B. L. Weber.   1977.   Ocean surface
    currents mapped by radar, *Science, 198:* 138-144.

Barrick, D. E. and B. L. Weber.   1977.   On the nonlinear theory of
    gravity waves on the ocean's surface.   Part II: Interpretation
    and applications, *J. Phys. Oceanogr.,* 7: 11-21.

Barrick, D. E.   1977a.   Extraction of wave parameters from measured
    HF sea-echo Doppler spectra, *Radio Science, 12:*   415-424.

Barrick, D. E.   1977b.   The ocean waveheight nondirectional spectrum
    from inversion of the HF sea-echo Doppler spectrum, *Remote Sens-
    ing of Environment, 6:*   201-227.

Barrick, D. E.   1978.   HF radio oceanography -- a review, *Boundary-
    Layer Meteor., 14:*   35-55.

Barrick, D. E. and R. J. Lipa. 1978. A compact transportable HF radar system for directional coastal wavefield measurements, In: *Ocean Wave Climate*, M. D. Earle and A. Malahoff (Eds.), Plenum, New York, (This volume).

Crombie, D. D. 1955. Doppler spectrum of sea echo at 13.56 Mc/s, *Nature*, *175*: 681-682.

Crombie, D. D. 1972. Resonant backscatter from the sea and its applications to physical oceanography, *Proceedings of the IEEE Ocean '72 Conference* (IEEE Publ. 72CHO 660-1 OCC), 173-179.

Hasselmann, K. 1971. Determination of ocean wave spectra from Doppler radio return from the sea surface, *Nature Physical Science*, *229*: 1617.

Lipa, B. J. 1977. Derivation of directional ocean-wave spectra by integral inversion of second-order radar echoes, *Radio Science*, *12*: 425-434.

Lipa, B. J. 1978. Inversion of second-order radar echoes from the sea, *J. Geophys. Res.*, *83*, 959-962.

Maresca, J. W. Jr. 1978. High-frequency skywave measurements of waves and currents associated with tropical and extra-tropical storms, In: *Ocean Wave Climate*, M. D. Earle and A. Malahoff (Eds.), Plenum, New York, (This volume).

Norton, K. A. 1941. The calculation of ground wave field intensity over a finitely conducting spherical earth, *Proc. IRE*, *29*: 623-639.

Ruck, G. T., D. E. Barrick, W. D. Stuart, and C. K. Krichbaum. 1970. *Radar Cross Section Handbook (Vols. I and II)*, Plenum, New York, 7-13, 671-676.

Sommerfeld, A. 1909. The propagation of waves in wireless telegraphy, *Ann. Physik*, *28*: 665-736.

Skolnik, N. I. 1962. *Introduction to Radar Systems*, McGraw-Hill Book Company: New York, 20-56, 521-534.

Stewart, R. H. and J. W. Joy. 1974. HF radio measurements of surface currents, *Deep Sea Res.*, *21*: 1039-1049.

Teague, C. C., G. L. Tyler, and R. H. Stewart. 1977. Studies of the sea using HF radio scatter, *IEEE Trans. on Antennas and Propagation*, *AP-25*: 12-19.

Teague, C. C. 1978. Synthetic aperture HF radar wave measurement experiments, In: *Ocean Wave Climate*, M. D. Earle and A. Malahoff (Eds.), Plenum, New York, (This volume).

Tyler, G. L. W. E. Faukerson, A. M. Peterson, and C. C. Teague. 1972. Second-order scattering from the sea: ten-meter observations of the Doppler continuum, *Science*, *177*: 349-351.

Wait, J. R. 1966. Theory of the HF ground wave backscatter from sea waves, *J. Geophys. Res.*, *71*: 4839-4842.

Weber, B. L. and D. E. Barrick. 1977. On the nonlinear theory of gravity waves on the ocean's surface. Part I: derivations, *J. Phys. Oceanogr.*, *7*: 3-10.

A Compact Transportable HF Radar System for Directional Coastal Wave Field Measurements

Donald E. Barrick[1] and Belinda J. Lipa[2]

[1]Wave Propagation Laboratory, National Oceanic and Atmospheric Administration and [2]SRI International

ABSTRACT

A low-powered transportable coastal radar system which can measure the first five angular Fourier coefficients of the wave height directional spectrum as a function of wave number is proposed and described. Operating at a single frequency in the upper HF region, the surface-wave radar employs a novel, stationary three-element receiving antenna to obtain angular information. The received signals from two crossed loop antennas and a monopole, all aligned along the same vertical axis and standing ~2 m high, are combined digitally to form and scan a broad cardioid beam. The second-order portion of the sea-echo Doppler spectrum is used to extract wave spectral information. This echo portion is described mathematically by a nonlinear integral equation. Trigonometric basis functions are used to represent the radar system output (both first and second order) as well as the wave height spectrum's angular dependence.

The first-order echo is used to linearize the integral equation in an approximation valid at upper HF for higher sea states (e.g., at 25 MHz for rms sea wave heights greater than 0.4 m). A Fredholm linear integral equation in which the radar data and the desired wave data are five-element vectors or tensors is then obtained. Different inversion methods are employed for the echo region between the first-order peaks and for the region beyond these peaks. Inversion error is examined based upon N-sample averaging of the random sea-echo voltage, and a stabilization technique is introduced to circumvent the problem of ill-conditioning. The standard deviation of the five coefficients is obtained from simulations using a

Phillips wave spectral model and error propagation theory. The accuracy of these radar-derived coefficients is compared with that obtained with a pitch-and-roll buoy over the same two-hour observing period and for the same frequency resolution. Near the spectral peak, typical radar inversion errors are 2-3% versus 13% buoy errors for the zero-order coefficient (i.e., the nondirectional wave height spectrum); the two first-harmonic coefficient accuracies are typically 2-4% for the radar, while they can be as high as 17% for the buoy. The less important second-harmonic coefficient comparisons are 2-4% for the radar and ∿4% for the buoy. These accuracies are generally consistent with the inverse square-root relation to the number of independent samples; for the same observation time and frequency resolution, the radar observes many more samples from area averaging.

BACKGROUND

The preceding paper reviews progress in the use of coastal HF radar systems for measurement of ocean surface features. In the upper portion of the HF band where the second-order region of the sea-echo Doppler spectrum is used to extract wave information, it is seen that conventional systems require large permanent antenna systems hundreds of meters long. There are many potential applications where such a system must be ruled out because of its size, cost, or adverse environmental impact.

The characteristics of a desirable coastal radar system for monitoring the wave height directional spectrum include the following: (i) The hardware should be compact, low-powered, and reliable; (ii) the radar system operation should be mini-computer controlled, with a minimum of required operator intervention; (iii) the data should be processed digitally so that final desired output results can be available in near-real time; (iv) the antenna system should be simple, non-moving, easily erectable, and unobtrusive on the beach; (v) the system should obtain its data by transmitting only a single frequency; it should not require a knowledge of system gains, path losses, and other amplitude factors that are difficult to measure and likely to vary from day to day. The NOAA current-mapping radar system, which is described in the previous paper, possesses all of these desirable attributes. The requirement for measuring directional wave fields instead of surface currents means, however, that items (iv) and (v) must be reconsidered.

In order to utilize data obtained at a single radar frequency, the preceding paper shows that the second-order portion of the Doppler spectrum (described mathematically by Equation (3) there) must be used to obtain wave information. Any practical inversion scheme employs the first-order echo (originating from short, fully

developed waves) as a signal normalizing factor, thus removing
unknown path losses and drifting system gains which would otherwise contaminate the desired data. The technique discussed here
will also employ this self calibration, thus satisfying item (v)
above.

     To provide wave field information, however, a different
antenna system from that used for the current-mapping radar is
needed. This is necessitated by the fact that the second-order
Doppler spectrum is required, whereas we used only the first-order
echo for currents. The former three or four antenna systems and
the resulting simple direction-finding equation will not work
for the present application, because that technique assumes that
no more than two signals arriving from different directions will
have the same Doppler shift. First-order sea echo in the absence
of any current returns at a single Doppler frequency, even though
the scattering patch may subtend 180° of sea (e.g., a semicircular
range gate). A current pattern will spread this discrete frequency into a band when the radial current velocity varies over
the resolution cell area; a realistic pattern should produce at
most two signals from different directions having the same radial
velocity and hence the same Doppler shift. However, the Doppler
spectrum of the second-order sea echo from a single (narrow-beam)
azimuthal direction is already spread over a continuum. Signals
arriving from many directions (e.g., the semicircular range gate)
will all have the same Doppler shift, and therefore the former
direction-finding algorithm will be unable to extract direction
for the second-order signals.

     As a result, we propose and analyze a new receiving antenna
system to circumvent this failure.   Three single antennas are
employed: two crossed (orthogonal) loop antennas and a monopole,
all aligned along the same vertical axis, as shown in Figure 1.
As with the current-mapping radar system, time-multiplex switching
at a 0.5 millisecond rate among the three is used to record the
three complex voltages, for all practical purposes simultaneously.
The antennas are not moved or rotated mechanically. However, the
signal voltages are combined digitally to cause a fairly broad
beam to be scanned in azimuth. This produces an integral equation
over angle which yields the first five Fourier angular coefficients
for the signal return. By expanding the wave height directional
spectrum over the same angular Fourier basis functions, a second
integral equation involving the second-order Doppler spectrum is
solved to obtain these five angular wave height coefficients as
a function of wave frequency. Thus, with this very simple antenna
system, precisely the same wave directional spectral information
as one obtains with a pitch-and-roll buoy (*Stewart*, 1977) is
recovered so that one-to-one comparisons of data from two different
wave sensor systems can be made.

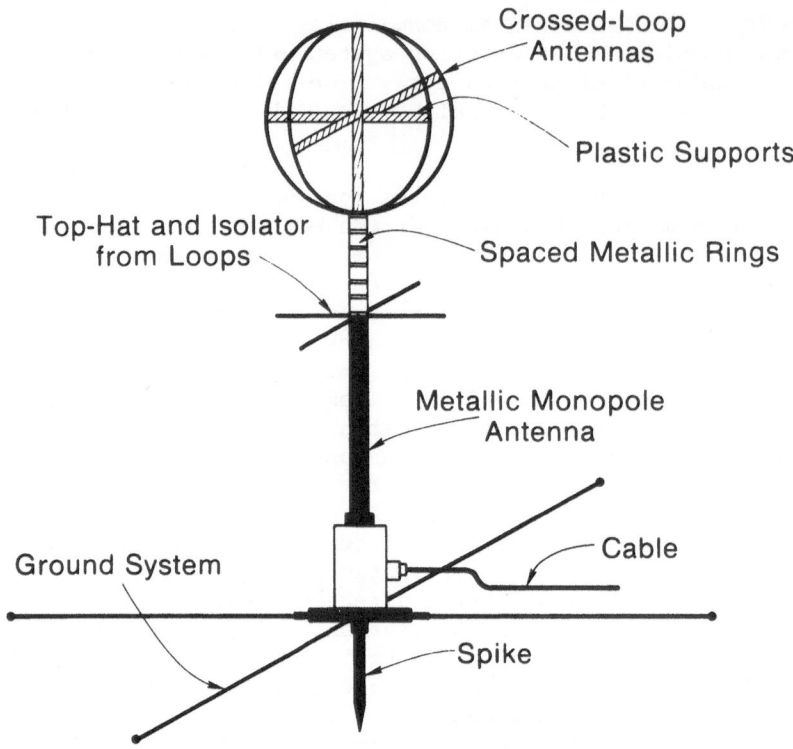

Fig. 1. Sketch of three-element (cross-loop/monopole) receiving antenna used for wave height directional spectrum measurements. The height is approximately 2 m.

THE ANTENNA SYSTEM AND BEAM SCANNING

The transmitting antenna consists of a simple monopole, or "whip," fed against a four-element radial ground plane; it radiates power isotropically in azimuth. It is tuned or matched to the 50Ω input from the transmitter and feed line, so as to effect maximum power transfer to the radiated field within the upper HF band of radar operation. The length of the antenna element and the four radial grounding elements are ∿2.6 m; these elements are available commercially from amateur radio shops. Figure 2 shows a sketch of the system as it has been operated in the past.

The ability to extract wave direction derives from the receiving antenna system. Two separate loop antennas, orthogonal in direction, are employed (i.e., "crossed" loops) along with a monopole. The signals from the three antennas are combined after

Fig. 2.   Sketch of NOAA coastal HF radar system for wave field measurements. The omnidirectional transmitting antenna is on the left and the receiving antenna is on the right.

analog-to-digital conversion following reception to form a broad cardioid beam with shape factor $\cos^4((\psi-\phi)/2)$. This beam is caused to scan in angle ψ, where ϕ is the direction of the originating signal, both with respect to any arbitrary reference. Figure 3 shows this beam pattern at the coast and the resulting integral which relates the continuous sea echo $g(\phi)$ originating from a patch at ϕ to the system output, $f(\psi)$, at a scan angle ψ.

Beam synthesis and scanning occurs in the following manner. A single small loop (whose total circumference is less than one-third the radar wave length, λ) has a voltage pattern which varies as $\sin\phi$, where ϕ is taken with respect to the loop axis. Larger loops whose electrical circumference exceeds $\lambda/3$ have a Bessel-function pattern (*Kraus*, 1950); this departure from the simple $\sin\phi$ pattern precludes their use in the simple manner suggested

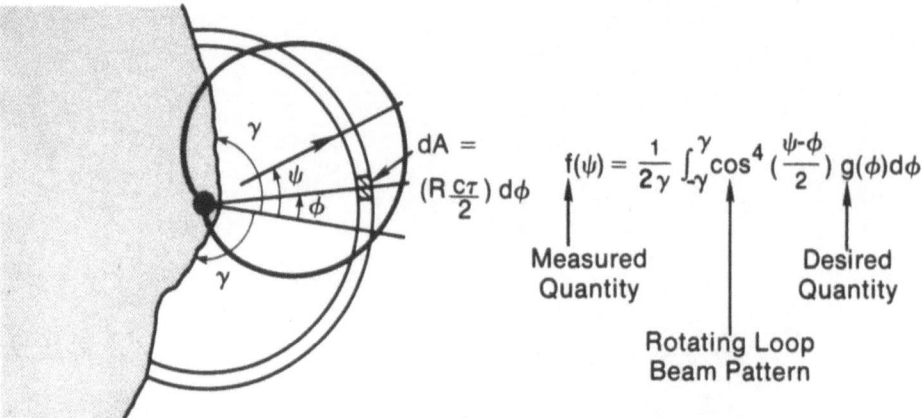

Fig. 3. Sketch of beam pattern and integral of radar-measured signal pattern, $f(\psi)$, vs actual echo pattern, $g(\phi)$, for loop/monopole scanning system.

here. Another identical loop at right angles to the first thus has a pattern $\cos\phi$ with respect to the same angle reference. Call the first loop "B" and the second loop "A." If the loop planes intersect along the same vertical line and the loops are electrically identical, the received signals can later be multiplied by factors proportional to $\cos\psi$ (for Loop A) and $\sin\psi$ (for Loop B) and summed, as shown in Figure 4. The digital sum voltage is then proportional to $\cos(\psi-\phi)$. Furthermore, if a monopole antenna lies along the same vertical axis as the loop-plane intersection, its omnidirectional azimuthal pattern produces a voltage independent of the signal direction, ϕ. If its amplitude is adjusted to be equal to that of the loops at their beam maxima (and assuming there is no phase mismatch due to unequal phase paths to and through the receiver), then the signal from this can be added digitally to the previous crossed-loop voltage sum to give $1 + \cos(\psi-\phi) = 2 \times \cos^2((\psi-\phi)/2)$. The power pattern is then proportional to the square of this quantity, giving rise to the $\cos^4$ beam pattern shown in Figure 3. The integral there is obtained due to the continuous contribution of sea echo with respect to azimuth angle ϕ. Thus we have created a broad beam and can cause it to be scanned, all by digital manipulation of the three received signals; no mechanical rotation of the antennas is required.

ϕ — Azimuth Angle From Loop B Axis;
ψ — Desired Direction of Beam Maximum;
$S(\phi)$ — Signal From Angle ϕ.

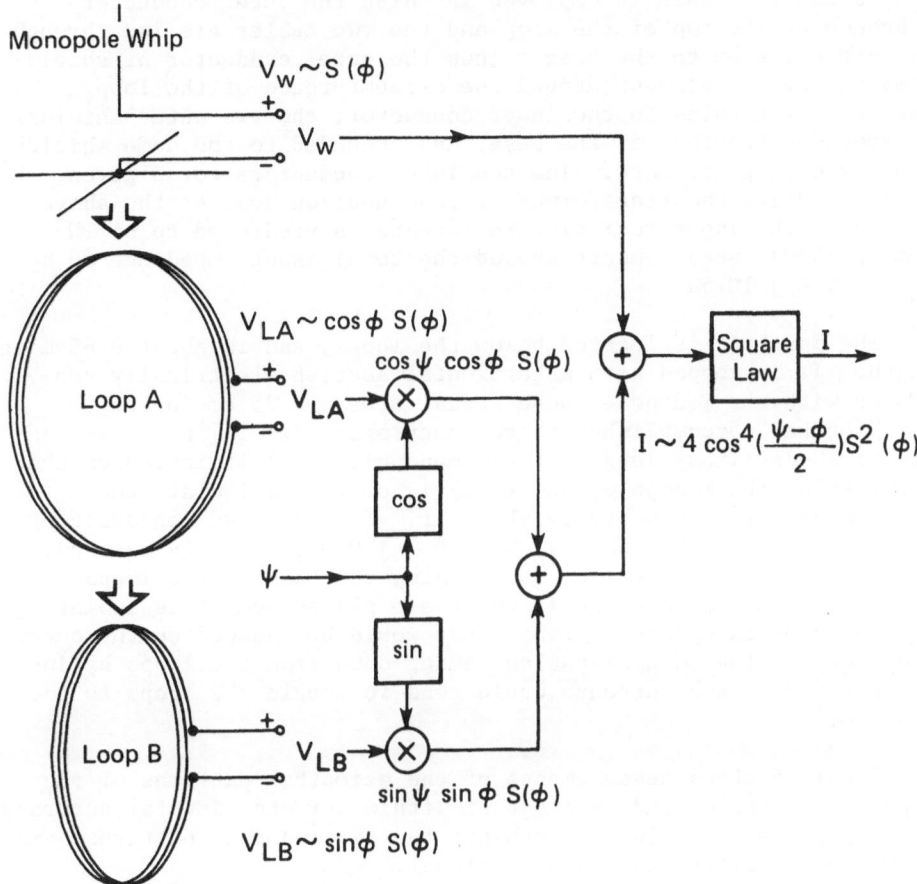

Fig. 4. Block diagram of mathematical combination of signals from the three antennas to produce $\cos^4((\psi-\phi)/2)$ beam pattern.

A sketch of the antenna system is shown in Figure 1. Based upon this sketch, we have designed and tested an antenna system, a photograph of which is shown in Figure 5. The rest of this section discusses the electrical properties of this system.

Each loop is 1.0 m in diameter; at 25 MHz, the total electrical length of the (single) wire in the hollow plastic circular tube is 0.266λ, meeting the required criterion for maintaining the $\sin\phi$ magnetic dipole pattern. To maintain electrical balance and isolation, a coaxial cable is employed in which the outer conductor is broken at the top of the loop and the two cables are led through the vertical stem to the base. Thus the outer conductor or shield forms the active element around the circumference of the loop, capacitively coupled to the inner conductor; the two outer shields are grounded together at the base, and grounded to the same shields from its orthogonal mate. The two inner conductors for a given loop then drive the transformer at the junction box. At the above frequency, the input radiation resistance is predicted to be 1Ω (*Kraus*, 1950); measurements showed the total input impedance to be $Z_{in} \simeq 0.9 - j\ 18.5\Omega$.

The monopole is located below the loops, and is about 0.85 m in length. It is topped with a horizontal section electrically continuous with the monopole whose radial arms are 25 cm in length. This "top hat" accomplishes three functions: (i) it increases the effective electrical length of the monopole; (ii) it increases the bandwidth of the monopole; and (iii) it serves to isolate the monopole from the loop system above it. The measured impedance of the monopole at 25 MHz was $Z_{in} \simeq 48.5 - j\ 0.5\Omega$, a nearly resonant condition. The top hat lies 50 cm below the base of the loops. Metallic bands spaced slightly apart are placed over this isolating section to break up any currents that would be induced on the outer conductors of the coaxial cables coming down from the loops by incoming fields; such currents would tend to couple the loops to the monopole.

Figure 6 shows measurements of the azimuthal patterns of the loops and monopole, indicating that within our experimental accuracy the loop patterns do in fact exhibit the desired $\cos^2$ patterns, and the monopole pattern is omnidirectional.

A switching and balancing network was designed and positioned at the antenna base, 1/2-wave length distant from the loops. Switching signals coming from the radar system minicomputer cause this junction box to sequence between the three antennas every 0.5 millisecond and pass the received consecutive echo voltages back along a single line to the receiver; this eliminates the possibility of introducing differential mismatch to the three signals along the path from the junction box back through the receiver. The difference

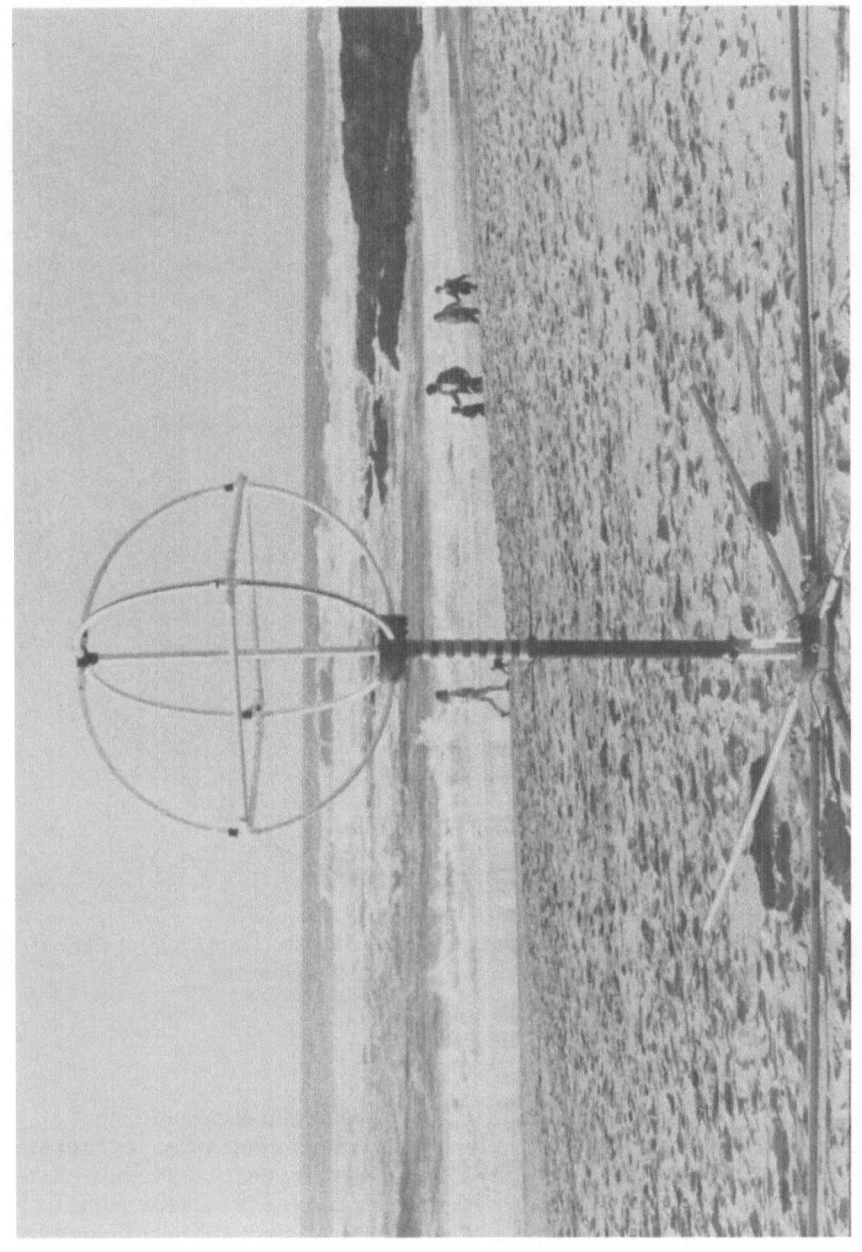

Fig. 5. Photograph of loop/monopole antenna system as it was operated in California, January 1978.

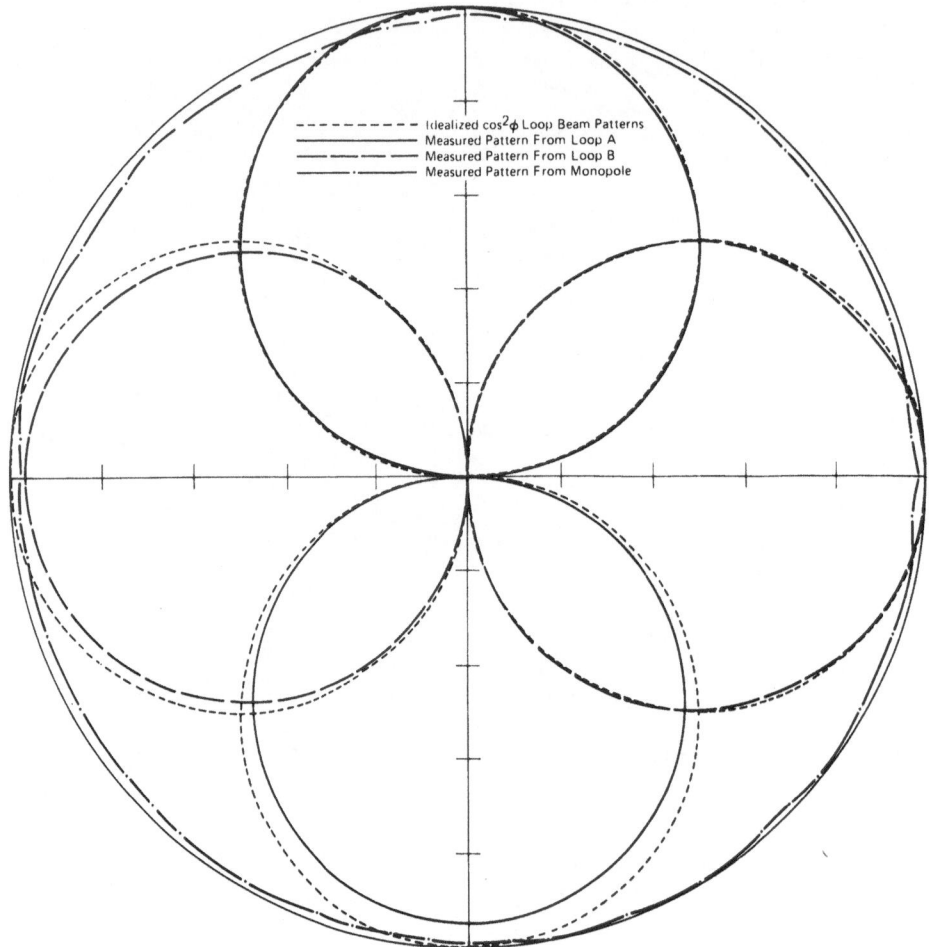

Fig. 6. Measured azimuth patterns (broken lines) vs desired patterns (solid lines) for the three antenna elements.

in path lengths from the antennas to the junction box and input impedances at the base of the three antennas produces two effects, a phase mismatch and a voltage amplitude disparity of ~21 dB. Both can be easily corrected, either by using matching transformers in the junction box and/or by appropriate adjustment within the digital processor following the receiver. A crude balancing is done at the junction box, matching the amplitudes to within one decibel and the phase to within two degrees; the remaining fine balancing is done later in the software.

The amplitude mismatch has a more serious consequence, however. It means that the loop antennas are 21 dB less efficient than the monopole. "Balancing" in effect is equivalent to introducing 21 dB attenuation into the path of the monopole signal. While this has the desirable attribute of making the three signal levels at the A/D convertor equivalent in magnitude (so that quantization error is the same for all three), it results in received signals lower by ∿21 dB than obtainable from the monopole. If the dominant additive system noise is not external (at 25 MHz, external noise is often near or below the level of internal noise), then the system signal-to-noise ratio is reduced accordingly. This translates into a decrease in maximum radar range to ∿45 km, down from 70 km. This decrease may not be of importance in most applications where one desires the wave height directional spectrum only within a distance ∿30 km from the radar. However, alternate loop designs are being examined with a goal to increase their efficiency.

The junction box removes gross amplitude and phase imbalances among the antennas. Fine imbalances are removed in the software based on data obtained during a "calibrate run"; in this run a weak continuous wave signal is transmitted several wave lengths from the antenna, usually along the axis of one of the loops. From the spectral output of the fast-Fourier transform (FFT), amplitude and phase differences observed between the monopole and loops are noted and later used to multiply the sea-echo FFT outputs to remove these remaining small imbalances, before forming the cardioid beam discussed earlier.

THE FOURIER BASIS FUNCTION FORMULATION
OF THE INVERSION PROBLEM

Sea-echo average spectral power is conveniently normalized and defined as the average radar backscattering cross section per unit surface area per radian/second (spectral) bandwidth; this is discussed before Equation (1) of the previous paper. Here, we denote this quantity as $\sigma(\omega,\phi)$, where it can represent either the first-order portion of the Doppler spectrum $\sigma_{(1)}(\omega,\phi)$, (for ω near $\pm\ \omega_B = \sqrt{2k_o g}$) or the second-order spectral contribution, $\sigma_{(2)}(\omega,\phi)$. Then one can conveniently define a corresponding quantity with the same dimensions representing the output of the broad-beam radar at scan angle ψ (see Figure 3) as follows

$$\tilde{\sigma}(\omega,\psi) \equiv \frac{1}{2\gamma} \int_{-\gamma}^{\gamma} \cos^4\left(\frac{\psi-\phi}{2}\right)\underset{\sim}{\sigma}(\omega,\phi)d\phi \ , \tag{1}$$

where 2γ is the total angle of sea surface within a given circular range cell subtended by the coast. We conveniently define ψ and ϕ

here with respect to the bisector of this partial ring of sea. The
kernel factor $\cos^4((\psi-\phi)/2)$ can be considered the antenna pattern
gain function; thus more complicated antenna patterns could be substituted for this quantity in the above integral and a solution
developed analogous to that shown below.

The kernel factor above has the particularly desirable properties that it is symmetric and expandable as a diagonal matrix of
trigonometric functions, i.e.,

$$\cos^4(\tfrac{\psi-\phi}{2}) = \sum_{n=-2}^{2} a_n tf_n(\psi) tf_n(\phi) \ , \tag{2}$$

where

$$tf_n(\alpha) \equiv \begin{cases} \cos(n\alpha) & \text{for } n \geq 0 \\ \sin(n\alpha) & \text{for } n < 0 \end{cases}$$

and

$$a_{-2} = a_2 \equiv \tfrac{1}{8}; \qquad a_{-1} = a_1 \equiv \tfrac{1}{2}; \qquad a_o \equiv \tfrac{3}{8} \ .$$

Hence Equation (1) can be expressed as an angular Fourier
series having five nonzero coefficients

$$\tilde{\sigma}(\omega,\psi) = \frac{1}{2\pi} \sum_{n=-2}^{2} b_n(\omega) tf_n(\psi) \ , \tag{3}$$

where

$$b_n(\omega) \equiv \frac{2}{\varepsilon_n} \int_{-\pi}^{\pi} \tilde{\sigma}(\omega,\psi) tf_n(\psi) d\psi \ , \text{ or} \tag{4}$$

$$b_n(\omega) = \frac{a_n \pi}{\gamma} \int_{-\gamma}^{\gamma} \underset{\sim}{\sigma}(\omega,\phi) tf_n(\phi) d\phi \ . \tag{5}$$

and $\qquad \varepsilon_n = \begin{cases} 2 \text{ for } n = 0 \\ 1 \text{ for } n \neq 0 \ . \end{cases}$

The coefficients $b_n(\omega)$ are obtained from experimental measurements of $\tilde{\sigma}(\omega,\psi)$ using Equation (4); thus the five coefficients b_n can be considered the observed quantities. The desired quantity is the wave height directional spectrum, $S(\bar{\kappa})$. Because the Doppler spectrum is expressed with Fourier angular coefficients, this representation is natural and useful for the directional spectrum also, i.e.,

$$S(\bar{\kappa}) = S(\kappa,\theta) = \frac{1}{2\pi} \sum_{n=-2}^{2} c_n(\kappa) t f_n(\theta) , \qquad (6)$$

where the series is shown truncated to correspond to the five Doppler coefficients and the bar indicates a vector.

A sample showing how a truncated spectrum of the form of Equation (6) might appear in three dimensions is provided in Figure 7. This isometric sketch is plotted at several discrete

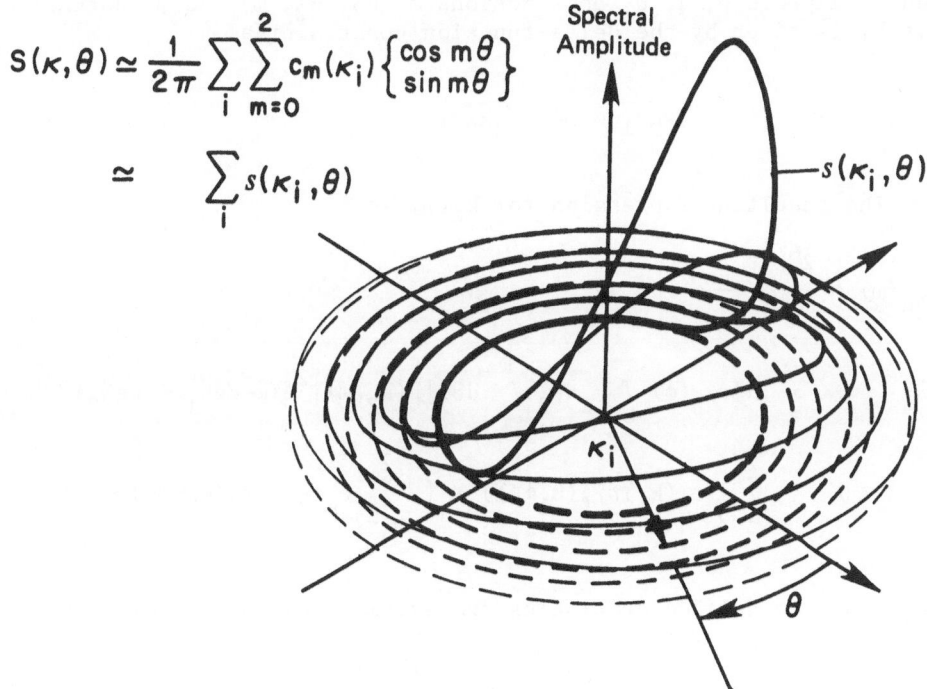

Fig. 7. Illustration of one possible graphical presentation of the five-coefficient inversion output vs angle, θ, for varying ocean wave number, κ_1.

wave numbers, κ_1, and illustrates the typical buildup of wind-wave energy at lower wave numbers, and angular skewing which can take place versus wave number.

An integral equation relating the desired $c_n(\kappa)$ to the observed $b_n(\omega)$ in the second-order region of the Doppler spectrum is next developed. The wave height directional spectrum in Equation (6) is used with Equation (3) of the preceding paper. Integration over ϕ, as indicated in Equation (5), gives an expression for $b_n(\omega)$ in the second-order region. The original double integral over p,q is simplified in the following manner. First, each of the four terms in the double summation over m_1, m_2 contribute in different regions of the Doppler spectrum. For example, in the region $\omega_B < \omega < 2\omega_B$, only the term with $m_1 = m_2 = +1$ contributes. Only the analysis involving this term is shown here for illustration. Second, the integrand is symmetric in the p-q plane; hence, we consider the contribution only for p > 0, multiplying by 2 to account for p < 0. Third, the delta function permits reduction of the double integral to a single integral. The remaining integration is to be done over k_2, which gives rise to a linear integral equation over small wave numbers, k_2, corresponding to the long ocean waves of interest. Then we express p, q, θ_1 as functions of k_2, θ_2, ω, and ϕ, noting that k_1 is given by the delta-function constraint as

$$\sqrt{gk_1} = \omega - \sqrt{gk_2} . \tag{7}$$

The resulting expression for $b_n(\omega)$ is

$$b_n(\omega) = \frac{2^6 k_o^4 \pi^2 a_n}{\gamma}$$

$$\times \int_{-\gamma}^{\gamma} d\phi\, tf_n(\phi) \int_0^\infty k_2 dk_2 \int_{-\pi}^{\pi} d\theta_2 |\Gamma(k_2,\omega)|^2 \delta(\omega - \sqrt{gk_1} - \sqrt{gk_2})$$

$$\times \left\{ \sum_{\ell=-2}^{2} c_\ell(k_1) tf_\ell(\theta_1+\phi) \right\} \times \left\{ \sum_{m=-2}^{2} c_m(k_2) tf_m(\theta_2+\phi) \right\} \tag{8}$$

where the sea surface statistics are assumed homogeneous over the angular coverage area (i.e., $-\gamma \leq \phi \leq \gamma$) so that $S(\kappa,\theta)$ observed from a different direction $\theta+\phi$ is merely $S(\kappa,\theta+\phi)$.

There are two sets of unknowns, $c_\ell(k_1)$ and $c_m(k_2)$, in the above equation. Because these sets are multiplicative, the integral equation is nonlinear. For higher sea states and/or radar frequencies such that $c_m(k_2)$ has its typically sharp spectral peak for $k_2 \ll 2k_o$, this integral can be linearized. For example, at a radar frequency of 25 MHz, the rms ocean wave height need exceed only 30 cm in order for this approximation to be satisfied. Then the only region of p-q space (for p>0) which effectively contributes to the integral is the small area near $k_2 = 0$ (see Figure 8). The constant-Doppler contours

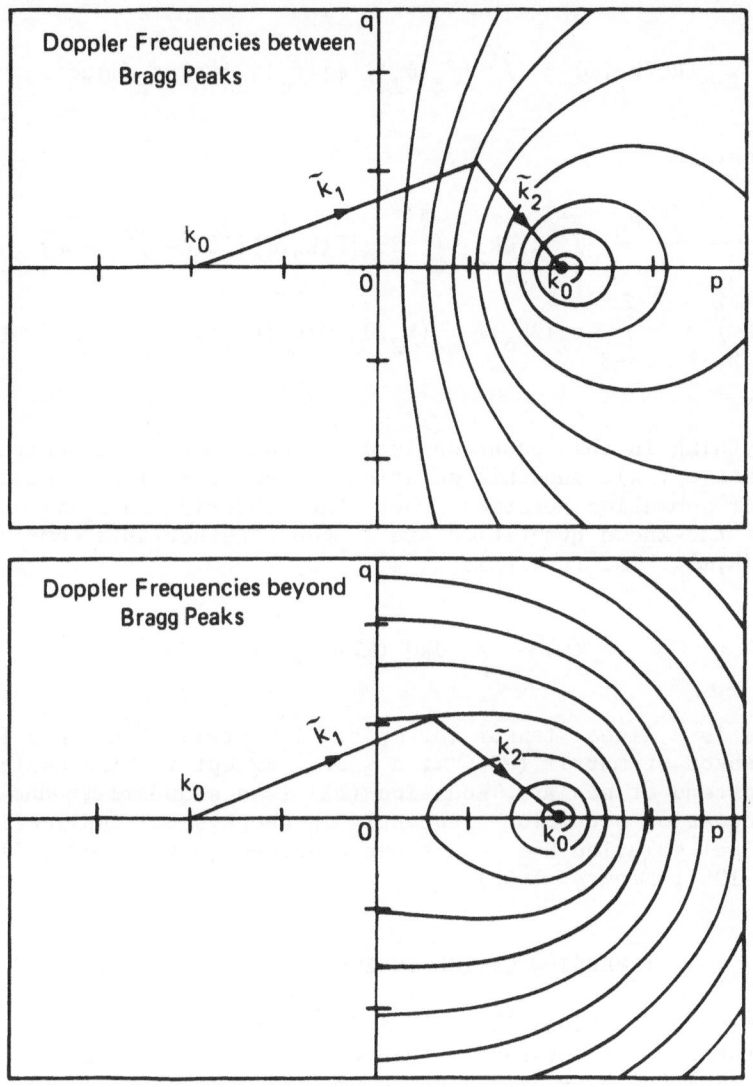

Fig. 8. Contours of constant Doppler frequency vs p,q; symmetry obtains for p<0. Contours near $p=k_o$ become nearly circular, and represent Doppler shifts near the Bragg frequency, $\sqrt{2gk_o}$.

in this region approximate circles centered on $k_2 = 0$ (i.e., $p = k_o$, $q = 0$), and in this region $k_1 \simeq 2k_o$. Thus $c_\ell(k_1) \simeq c_\ell(2k_o)$ and the latter are no longer unknown, but are determined from the first order echo (as shown in Appendix A) and substituted into the above equation. This leaves only the $c_m(k_2)$ as unknowns, with the remaining quantities k_1 and θ_1 being known or determined in terms of k_2, ω, and ϕ. Finally the known ϕ dependence is integrated out (either exactly or numerically, whichever is more convenient) giving the known constants

$$I_{\ell mn}(k_2, \theta_2, \omega) \equiv \int_{-\gamma}^{\gamma} tf_\ell(\theta_1 + \phi) tf_m(\theta_2 + \phi) tf_n(\phi) d\phi . \qquad (9)$$

Thus, we obtain

$$b_n(\omega) = \frac{2^6 k_o^4 \pi^2 a_n}{\gamma} \int_0^\infty k_2 dk_2 \int_{-\pi}^{\pi} d\theta_2 |\Gamma(k_2, \omega)|^2 \delta(\omega - \sqrt{gk_1} - \sqrt{gk_2})$$

$$\times \sum_{\ell=-2}^{2} \sum_{m=-2}^{2} c_\ell(2k_o) I_{\ell mn}(k_2, \theta_2, \omega) c_m(k_2) . \qquad (10)$$

Everything in this equation is now known except the desired coefficients $c_m(k_2)$, and this equation can be rewritten in simplified tensor-summation notation, where the subscript on k_2 is dropped and all of the known quantities are lumped together in a common kernel $G_n^m(k, \omega)$. The result is

$$b_n(\omega) = \int_0^\infty dk G_n^m(k, \omega) c_m(k), \qquad (11)$$

where this is a linear tensor (or matrix) integral, G_n^m being a five-by-five tensor or matrix ($-2 \leq m, n \leq 2$). Except for the use of tensors instead of scalars, Equation (11) is a standard Fredholm integral equation occurring frequently in geophysical applications, its numerical solution has been treated extensively (*Lipa*, 1977; *Phillips*, 1962; *Twomey*, 1963).

INVERSION OF THE INTEGRAL EQUATION

In solving Equation (10) numerically, the desired coefficient $c_m(k)$ is taken to be constant within wave number bands, such that $c_m^m(k) \simeq c_m(k_j)$ for $k_j - \Delta_j/2 < k < k_j + \Delta_j/2$, with $j = 1, 2, \ldots J$. The integral equation is then converted to a linear algebraic equation,

$$b_n(\omega) = \sum_{j=1}^{J} \sum_{m=-2}^{2} a_{jmn}(\omega) c_m(k_j) \ , \tag{12}$$

where

$$a_{jmn}(\omega) \equiv \int_{k_j - \Delta_j/2}^{k_j + \Delta_j/2} G_n^m(k,\omega)\,dk =$$

$$\frac{2^6 k_o^4 \pi^2 a_n}{\gamma} \sum_{\ell=-2}^{2} c_\ell(2k_o) \int_{k_j - \Delta_j/2}^{k_j + \Delta_j/2} k\,dk$$

$$X \int_{-\pi}^{\pi} d\theta \left| \Gamma(k,\omega) \right|^2 \delta(\omega - \sqrt{gk_1} - \sqrt{gk}) I_{\ell mn}(k,\theta,\omega) \ . \tag{13}$$

The coefficients $I_{\ell mn}(k,\theta,\omega)$ defined in Equation (9) are evaluated numerically first, using identities to express $tf(\theta+\phi)$ in terms of $tf(\theta)$ and $tf(\phi)$ (e.g., $\cos(\theta+\phi) = \cos\theta\,\cos\phi - \sin\theta$ X $\sin\phi$). Then the remaining double integral in Equation (13) is solved using a method described by *Lipa* (1977). This consists of integrating out one variable in closed form by making use of the sifting property of the Dirac-delta function, and then performing the remaining integration numerically.

Equation (12) is solved for the c's in terms of the b's by use of a method that is based on the shape of the frequency contours and the contribution to the radar cross section from different wave lengths. In Figure 9 circles are drawn on the p-q plane about the point $k_o, 0$ defining bands in wave number space (k_2 or simply k). Consider a contour defined by a frequency ω_1 slightly greater in absolute value than the Bragg frequency $\sqrt{2k_o g}$, which lies entirely within the first wave number band. For this contour, Equation (12) becomes

$$b_n(\omega_1) = \sum_{m=-2}^{2} a_{1mn}(\omega_1) c_m(k_2) \ . \tag{14}$$

When the other Doppler sideband (i.e., $\omega_1 = -\left|\omega_1\right| < -\sqrt{2k_o g}$) is included, there are ten equations (i.e., $b_n(\pm\left|\omega_1\right|)$ for $-2 \leq n \leq 2$) in five unknowns (i.e., $c_m(k_1)$, $-2 \leq m \leq 2$). These are solved using the methods outlined in Appendix B. To obtain values for $c_m(k_2)$ in the second wave number band, a frequency contour ω_2 which intersects the first two

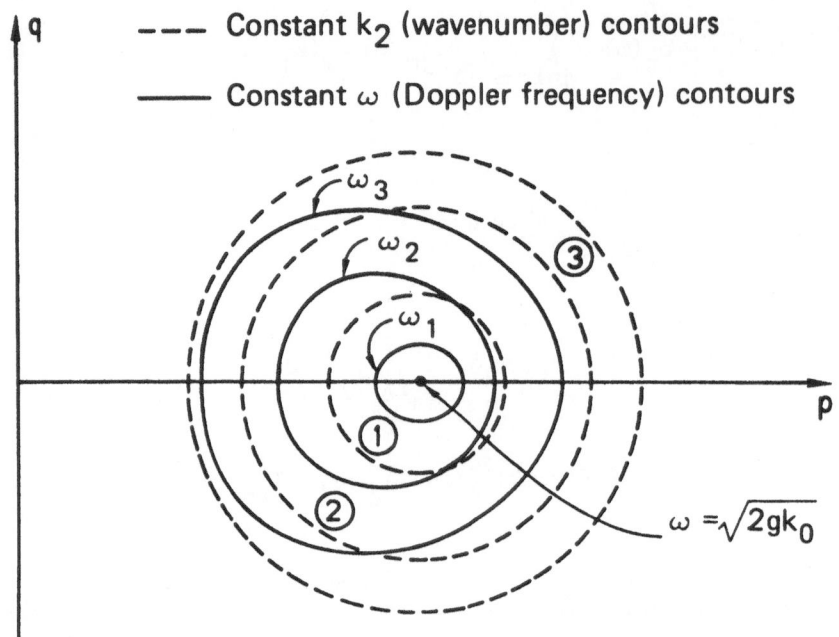

Fig. 9. Details of constant Doppler contours and constant wave number bands near the Bragg frequency, $\sqrt{2gk_o}$.

wave number bands is used. Then

$$b_n(\omega_2) = \sum_{m=-2}^{2} [a_{1mn}(\omega_2)c_m(k_1) + a_{2mn}(\omega_2)c_m(k_2)] , \qquad (15)$$

with an analogous equation for $b_n(-|\omega_2|)$. Values for $c_m(k_2)$ are obtained by substitution of the values for $c_m(k_1)$ derived from Equation (14). This procedure is followed for increasing frequencies and wave numbers until the desired profile of $c_m(k)$ is found for all wave numbers of interest.

For this method to be successful with noisy data, it is essential that the contribution to the echo signal at the Doppler frequency corresponding to j be dominated by the wave number corresponding to j rather than the wave number corresponding to j+1. This requirement then limits the obtainable wave number resolution. When $|\omega| > \sqrt{2k_og}$ (i.e., outside the Bragg lines), the effect of the smaller wave numbers dominates the contribution at a given ω, and hence both good resolution and stability result from the use of the procedure described in Equations (14) and (15).

Similar success can be obtained for Doppler frequencies between the Bragg lines ($|\omega| < \sqrt{2k_0 g}$). In this region, however, it is the higher wave numbers whose contributions dominate for a given ω. For this region, therefore, one starts at the higher saturated ocean wave numbers and iterates as described above to obtain the profile of $c_m(k_j)$ for decreasing values of k_j. In this saturation region, $k_j = 2k_0$ is used as the starting wave number, and starting values for the coefficients $c_m(2k_0)$ are obtained from the first-order echoes, as described in Appendix A. Thus two independent estimates for $c_m(k)$ are derived: one from Doppler frequencies beyond the Bragg lines and the other from Doppler frequencies between the Bragg lines. These can be averaged to obtain a final value for $c_m(k)$.

FINITE-SAMPLE ERROR ANALYSIS

All of the mathematics employed up to now assumes infinite ensemble averaging. Thus $b_n(\omega)$, $c_n(k)$, $g(\omega,\phi)$, and $\tilde{\sigma}(\omega,\psi)$ all imply infinite ensemble averages. In practice, however, one employs a "sample average" consisting of the sum of a finite number, N, of samples. These N samples, which are assumed to be statistically independent, can come from spectra taken over different short segments of time, from performing running frequency averaging on contiguous points of the Doppler spectra (these two are essentially identical operations, merely referred to different domains), or can be obtained from contiguous but nonoverlapping range cells. In practice, the N independent samples are collected from all three of the above sources. One must merely ascertain that the sea surface statistics are homogeneous or stationary over the time and space scales during which the N samples are gathered.

Barrick and Snider (1977) have shown that sea-echo voltage spectra are Gaussian distributed. Furthermore, Doppler spectra become uncorrelated (and hence statistically independent) for time intervals greater than ~25 seconds at frequencies above 10 MHz. Finally, they observed that contiguous range cells of sea echo are statistically independent to spatial separations as small as 3 km (20 μs pulse width); no experimental data were gathered for finer spatial separations. All of these facts can be used in determining how to optimally perform finite-sample spectral averaging over space, time, and frequency, and then how to estimate N, the number of independent samples.

Because a finite-sample average is still a random variable, the observables $b_n(\omega)$ will be somewhat random, rather than smooth as they would be if the sample size were infinite. Thus the procedure outlined in the preceding section for determining the sea wave height spectral coefficient, $c_n(k)$, from the $b_n(\omega)$ will produce values which also vary randomly about their true mean. The purpose of this section is to determine this standard deviation (which is called the rms statistical error) of the resulting $c_n(k)$ from the mean. To do this, we must first find the covariance matrix of the $b_n(\omega)$ for N-sample averages. This is defined as

$$\text{Cov}[{}_N b_m \cdot {}_N b_n] \equiv \langle {}_N b_m \cdot {}_N b_n \rangle - \langle {}_N b_m \rangle \langle {}_N b_n \rangle , \qquad (16)$$

where the braces $\langle ... \rangle$ denote infinite ensemble averages, and ${}_N b_n$ denotes an N-sample average corresponding to the input $b_n(\omega)$. By definition, $\lim_{N \to \infty} {}_N b_n = b_n(\omega)$.

In Appendix C the above covariance matrix is derived. The final result is

$$\text{Cov}[{}_N b_m \cdot {}_N b_n] = \frac{1}{N} \sum_{m,n} \sum \delta^{ij}_{mn} b_i(\omega) b_j(\omega) = \frac{1}{N} \delta^{ij}_{mn} b_i(\omega) b_j(\omega), \quad (17)$$

where the summation convention from tensor analysis is employed in the latter representation. The tensor elements δ^{ij}_{mn} are derived in Appendix C and given in Table 1.

In addition, the covariance matrix for the second-order Doppler spectral power at different frequencies is needed. Because the linearized integral equation, Equation (11), is itself of matrix rather than scalar form, this second covariance matrix will have 5 x 5 matrices as its individual elements. Appendix D shows that the second-order Doppler voltages at different frequencies are uncorrelated. This implies that the covariance matrix at the 2J frequency points called for in the preceding section is itself diagonal. Hence the final covariance matrix is

$$\begin{bmatrix} \text{Cov}[{}_N b_m \cdot {}_N b_n]_{\omega_1} & 0 & 0 & \cdots & 0 \\ 0 & \text{Cov}[{}_N b_m \cdot {}_N b_n]_{\omega_2} & 0 & \cdots & 0 \\ 0 & 0 & \cdots & & \vdots \\ \vdots & \vdots & & & \\ 0 & 0 & & \cdots & \text{Cov}[{}_N b_m \cdot {}_N b_n]_{\omega_{2J}} \end{bmatrix}$$

$$(18)$$

TABLE 1. Tensor elements of covariance matrix in Equation (17). All δ_{mn}^{ij} for i,j different from those indices shown are identically zero. Also, these matrix elements are symmetric in m,n.

m,n = -2, -2 $\quad \delta_{mn}^{00} = \frac{1}{18}; \; \delta_{mn}^{-2-2} = \frac{1}{2}; \; \delta_{mn}^{22} = -\frac{1}{2}$

m,n = -2, -1 $\quad \delta_{mn}^{01} = \delta_{mn}^{10} = \frac{1}{12}; \; \delta_{mn}^{-1-2} = \delta_{mn}^{-2-1} = \frac{1}{4}; \; \delta_{mn}^{12} = \delta_{mn}^{21} = -\frac{1}{4}$

m,n = -2,0 $\quad \delta_{mn}^{1-1} = \delta_{mn}^{-11} = \frac{1}{8}; \; \delta_{mn}^{0-2} = \delta_{mn}^{-20} = \frac{1}{6}$

m,n = -2,1 $\quad \delta_{mn}^{0-1} = \delta_{mn}^{-10} = \frac{1}{12}; \; \delta_{mn}^{-12} = \delta_{mn}^{2-1} = \frac{1}{4}; \; \delta_{mn}^{1-2} = \delta_{mn}^{-21} = \frac{1}{4}$

m,n = -2,2 $\quad \delta_{mn}^{2-2} = \delta_{mn}^{-22} = \frac{1}{2}$

m,n, = -1, -1 $\quad \delta_{mn}^{00} = \frac{4}{9}; \; \delta_{mn}^{02} = \delta_{mn}^{20} = -\frac{2}{3}; \; \delta_{mn}^{-1-1} = \frac{1}{2}$

m,n = -1,0 $\quad \delta_{mn}^{-10} = \delta_{mn}^{0-1} = \frac{5}{12}; \; \delta_{mn}^{1-2} = \delta_{mn}^{-21} = \frac{1}{4}; \; \delta_{mn}^{2-1} = \delta_{mn}^{-12} = -\frac{1}{4}$

m,n = -1,1 $\quad \delta_{mn}^{1-1} = \delta_{mn}^{-11} = \frac{1}{4}; \; \delta_{mn}^{-20} = \delta_{mn}^{0-2} = \frac{2}{3}$

m,n = -1,2 $\quad \delta_{mn}^{-10} = \delta_{mn}^{0-1} = -\frac{1}{12}; \; \delta_{mn}^{1-2} = \delta_{mn}^{-21} = \frac{1}{4}; \; \delta_{mn}^{-12} = \delta_{mn}^{2-1} = \frac{1}{4}$

m,n = 0,0 $\quad \delta_{mn}^{00} = \frac{1}{2}; \; \delta_{mn}^{11} = \frac{1}{4}; \; \delta_{mn}^{-1-1} = \frac{1}{4}; \; \delta_{mn}^{22} = \frac{1}{2}; \; \delta_{mn}^{-2-2} = \frac{1}{2}$

m,n = 0,1 $\quad \delta_{mn}^{01} = \delta_{mn}^{10} = \frac{5}{12}; \; \delta_{mn}^{12} = \delta_{mn}^{21} = \frac{1}{4}; \; \delta_{mn}^{-1-2} = \delta_{mn}^{-2-1} = \frac{1}{4}$

m,n = 0,2 $\quad \delta_{mn}^{11} = \frac{1}{8}; \; \delta_{mn}^{-1-1} = -\frac{1}{8}; \; \delta_{mn}^{20} = \delta_{mn}^{02} = \frac{1}{6}$

m,n = 1,1 $\quad \delta_{mn}^{00} = \frac{4}{9}; \; \delta_{mn}^{02} = \delta_{mn}^{20} = \frac{2}{3}; \; \delta_{mn}^{11} = \frac{1}{2}$

m,n = 1,2 $\quad \delta_{mn}^{01} = \delta_{mn}^{10} = \frac{1}{12}; \; \delta_{mn}^{12} = \delta_{mn}^{21} = \frac{1}{4}; \; \delta_{mn}^{-1-2} = \delta_{mn}^{-2-1} = -\frac{1}{4}$

m,n = 2,2 $\quad \delta_{mn}^{00} = \frac{1}{18}; \; \delta_{mn}^{22} = \frac{1}{2}; \; \delta_{mn}^{-2-2} = -\frac{1}{2}$

where matrix algebra instead of scalar algebra is implied in the subsequent manipulation of the diagonal elements in Equation (18).

Values of $b_n(\omega)$ are calculated using a Phillips wind-wave model for the wave height directional spectrum in Equation (3) of the preceding paper, and then Equation (5) of this paper is used to find the $b_n(\omega)$. Using Equation (17) above, the input signal covariance matrix, Equation (18), is then obtained. The Phillips model has the form

$$S(\kappa,\theta) = f(\kappa)g(\theta), \qquad \text{where} \tag{19a}$$

$$f(\kappa) = \begin{cases} \dfrac{.01}{2\kappa^3} & \text{for} \quad \kappa > \kappa_{co} \ (=g/u^2) \\[3ex] 0 & \text{for} \quad \kappa < \kappa_{co} \end{cases}, \tag{19b}$$

and

$$g(\theta) = \frac{4}{3\pi} \cos^4(\frac{\theta-\theta_o}{2}). \tag{19c}$$

Here, $g(\theta)$ is a directional factor, g is the acceleration due to gravity, u is the wind speed in m/s, and θ_o is the wind direction taken with respect to any desirable angular reference from shore. The spectrum is defined and normalized such that

$$S(\overline{\kappa})d^2\overline{\kappa} = S(\kappa_x,\kappa_y)d\kappa_x d\kappa_y = S(\kappa,\theta)d\kappa d\theta , \tag{19d}$$

and hence

$$h^2 = \int_o^\infty d\kappa \int_{-\pi}^\pi d\theta S(\kappa,\theta) , \tag{19e}$$

$$\text{or} \quad h^2 = \int_o^\infty f(\kappa) \, d\kappa , \tag{19f}$$

$$1 = \int_{-\pi}^\pi g(\theta)d\theta \tag{19g}$$

where mean-square wave height is defined as h^2 and the nondirectional wave height spectrum is given by $f(\kappa)$.

The techniques of Appendix B, where the input "signal covariance mattix," C_b is as calculated above, is used to obtain the output "parameter covariance matrix" for the directional coefficients through the transformation given in Equation (B-4). In particular, the square roots of the diagonal elements (the variances) are wanted. These are the standard deviations of the N-sample wave height spectral coefficients. This is done for the above model where three wind directions, $\theta_0 = 0°$, $45°$, $90°$ are employed. The standard deviations are the same (respectively) for $\theta_0 = 180°$, $135°$, $90°$; $180°$, $225°$, $270°$; and $360°$, $315°$, $270°$. A cutoff κ_{co} is chosen such that $\kappa_{co}/k_0 = 0.1$; at 20 MHz, this corresponds to an rms sea wave height of 1.2 m and wave period of 10 seconds. The wave numbers k_i at which we wish to retrieve the $c_n(k_i)$ are chosen such that the first value, k_1, lies near but below the cutoff, κ_{co}, and the other nine define the spectrum for $\kappa_{co} < \kappa < 3\kappa_{co}$. The $\phi=0$ direction of the wave height spectra is selected to lie along the shore, and a straight coastline is specified. Typical retrieved values are shown in Figure 10. This figure gives values of $c_n(k_i)$, $-2 < n < 2$, for $\phi_0 = 45°$. The dots are the mean coefficient values, while the vertical error bars are the standard deviations obtained using the transformation (Equation (B-4)).

The results shown in Figure 10 include the average values of c_n obtained from inverting both the outer sides ($|\omega| > \omega_B$) and the inner sides ($|\omega| < \omega_B$) of the Doppler spectrum. Three hundred such Doppler spectra are assumed to be averaged, which include 60 independent spectra sequential in time for five contiguous semi-circular range cells, each 1.2 km wide (corresponding to our 8 μs radar pulse-width). Thus the patch of sea from which radar data are collected and averaged is ∿7.2 km in radius. In this analysis the sea statistics are assumed homogeneous over this observation area. This assumption will not be valid for offshore wind and wave conditions where wave conditions vary with fetch. The latter situation requires a different analysis which will be presented in a subsequent publication. Requiring a temporal (Doppler) frequency resolution of 0.0064 Hz (compatible with the desired wave number resolution in k_i), this implies that the observation of 60 sequential spectral would require ∿2 hours. Hence, there are (60 time samples) X (5 range samples) X (2 sideband regions) = 600 total independent samples.

In Appendix B a method is described for stabilizing the solution of equations by minimizing the deviation of the solution from an initial guess. The value of the Lagrange multiplier, γ, used for stabilization is determined by the statistical fluctuation in the input data, b_n. If γ is zero, the random data fluctuations can cause unstable, ill-conditioned solutions. On the other hand, if γ is too large, then the retrieved value for the desired parameters, c_m, will not be significantly different from the initial guess, c_m^i; furthermore, when the retrieved values, c_m, are then

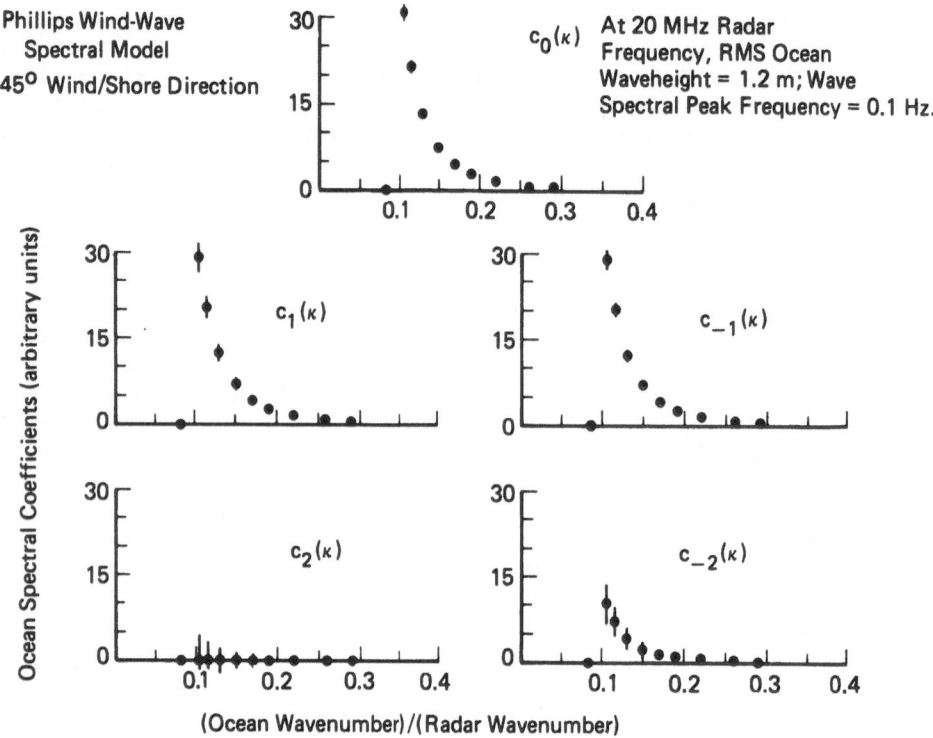

Fig. 10.   Coefficients retrieved from the inversion technique,
including error-bars, for 45° wind/shore direction with Phillips
spectral model input.

substituted into Equation (B-1) to recover the input data, b_n, the
latter tend to fall outside the region of fluctuation of the original
input data. In the following, γ is selected so that the retrieved
values of c_m, obtained from Equation (B-3), fall within the error
bars (or diagonal element standard deviations) obtained from Equation (B-4). The top radar entries in Table 2 thus represent the
worst case of no stabilization ($\gamma=0$) and no initial guess. The middle radar entries use stabilization, but with a poor initial guess
(i.e., 100% in error). The bottom entries use stabilization with a
good initial guess (within 20%). Figure 10 is for the second
condition.

TABLE 2. Standard deviation of directional spectrum
coefficients at spectral peak. The standard
deviations are expressed as a percentage of c_0.
For each coefficient, the left column is for
radar observations and the right is for buoy
observations. The top radar entries correspond
to no stabilization and no initial guess; the
middle entries are for stabilization with a
poor initial guess; the bottom entries are for
stabilization with a good initial guess (within
20%). The results are for two hours of wave
observations at 0.0064 Hz frequency resolution.

Wind Angle	c_{-2}		c_{-1}		c_0		c_1		c_2	
	12.6		15.8		8.6		4.3		20.4	
90°	6.5		6.1		4.2		4.1		9.4	
	2.8;	0.0	2.6;	17.2	2.2;	12.9	2.6;	0.0	3.0;	4.3
	16.0		12.2		9.0		13.0		11.2	
45°	10.0		5.2		3.8		8.1		10.8	
	3.5;	4.3	2.8;	12.2	2.3;	12.9	2.8;	12.2	3.0;	0.0
	12.2		7.8		8.6		18.6		16.5	
0°	10.2		6.1		3.3		6.6		7.8	
	3.0;	0.0	4.0;	0.0	2.9;	12.9	1.8;	17.2	1.9;	4.3

The entries in Table 2 are standard deviations of the five
coefficients near the spectral peak (the most important region)
for the three wind directions. They are given as percentages of
c_0 at the spectral peak. Also shown as the rightmost entries are
the standard deviations of these coefficients obtained from the
tilt buoy described in *Tyler et al.* (1974) and *Stewart* (1977). The
same wave spectral model, Equation (19), is used for the buoy
simulations as for the radar. Appendix E shows how these coefficients are obtained from buoy data. Two hours of buoy data are
assumed to be used (just as with the radar) with the same frequency
resolution of 0.0064 Hz. Since the buoy is a point measurement, no
area averaging is possible as with the radar, and there is only one
sideband with the buoy. Hence there are only 60 independent samples
possible with the buoy in the same two-hour observation period, and
one obtains the usual chi-squared distribution for the variances,
with 120 degrees of freedom. The percentage error for $_Nc_o$, for

example, is merely $100\%/\sqrt{N} = 12.9\%$. This represents a fair comparison because both buoy and radar observations are taken over the same two-hour period with the same 0.0064 Hz frequency resolution.

Statistical sampling normally produces greater errors for buoy-measured data, especially for the lowest three coefficients, c_0, c_1, and c_{-1}. These are the most important coefficients, for they contain the essential information about wave height, the nondirectional spectrum, and mean wave direction, as shown in the next section. Furthermore, for realistic directional wave spectra, c_2 and c_{-2} are considerably smaller than the first three coefficients (e.g., for the model of Equation (19c), c_{-2} and c_2 are 1/3 of c_0 and 1/4 of c_1, c_{-1}), so that they are generally less important.

DETERMINATION OF MODEL PARAMETERS

The five lowest-order Fourier angular coefficients vs wave number, $c_n(\kappa)$, are the most basic representation of the wave height spectral directional dependence; from the truncated series one can plot a smoothed estimate of the angular pattern. However, the coefficients as they stand do not admit to a quick interpretation of important directional parameters, such as the dominant or mean wave direction.

A wave directional model for wind-driven seas which has become popular over the past several years represents the angular dependence at a given wave length as $\cos^s((\theta-\theta_o)/2)$ (*Tyler et al.*, 1974; *Stewart and Barnum*, 1975), where θ is the wave angle from a selected direction. In this expression, θ_o can be interpreted as the mean wave direction, and s as a spreading factor relatable to the wave beamwidth. Quantitatively, one can equate the truncated Fourier series wave spectrum of Equation (6) to a three-parameter version of this model as follows

$$\frac{1}{2\pi} A(s)\cos^s(\frac{\theta-\theta_o}{2}) = \frac{1}{2\pi} \sum_{n=-2}^{2} c_n tf_n(\theta) , \qquad (20)$$

where it is understood that s, θ_o, and the c_n are functions of wave number, κ, and that $s \geq 0$. *Longuet-Higgens et al.* (1963) first represented the pitch-and-roll buoy output in terms of the truncated, five-coefficient Fourier angular representation. They showed that truncation could result in a spectrum which was slightly negative at certain angles, and suggested the model of Equation (20) with s=4 as a step to correct this situation. *Tyler et al.*(1974) obtained some of the relationships between the model parameters and moments

of the directional spectrum; the latter are the natural outputs
of the buoy. Here a summary of all of the pertinent relationships
between the angular coefficients, c_n, employed in this paper and
the model parameters is presented. Simplified derivations of these
results are presented in Appendix F. Thus, for $s \geq 0$ the following
equations relate the c_n to the model parameters:

$$\frac{c_1}{c_0} = \frac{2s}{s+2} \cos\theta_o \equiv C_1 \quad ; \quad \frac{c_{-1}}{c_0} = \frac{2s}{s+2} \sin\theta_o \equiv C_{-1} \quad ; \quad (21)$$

$$\frac{c_2}{c_0} = \frac{2s(s-2)}{(s+2)(s+4)} \cos 2\theta_o \equiv C_2 \; ; \; \frac{c_{-2}}{c_0} = \frac{2s(s-2)}{(s+2)(s+4)} \sin 2\theta_o \equiv C_{-2};$$
$$(22)$$

$$\frac{A(s)}{c_0} = 2\pi / \int_{-\pi}^{\pi} \cos^s \frac{\theta}{2} \, d\theta = \frac{\sqrt{\pi}\,\Gamma(s/2+1)}{\Gamma(s/2+1/2)} \, , \quad (23)$$

where $\Gamma(x)$ is the Gamma function of argument x (see *Abramowitz and
Stegun*, 1964, Eq. 6.1.49). These equations exhibit the proper behavior in the limit of swell; i.e., as $s \to \infty$, the left side of Equation (20) becomes $c_0 \delta(\theta - \theta_o)$, and Equations (21), (22), and (23)
tend to the proper values in this limit.

Having c_n for $-2 \leq n \leq 2$, one can readily determine s and θ_o;
to see this, define $\mu_{1,2}$ as the modulus of the C's, i.e.,

$$\mu_1 \equiv \sqrt{C_1^2 + C_{-1}^2} \quad ; \quad \mu_2 \equiv \sqrt{C_2^2 + C_{-2}^2} \, . \quad (24)$$

Then from Equation (21) we have

$$\theta_o = \tan^{-1}(C_{-1}/C_1) \; ; \quad s = \frac{2\mu_1}{1-\mu_1} \, , \quad (25)$$

or if we use Equation (21) we obtain

$$\theta_o = \frac{1}{2} \tan^{-1}(C_{-2}/C_2) \quad ; \quad s = \frac{1 - 3\mu_2 + \sqrt{1-6\mu_2 + 17\mu_2^2}}{1 - \mu_2} \, . \quad (26)$$

Hence one first determines θ_o and s from either Equation (25)
or Equation (26) (or by averaging the values of s and θ_o obtained
from both sets of equations). Then Equation (23) is employed to

find A(s). Check values of $A(s)/c_0$ are 2, 8/3, and 128/35 for
s = 2, 4, 8 respectively.

A physically meaningful parameter of the model represented by
the left side of Equation (20) is the half-power beamwidth. This
is readily expressed in terms of the spread coefficient, s, as

$$\Delta\theta = 4\,\cos^{-1}[(1/2)^{1/s}] \quad . \tag{27}$$

The quantity $c_0(\kappa)$ is of itself meaningful; it is the wave
height nondirectional spectrum, $f(\kappa)$, as defined in Equation (19),
as can be seen from Equation (6) by integrating θ over 2π. In
other words,

$$S(\kappa) = \int_0^{2\pi} S(\kappa,\theta)d\theta = c_0(\kappa) = f(\kappa), \tag{28}$$

from which the mean-square wave height is obtained as

$$h^2 = \int_0^\infty S(\kappa)d\kappa = \int_0^\infty c_0(\kappa)d\kappa \quad . \tag{29}$$

SUMMARY AND CONCLUSIONS

An HF radar remote-sensing technique for measuring the ocean
wave height directional spectrum in coastal waters has been developed
and analyzed. The shore-based radar is small, inexpensive, transportable, and minicomputer controlled. The second-order portion of
the sea-echo Doppler spectrum is employed, and by normalizing with
respect to the first-order echo, a linearized integral equation is
developed and solved.

The unique feature of the system which allows the wave
directional characteristics to be extracted is the receiving
antenna system. It consists of a monopole and two loops crossed
along the same vertical axis. This antenna system has been built,
and both laboratory and field testing show that it performs electrically as expected. By recording the Fourier-transformed complex
voltages on these three antennas, a cardioid beam pattern is formed
and rotated electrically, all by manipulation of the received data
within the minicomputer software. The antenna beam pattern is
resolvable into the lowest five Fourier angular coefficients; this
in turn permits solving for coefficients of the wave height directional spectrum at a number of ocean wave numbers (or wave lengths).

The number of coefficients available depends on the number of
incoherent averages used and the accuracy desired. It is shown
that accurate estimates of the first five angular coefficients
may be obtained with this radar system. It is natural to use the
pitch-and-roll buoy for comparisons, since it also measures the
first five angular coefficients. In general, other antenna schemes
can be implemented to yield more wave directional information. For
example, by using the antennas and beam-forming system here for
both transmit and receive (instead of just receive), as many as
nine angular coefficients should be obtainable.

While awaiting actual comparisons of radar vs buoy measurements,
we have developed and numerically tested the radar inversion technique against simulated model data. In particular, errors in
retrieving the wave field coefficients due to finite sample-size
averaging are studied. Typical accuracies obtainable by the radar
over a two-hour observation period are 2-3% for c_o, c_1, and c_{-1}.
Equivalent buoy accuracies for these quantities are 13% for c_o and
$0-17\%$ for c_1 and c_{-1}. Radar accuracies for c_2 and c_{-2} are
equivalent to buoy accuracies; these coefficients are less important
for reconstructing wind-wave directional spectra. The lowest three
coefficients, for example, give sufficient information to retrieve
the parameters for the popular $\cos^s((\theta-\theta_o)/2)$ wave beam model.

A summary of the radar digital signal processing steps in the
production of a wave height directional spectrum is shown in Figure
11. This chart begins with the output time series from the analog-
to-analog converter for the three antennas and ends with various
graphical terminal displays of the desired wave field quantities.

There are three additional types of errors or inaccuracies
which must be examined further. These are, of course, in addition
to hardware inaccuracies or failures which afflict any system. The
first is the error inherent in the linearization assumptions used to
obtain Equations (11) and/or (12); this error will decrease with
increasing sea state and/or radar frequency. It is presently being
analyzed and will be discussed in a subsequent paper. The second
error is the effect of nonuniform illumination of the semi-circular
ring of sea due to path-loss differences vs angle; these effects can
arise at angles which graze the coast due to hills, cliffs, or other
terrain features which hinder radio propagation. The effect of this
will be studied experimentally. This error can be minimized by
exercising care in site selection. The third error results from our
assumption that the sea state or wave field statistics are homogeneous
within the area over which sea-echo data are gathered. This will
be true for onshore wind and wave conditions, where the maximum radar
range is kept below ∿20 km. Under conditions of offshore winds when

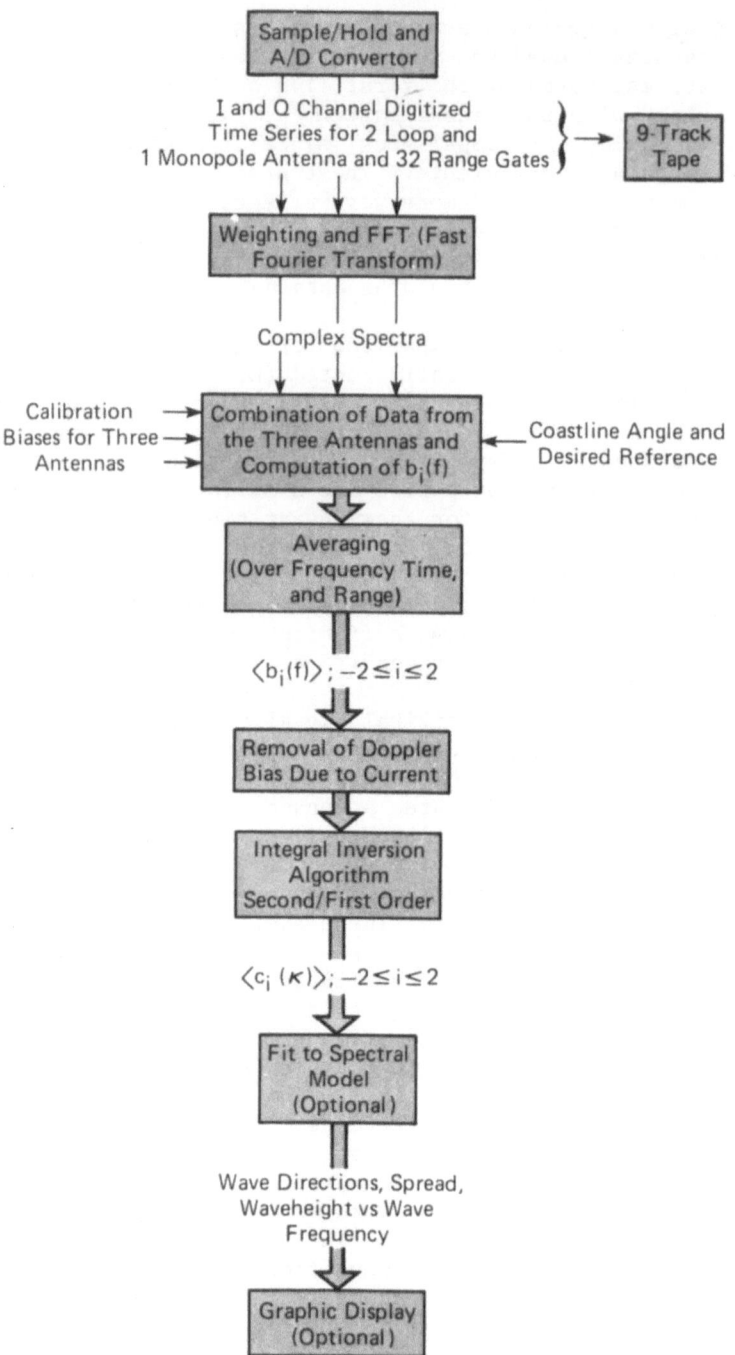

Fig. 11. Steps in the software production of a wave height directional spectrum for loop/monopole radar system.

the sea state varies with distance from shore (fetch), the present technique must be extended to account for and measure this variation This problem is the next major subject we shall analyze in the evolution of this radar technique for coastal wave-field monitoring.

APPENDIX A

By restricting attention to the first-order region of the Doppler spectrum (i.e., ω near $\pm\omega_B$) we obtain simple linear relationships between the $b_n(\omega)$ and the $c_n(2k_o)$; this is readily established by substituting Equation (1) of the preceding paper into Equation (5) here. Ideally, the first-order spectral shapes are impulse functions at $\pm\omega_B$; in radar observations they occur as narrow spectral spikes. Integration over the narrow frequency region of width Δ which comprises this spike at ω_B, provides

$$b_n \equiv \int_{\omega_B-\Delta/2}^{\omega_B+\Delta/2} b_n(\omega)\,d\omega = \frac{a_n}{\gamma}\pi \int_{-\gamma}^{\gamma} d\phi \int_{\omega_B-\Delta/2}^{\omega_B+\Delta/2} d\omega \underset{\sim}{\sigma}(\omega,\phi)\,tf_n(\phi)$$

$$= \frac{2^6\pi^2 a_n k_o^4}{\gamma} \int_{-\gamma}^{\gamma} S(2k_o,\phi)\,tf_n(\phi)\,d\phi, \quad\text{and using Equation (6),}$$

$$b_n = \frac{2^6\pi^2 a_n k_o^4}{\gamma} \sum_{\ell=-2}^{2} c_\ell(2k_o) \int_{-\gamma}^{\gamma} tf_\ell(\phi)\,tf_n(\phi)\,d\phi . \tag{A-1}$$

If the above integration is extended over 2π, (i.e., water completely surrounding the radar), then there would be a one-for-one correspondence between the b_n and $c_n(2k_o)$. For an arbitrary angle 2γ, however, there are generally five terms left in the summations (because of the nonorthogonality of the basis functions over any angle than $2\gamma = 2\pi$). This means that there is a linear system of five equations in five unknowns to be solved for the $c_n(2k_o)$. The solution is straightforward; in the presence of noise or random errors in the b_n, however, one must employ the techniques discussed in the paper to estimate the errors in the recovered $c_n(2k_o)$.

The case of a straight coastline is very common ($\gamma = \pi/2$), and the solutions for the c_n are given for that case. Simplification is possible because the above integral is zero for the following (ℓ,n) subscript pairs: (2,-2); (-2,2); (1,-1); (-1,1); (-2,1); (1,-2); (2,-1); (-1,2); (0,-1); (-1,0); (0,-2); (-2,0); (0,2); (2,0). The resulting solutions for the unknown c_ℓ are:

$$2^6 \pi k_o^4 c_{-2}(2k_o) = 4\xi_1 b_{-2} - \xi_2 b_{-1} \quad ;$$

$$2^6 \pi k_o^4 c_{-1}(2k_o) = \xi_1 b_{-1} - 4\xi_2 b_{-2} \quad ;$$

$$2^6 \pi k_o^4 c_o(2k_o) = \frac{4}{3}\xi_3 b_o - \xi_4 b_1 + 4\xi_5 b_2 \quad ; \qquad (A-2)$$

$$2^6 \pi k_o^4 c_1(2k_o) = \frac{4}{3}\xi_4 b_o + \xi_6 b_1 - 4\xi_7 b_2 \quad ;$$

$$2^6 \pi k_o^4 c_2(2k_o) = \frac{4}{3}\xi_5 b_o - \xi_7 b_1 + 4\xi_8 b_2 \quad ;$$

where

$$\xi_1 = \frac{3/4}{3\pi/8 - 8/3\pi} \quad ; \quad \xi_2 = \frac{2/\pi}{3\pi/8 - 8/3\pi} \quad ;$$

$$\xi_3 = \frac{1/2 - 8/9\pi^2}{\pi/2 - 44/9\pi} \quad ; \quad \xi_4 = \frac{2/\pi}{\pi/2 - 44/9\pi} \quad ;$$

$$\xi_5 = \frac{8/3\pi^2}{\pi/2 - 44/9\pi} \quad ; \quad \xi_6 = \frac{1}{\pi/2 - 44/9\pi} \quad ;$$

$$\xi_7 = \frac{4/3\pi}{\pi/2 - 44/9\pi} \quad ; \quad \xi_8 = \frac{1 - 8/\pi^2}{\pi/2 - 44/9\pi}$$

APPENDIX B

This appendix describes the inversion of linear matrix equations and the propagation of errors through this inversion process when the input contains noise. Equation (12) of the text is an example of a set of linear equations to be inverted, which we write in matrix form here as

$$b = ac, \qquad (B-1)$$

where $b = (\ldots b_n(\omega_i)\ldots)$ represents the input or measured data, $c = (\ldots c_m(k_j)\ldots)$ represents the parameters to be derived, and "a" is the known transformation matrix. In this case, the matrix "a" is IxJ ($1 \leq i \leq I$ and $1 \leq j \leq J$), but the elements themselves are 5 x 5 matrices over the m,n. Hence the appropriate matrix arithmetic operations are implied in all subsequent operations. For the situations we are considering, J is equal to or greater than I and the system is overdetermined. The solution, obtained by minimizing the sum of squared deviations $\sum_i |b(\omega_i) - a_{ij}c(k_j)|^2$ is

$$c = (a^Ta)^{-1}a^Tb \quad . \tag{B-2}$$

When the signal contains noise, Equation (B-1) becomes $b + \varepsilon$ = ac (where ε represents random fluctuation about the mean, b) and there is a range of possible solutions for c. The inversion problem is often ill-posed; that is, small changes in the data due to random noise can lead to severe distortion in the solution because the matrix a^Ta has some small eigenvalues. To infer values for $c(k_j)$, the solution must be stabilized to eliminate the spurious contribution from random noise while preserving the information content of the signal. The stabilizing method we use, due to *Phillips* (1962) and *Twomey* (1963), includes mathematical constraints in the solution of the matrix equation. An initial guess is made for the values of c and the deviation of the ultimate solution from this approximation is minimized, with Equation (B-1) as auxiliary constraints, to give

$$c = (a^Ta + \gamma I)^{-1} (a^Tb + \gamma c^I) \quad , \tag{B-3}$$

where I is the identity matrix, c^I is the initial solution, and γ is a Lagrange multiplier. Thus the magnitude of the smallest eigenvalues of the modified transformation matrix $a^Ta + \gamma I$ is increased for nonzero values of γ, resulting in stabilization of the solution. Too large a value for γ, however, produces a solution for c from Equation (B-3) which when substituted into Equation (B-1) gives values of b that exhibit large deviations from the input data. In practice, γ is chosen as large as possible, subject to the constraint that these deviations are within the statistical fluctuations of b.

The standard deviation in the derived parameters may be determined from the statistical uncertainty in the data using the theory of linear propagation of errors (*Brandt*, 1970). These standard deviations are the square roots of the parameter covariance matrix diagonal elements. The parameter covariance matrix, C_c, is given in terms of the input or data covariance matrix, C_b, as

$$C_c = [(a^Ta + \gamma I)^{-1}a^T] C_b [(a^Ta + \gamma I)^{-1}a^T]^T \tag{B-4}$$

APPENDIX C

Corresponding to Equation (1) for infinite ensemble-averaged power, we define the unaveraged voltage FFT (omitting but understanding the ω-dependence) as

$$\overset{\circ}{v}(\psi) = \frac{1}{2\gamma} \int_{-\gamma}^{\gamma} \cos^2(\frac{\psi-\phi}{2}) \, \underset{\circ}{v}(\phi) d\phi \quad . \tag{C-1}$$

The angular sea-echo received spectral voltage $v(\phi)$ is complex, whose real and imaginary parts are zero-mean Gaussian random variables; $v(\phi)$ is the output of a narrow-beam scanning system in the limit where the antenna gain pattern becomes infinitesimally narrow, i.e., a delta function. The above linear operation on $\underset{\circ}{v}(\phi)$ implies that $\overset{\circ}{v}(\psi)$ is also a complex Gaussian process.

The input $v(\phi)$ is random over time and space, but more importantly, we assume it to be random over angle, ϕ. Specifically, we take $\underset{\circ}{v}(\phi)$ to be uncorrelated with $\underset{\circ}{v}^*(\phi')$ for $\phi \neq \phi'$. Experimental sea-echo data (both first and second order) measured with a NOAA HF radar system (*Barrick et al.*, 1977) show that angular voltage samples are uncorrelated for $|\phi-\phi'|$ as small as 0.2°, so that this assumption is quite reasonable for the present analysis. Mathematically we express this as

$$<\underset{\circ}{v}(\phi)\underset{\circ}{v}^*(\phi')> \equiv \underset{\sim}{V}(\phi')\delta(\phi-\phi') = \underset{\sim}{V}(\phi)\delta(\phi'-\phi) \quad , \tag{C-2}$$

where angular braces denote infinite ensemble averages. We also define

$$\tilde{V}(\psi) \equiv 2\gamma \cdot <\overset{\circ}{v}(\psi)\overset{\circ}{v}^*(\psi)> \tag{C-3}$$

Hence using Equations (C-2) and (C-3) in Equation (C-1), we obtain

$$\tilde{V}(\psi) = \frac{1}{2\gamma} \int_{-\gamma}^{\gamma} \cos^4(\frac{\psi-\phi}{2}) \underset{\sim}{V}(\phi) d\phi \quad , \tag{C-4}$$

which is identical to Equation (1), i.e., $\underset{\sim}{V}(\phi)$, $\tilde{V}(\psi)$ can represent $\underset{\sim}{\sigma}(\omega,\phi)$, $\tilde{\sigma}(\omega,\psi)$.

Now we define N-sample spectral averages of the broad-beam system output as

$$_N\tilde{\sigma}(\psi) = \frac{2\gamma}{N} \sum_{i=1}^{N} \overset{\circ}{v}_i(\psi) \overset{\circ}{v}_i^*(\psi) \quad , \tag{C-5}$$

where we now subscript the random voltages to denote that they are samples from an ensemble.

Analogous to Equation (4), we also define finite-sample averages of the Fourier-coefficient observables as

$$_N b_n = \frac{2}{\varepsilon_n} \int_{-\pi}^{\pi} {}_N\tilde{\sigma}(\psi)\, tf_n(\psi)\, d\psi \quad , \tag{C-6}$$

where we imply the following limiting processes:

$$\tilde{\sigma}(\omega,\psi) = \operatorname*{Lim}_{N\to\infty} {}_N\tilde{\sigma}(\psi) \qquad \text{and} \qquad b_n(\omega) = \operatorname*{Lim}_{N\to\infty} {}_N b_n \tag{C-7}$$

Employing Equations (C-1), (C-5), and (C-6) in $\langle {}_N b_m \cdot {}_N b_n \rangle$ we obtain

$$\langle {}_N b_m \cdot {}_N b_n \rangle =$$

$$\int_{-\pi}^{\pi} d\psi \int_{-\pi}^{\pi} d\psi'\, tf_m(\psi)\, tf_n(\psi')\, \frac{1}{(2\gamma)^4} \int_{-\gamma}^{\gamma} d\phi \int_{-\gamma}^{\gamma} d\phi' \int_{-\gamma}^{\gamma} d\phi'' \int_{-\gamma}^{\gamma} d\phi'''$$

$$\text{x } \cos^2\left(\frac{\psi-\phi}{2}\right) \cos^2\left(\frac{\psi-\phi'}{2}\right) \cos^2\left(\frac{\psi'-\phi''}{2}\right) \cos^2\left(\frac{\psi'-\phi'''}{2}\right) \tag{C-8}$$

$$\text{x } \left(\frac{2\gamma}{N}\right)^2 \sum_{i=1}^{N} \sum_{j=1}^{N} \langle v_{oi}(\phi) v_{oi}^*(\phi'') v_{oj}(\phi') v_{oj}^*(\phi''') \rangle \quad .$$

Because the voltages $v_{o\ell}(\phi)$ are Gaussian, we can simplify this by noting that

$$\langle v_{oi}(\phi) v_{oi}^*(\phi'') v_{oj}(\phi') v_{oj}^*(\phi''') \rangle = \langle v_{oi}(\phi) v_{oi}^*(\phi'') \rangle \langle v_{oj}(\phi') v_{oj}^*(\phi''') \rangle$$

$$+ \langle v_{oi}(\phi) v_{oj}(\phi') \rangle \langle v_{oi}^*(\phi'') v_{oj}^*(\phi''') \rangle \tag{C-9}$$

$$+ \langle v_{oi}(\phi) v_{oj}^*(\phi''') \rangle \langle v_{oj}(\phi') v_{oi}^*(\phi''') \rangle \quad .$$

The second term is zero because the average of the nonconjugated product of any two complex voltages is zero by orthogonality requirements (i.e., the real and imaginary voltage parts are uncorrelated and have the same variances). Equation (C-2) is next employed for the first and third terms. The first term upon simplification yields identically $\langle_N b_n\rangle\langle_N b_n\rangle$. The third term further simplifies because "independent" samples implies that the average is nonzero only for i=j. Hence we obtain

$$\text{Cov}[_N b_m \cdot _N b_n] = \frac{4}{N\varepsilon_m\varepsilon_n} \int_{-\pi}^{\pi} d\psi \int_{-\pi}^{\pi} d\psi' \, tf_m(\psi) tf_n(\psi')$$

$$\times \frac{1}{(2\gamma)^2} \int_{-\gamma}^{\gamma} d\phi \int_{-\gamma}^{\gamma} d\phi' \cos^2(\frac{\psi-\phi}{2}) \cos^2(\frac{\psi'-\phi}{2})$$

$$\text{(C-10)}$$

$$\times \cos^2(\frac{\psi-\phi'}{2}) \cos^2(\frac{\psi'-\phi}{2}) \underset{\sim}{g}(\omega,\phi) \underset{\sim}{g}(\omega,\phi') \quad .$$

This equation will be simplified by integration over ψ, ψ' after use of the following definitions:

$$\cos^2(\frac{\psi-\phi}{2}) \cos^2(\frac{\psi-\phi'}{2}) \equiv \sum_{i=-2}^{2} d_i(\phi,\phi') tf_i(\psi) \qquad \text{(C-11)}$$

with an analogous expression for $\cos^2(\frac{\psi'-\phi}{2})\cos^2(\frac{\psi'-\phi'}{2})$, where

$$4d_0 \equiv 1 + \frac{1}{2}(\cos\phi\cos\phi' + \sin\phi\sin\phi');$$

$$4d_1 \equiv \cos\phi + \cos\phi';$$

$$4d_{-1} \equiv \sin\phi + \sin\phi'; \qquad \qquad \text{(C-12)}$$

$$4d_2 \equiv \frac{1}{2}(\cos\phi\cos\phi' - \sin\phi\sin\phi');$$

$$4d_{-2} \equiv \frac{1}{2}(\cos\phi\sin\phi' + \sin\phi\cos\phi') \quad .$$

Upon integrating out ψ and ψ' we now obtain

$$\text{Cov}[_N b_m \cdot _N b_n] = (\frac{2\pi}{2\gamma})^2 \frac{1}{N} \int_{-\gamma}^{\gamma} d\phi \int_{-\gamma}^{\gamma} d\phi' d_m(\phi,\phi') d_n(\phi,\phi') \underset{\sim}{g}(\omega,\phi) \underset{\sim}{g}(\omega,\phi') \quad .$$

$$\text{(C-13)}$$

In this equation, $d_m \cdot d_n$ is expanded as follows

$$d_m(\phi,\phi')d_n(\phi,\phi') \equiv \sum_{i=-2}^{2} \sum_{j=-2}^{2} a_i a_j \delta_{mn}^{ij} tf_i(\phi) tf_j(\phi) , \qquad (C-14)$$

where δ_{mn}^{ij} are constants readily obtained from the expansion; they are given in Table 1. Then we employ Equation (5) to simplify, obtaining

$$\text{Cov}[_N b_m \cdot {}_N b_n] = \frac{1}{N} \sum_{i=-2}^{2} \sum_{j=-2}^{2} \delta_{mn}^{ij} b_i(\omega) b_j(\omega) . \qquad (C-15)$$

APPENDIX D

The purpose of this appendix is to show that the complex fast Fourier transform (FFT) voltages of the received second-order sea-echo signal at two different Doppler frequencies are uncorrelated (and hence also statistically independent since *Barrick and Snider* (1977) demonstrated that these signals are Gaussian). Analogous to Equation (3) of *Barrick and Snider* (1977), one can express this complex voltage as being directly proportional to the second-order spatial-temporal wave height coefficient, $H_2(\overline{K},\Omega)$, where $\overline{K} \equiv \overline{K}_r = 2k_0\hat{x}$ for surface-wave backscatter, and Ω is the Doppler frequency being observed. Using the notation of *Barrick and Weber* (1977), we write

$$\overset{\circ}{v}_2(\Omega) \sim H_2(\overline{K},\Omega) , \text{ where}$$

$$H_2(\overline{K},\Omega) = \sum_{\overline{k},\omega} \sum_{\overline{k}',\omega'} A(\overline{k},\omega,\overline{k}',\omega') H_1(\overline{k},\omega) H_1(\overline{k}',\omega')$$

$$\hspace{9cm} (D-1)$$

$$\times \delta_{\overline{K}}^{\overline{k}+\overline{k}'} \delta_{\Omega}^{\omega+\omega'} .$$

Hence, $H_1(\overline{k},\omega)$ is the first-order spatial-temporal surface height Fourier series coefficient (assumed to be a complex Gaussian random variable). The δ's here are Kronecker deltas. $A(...)$, as derived in *Barrick and Weber* (1977), is the second-order wave height coupling coefficient, but is understood here to include the second-order electromagnetic coupling contribution also.

We show here that $\langle \overset{\circ}{v}_2(\Omega_1) \overset{\circ *}{v}_2(\Omega_2) \rangle = 0$ for $\Omega_1 \neq \Omega_2$; this implies that the FFT output at any two different Doppler frequencies is zero. First we rewrite Equation (D-1) using the first-order dispersion relation to express ω in terms of k, and ω' in terms of k'. Then a two-sided representation for H_1 is employed so that waves travelling over 360° of space can be included. These steps are detailed in *Barrick and Weber* (1977) following their Equation (1). We employ the Kronecker deltas to eliminate the series on ω and ω', and redefine the summation indices over $\bar{k}$, $\bar{k}'$ in certain cases to obtain (dropping the subscript on H_1):

$$H_2(\bar{K},\Omega) = \sum_{\bar{k}} \sum_{\bar{k}'} [A(\bar{k},\omega_g,\bar{k}',\omega_g')H(\bar{k})H(\bar{k}')\delta^{\bar{k}+\bar{k}'}_{\bar{K}}\delta^{\omega_g+\omega_g'}_{\Omega}$$

$$+ \quad A(-\bar{k},-\omega_g,\bar{k}',\omega_g')H*(\bar{k})H(\bar{k}')\delta^{-\bar{k}+\bar{k}'}_{\bar{K}}\delta^{-\omega_g+\omega_g'}_{\Omega}$$

$$+ \quad A(\bar{k},\omega_g,-\bar{k}',-\omega_g')H(\bar{k})H*(\bar{k}')\delta^{\bar{k}-\bar{k}'}_{\bar{K}}\delta^{\omega_g-\omega_g'}_{\Omega} \qquad \text{(D-2)}$$

$$+ \quad A(-\bar{k},-\omega_g,-\bar{k}',-\omega_g')H*(\bar{k})H*(\bar{k}')\delta^{-\bar{k}-\bar{k}'}_{\bar{K}}\delta^{-\omega_g-\omega_g'}_{\Omega}],$$

where $\omega_g \equiv \sqrt{gk}$ and $\omega_g' \equiv \sqrt{gk'}$.

At this point the correlation function of $H_2(\bar{K},\Omega_1)$ with $H_2^*(\bar{K},\Omega_2)$ is formed. In order to simplify notation, we define $A_m^n \equiv A(\bar{k}_n,\omega_{gn},\bar{k}_m,\omega_{gm})$, $A_m^{-n} \equiv A(-\bar{k}_n,-\omega_{gn},\bar{k}_m,\omega_{gm})$, etc. There will be sixteen terms in the correlation function because of four each in $H_2(\bar{K},\Omega_1)$ and $H_2(\bar{K},\Omega_2)$. We immediately eliminate terms which will obviously be zero, such as $\langle H(\bar{k}_1)H(\bar{k}_2)H(\bar{k}_3)H(\bar{k}_4) \rangle$, $\langle H(\bar{k}_1)H(\bar{k}_2)H(\bar{k}_3)H*(\bar{k}_4) \rangle$, etc.; we must always have two complex conjugate factors and two nonconjugated factors in each average for the result to be nonzero. This is a consequence of the definition of Equation (8) in *Barrick and Weber* (1977), because $\langle H(\bar{k})H(\bar{k}') \rangle \equiv 0$ even for $\bar{k} = \pm\bar{k}'$. We also drop the "g" subscript henceforth on the ω's. Six terms remain.

$$\langle H_2(\bar{K},\Omega_1)H_2^*(\bar{K},\Omega_2) \rangle =$$

$$\sum_{\bar{k}_1,\bar{k}_2,\bar{k}_3,\bar{k}_4} A_2^1 A_4^3 \langle H(\bar{k}_1)H(\bar{k}_2)H*(\bar{k}_3)H*(\bar{k}_4) \rangle \delta^{\bar{k}_1+\bar{k}_2}_{\bar{K}}\delta^{\omega_1+\omega_2}_{\Omega_1}\delta^{\bar{k}_3+\bar{k}_4}_{\bar{K}}\delta^{\omega_3+\omega_4}_{\Omega_2}$$

$$+ A_2^{-1} A_4^{-3} \langle H^*(\overline{k}_1) H(\overline{k}_2) H(\overline{K}_3) H^*(\overline{k}_4) \rangle \delta_{\overline{K}}^{-k_1+\overline{k}_2} \delta_{\Omega_1}^{-\omega_1+\omega_2} \delta_{\overline{K}}^{-\overline{k}_3+\overline{k}_4} \delta_{\Omega_2}^{-\omega_3+\omega_4}$$

$$+ A_2^{-1} A_{-4}^{3} \langle H^*(\overline{k}_1) H(\overline{k}_2) H^*(\overline{k}_3) H(\overline{k}_4) \rangle \delta_{\overline{K}}^{-\overline{k}_1+\overline{k}_2} \delta_{\Omega_1}^{-\omega_1+\omega_2} \delta_{\overline{K}}^{\overline{k}_3-\overline{k}_4} \delta_{\Omega_2}^{\omega_3-\omega_4}$$

$$+ A_{-2}^{1} A_4^{-3} \langle H(\overline{k}_1) H^*(\overline{k}_2) H(\overline{k}_3) H^*(\overline{k}_4) \rangle \delta_{\overline{K}}^{\overline{k}_1-\overline{k}_2} \delta_{\Omega_1}^{\omega_1-\omega_2} \delta_{\overline{K}}^{-\overline{k}_3+\overline{k}_4} \delta_{\Omega_2}^{-\omega_3+\omega_4} \qquad (D-3)$$

$$+ A_{-2}^{1} A_{-4}^{3} \langle H(\overline{k}_1) H^*(\overline{k}_2) H^*(\overline{k}_3) H(\overline{k}_4) \rangle \delta_{\overline{K}}^{\overline{k}_1-\overline{k}_2} \delta_{\Omega_1}^{\omega_1-\omega_2} \delta_{\overline{K}}^{\overline{k}_3-\overline{k}_4} \delta_{\Omega_2}^{\omega_3-\omega_4}$$

$$+ A_{-2}^{-1} A_{-4}^{-3} \langle H^*(\overline{k}_1) H^*(\overline{k}_2) H(\overline{k}_3) H(\overline{k}_4) \rangle \delta_{\overline{K}}^{-\overline{k}_1-\overline{k}_2} \delta_{\Omega_1}^{-\omega_1-\omega_2} \delta_{\overline{K}}^{-\overline{k}_3-\overline{k}_4} \delta_{\Omega_2}^{-\omega_3-\omega_4} \quad .$$

We now employ the property of Gaussian process given in Equation (C-9); this permits us to express the quadruple moments above as products of double moments. Again all terms which do not have a conjugate/nonconjugate pair in each average are eliminated. We eliminate still more terms by noting that factors such as $\langle H(\overline{k}_1) H^*(\overline{k}_2) \rangle$ are nonzero when $\overline{k}_1 = \overline{k}_2$; however, this requires the corresponding coupling coefficient be $A_{-2}^{1} = A_{-1}^{1} = A(\overline{k}_1, \omega_1, -\overline{k}_1, -\omega_1)$, and we proved in *Weber and Barrick* (1977) that this is identically zero. Then using the definition of the wave height directional spectrum from the above reference, i.e., $\langle H(\overline{k}) H^*(\overline{k}) \rangle = [(2\pi)^2/2L_x L_z] \times S(\overline{k})$, we obtain

$$[\frac{2L_x L_y}{(2\pi)^2}] \langle H_2(\overline{K}, \Omega_1) H^*(\overline{K}, \Omega_2) \rangle =$$

$$\sum_{\overline{K}_1} \sum_{\overline{k}_2} [2(A_2^1)^2 S(\overline{k}_1) S(\overline{k}_2) \delta_{\overline{K}}^{\overline{k}_1+\overline{k}_2} \delta_{\Omega_1}^{\omega_1+\omega_2} \delta_{\Omega_2}^{\omega_1+\omega_2} \qquad \text{(from first term)}$$

$$+ (A_2^{-1})^2 S(\overline{k}_1) S(\overline{k}_2) \delta_{\overline{K}}^{-\overline{k}_1+\overline{k}_2} \delta_{\Omega_1}^{-\omega_1+\omega_2} \delta_{\Omega_2}^{-\omega_1+\omega_2} \qquad \text{(from second term)}$$

$$+ (A_2^{-1})^2 S(\overline{k}_1) S(\overline{k}_2) \delta_{\overline{K}}^{-\overline{k}_1+\overline{k}_2} \delta_{\Omega_1}^{-\omega_1+\omega_2} \delta_{\Omega_2}^{-\omega_1+\omega_2} \qquad \text{(from third term)}$$

$$+ (A_{-2}^1)^2 S(\overline{k}_1) S(\overline{k}_2) \delta_{\overline{K}}^{\overline{k}_1 - \overline{k}_2} \delta_{\Omega_1}^{\omega_1 - \omega_2} \delta_{\Omega_2}^{\omega_1 - \omega_2} \qquad \text{(from fourth term)}$$

$$+ (A_{-2}^1)^2 S(\overline{k}_1) S(\overline{k}_2) \delta_{\overline{K}}^{\overline{k}_1 - \overline{k}_2} \delta_{\Omega_1}^{\omega_1 - \omega_2} \delta_{\Omega_2}^{\omega_1 - \omega_2} \qquad \text{(from fifth term)}$$

$$+ 2(A_{-2}^{-1})^2 S(\overline{k}_1) S(\overline{k}_2) \delta_{\overline{K}}^{-\overline{k}_1 - \overline{k}_2} \delta_{\Omega_1}^{-\omega_1 - \omega_2} \delta_{\Omega_2}^{-\omega_1 - \omega_2}] \quad \text{(from sixth term).}$$

$$\text{(D-4)}$$

The above expression could be simplified by combining the second with the third and the fourth with the fifth terms. However, this is not necessary to illustrate that all terms are zero. Take the first for example. The two final Kronecker deltas in this equation require that

$$\Omega_1 = \omega_1 + \omega_2 \qquad \text{and} \qquad \Omega_2 = \omega_1 + \omega_2 . \qquad \text{(D-5)}$$

The only way this can be satisfied is for $\Omega_1 = \Omega_2$. Otherwise one of the two Kronecker deltas is always zero when the other is nonzero. Hence we have proved that the voltages at two different Doppler frequencies Ω_1 and Ω_2 for second-order sea echo are uncorrelated.

APPENDIX E

Here we examine the output of the pitch-and-roll buoy, with the aim of deriving the standard deviations of the five Fourier angular coefficients obtained from N-sample averaging. The final result is that the N-sample variance of c_n is c_n^2/N, implying that the coefficients are not coupled.

These coefficients (infinite-ensemble-averaged) were first examined by *Longuet-Higgins et al.* (1963), whereas *Tyler et al.* (1974) preferred to express the buoy-derived outputs as angular moments, i.e.,

$$N_{pq}(\kappa) = \int_{-\pi}^{\pi} \cos^p\theta \sin^q(\theta) S(\kappa,\theta) d\theta , \qquad \text{(E-1)}$$

while our definition, Equation (6), produces

$$c_n(\kappa) = \frac{2}{\varepsilon_n} \int_{-\pi}^{\pi} tf_n(\theta)S(\kappa,\theta)d\theta \quad . \qquad (E-2)$$

The relationships between the moments and coefficients are thus readily deduced:

$$c_0(\kappa) = N_{00}(\kappa); \quad c_1(\kappa) = 2N_{10}(\kappa); \quad c_{-1}(\kappa) = 2N_{01}(\kappa);$$

$$\qquad (E-3)$$

$$c_2(\kappa) = 2N_{20}(\kappa) - 2N_{02}(\kappa); \quad c_{-2}(\kappa) = 4N_{11}(\kappa) \quad .$$

There is also the following auxiliary relation among the six moments, implying that only five of them are independent:

$$N_{00}(\kappa) = N_{20}(\kappa) + N_{02}(\kappa) \quad .$$

Omitting hardware details, the buoy produces three voltage signals vs time at its mooring point x,y which are proportional to surface vertical acceleration and the x,y slopes. The acceleration is time-integrated twice to give surface height. We assume that the appropriate proportionality constants are known, so that we may call the buoy output signals $\zeta(x,y,t)$, $\partial\zeta(x,y,t)/\partial x$, and $\partial\zeta(x,y,t)/\partial y$. In order to examine the N-sample statistics of the c_n, we employ the following Fourier-integral definitions:

$$\zeta(x,y,t) = \zeta(\overline{r},t) \equiv \int\int\int_{-\infty}^{\infty} H(\overline{\kappa},\omega)e^{-i\overline{\kappa}\cdot\overline{r}+i\omega t}d^2\overline{\kappa}d\omega \quad , \quad (E-4)$$

where the wave height directional spectrum is defined as

$$\langle H*(\overline{\kappa},\omega)H(\overline{\kappa}',\omega')\rangle \equiv S(\overline{\kappa}',\omega')\delta(\overline{\kappa}'-\overline{\kappa})\delta(\omega'-\omega) \quad , \qquad (E-5)$$

where, because ζ is real, we must have $H*(\overline{\kappa},\omega) = H(-\overline{\kappa},-\omega)$. Analogously, we can define cross spectra between any two quantities ζ_i, ζ_j (e.g., ζ_i could be $\partial\zeta(x,y,t)/\partial x$, etc.) as $S_{ij}(\overline{\kappa},\omega)$ in terms of the Fourier transforms H_i^* and H_j.

The deep-water dispersion relation to lowest order (*Barrick and Weber*, 1977) permits the following form for the wave height spatial-temporal spectrum:

$$S(\overline{\kappa},\omega) = \frac{1}{2} S(\overline{\kappa})\delta(\omega-\sqrt{g\kappa}) + \frac{1}{2} S(-\overline{\kappa})\delta(\omega+\sqrt{g\kappa}) \quad . \tag{E-6}$$

We define $S(\overline{\kappa})d^2\overline{\kappa} = S(\kappa_x,\kappa_y)d\kappa_x d\kappa_y = S(\kappa,\theta)d\kappa d\theta$. Then we can extend Equation (E-2) to obtain

$$\frac{2\omega}{g} c_n(\frac{\omega^2}{g}) = \frac{2}{\varepsilon_n} \int_0^\infty \kappa d\kappa \int_{-\pi}^{\pi} tf_n(\theta)S(\overline{\kappa},\omega)d\theta \equiv c_n^t(\omega) =$$

$$\frac{2}{\varepsilon_n} <\iint d^2\overline{\kappa} \iiint d^2\overline{\kappa}'d\omega'H(\overline{\kappa},\omega)H(\overline{\kappa}',\omega')tf_n(\theta)e^{-i(\overline{\kappa}+\overline{\kappa}')\cdot\overline{r}}> , \tag{E-7}$$

where the left-hand side is the temporal Fourier angular coefficient, as employed by *Longuet-Higgins et al.* (1963). We now define an N-sample average:

$$_N c_n(\omega) \equiv \frac{2}{\varepsilon_n N} \sum_{i=1} \iint d^2\overline{\kappa} \iiint d^2\overline{\kappa}'d\omega'H^i(\overline{\kappa},\omega)H^i(\overline{\kappa}',\omega')tf_n(\theta)e^{i(\overline{\kappa}+\overline{\kappa}')\cdot\overline{r}} \tag{E-8}$$

First we show that $c_n^t(\omega)$ is formed from the averaged temporal cross spectra of the heights and slopes. Then we show that the N-sample variance of the coefficient c_n is merely c_n^2/N. We do this by concentrating on c_{-2}, as it illustrates adequately the procedure.

We start with Equation (E-7), employing the fact that $tf_{-2}(\theta) = \sin2\theta = 2\sin\theta\cos\theta$, and obtain

$$c_{-2}^t(\omega) =$$

$$4<\iint d^2\overline{\kappa} \iiint d^2\overline{\kappa}'d\omega'\frac{(-i\kappa\cos\theta)(-i\kappa\sin\theta)}{(-i\kappa)^2} H(\overline{\kappa},\omega)H(\overline{\kappa}',\omega')e^{-i(\overline{\kappa}+\overline{\kappa}')\cdot\overline{r}}> . \tag{E-9}$$

Notice that $\kappa\cos\theta = \kappa_x$ and $\kappa\sin\theta = \kappa_y$, where $\overline{\kappa} = (\kappa_x,\kappa_y)$. By differentiating Equation (E-4) with respect to x, we see that the Fourier transform of $\zeta_x(x,y,t)$ is $-i\kappa_x H(\overline{\kappa},\omega) = -\kappa\cos\theta H(\overline{\kappa},\omega)$. We now plan to change variables in Equation (E-9) from κ',ω to

$-\overline{\kappa}',-\omega'$. Then we note that when we average $<H(\overline{\kappa},\omega)H(-\overline{\kappa}',-\omega')> =$ $<H(\overline{\kappa},\omega)H*(\overline{\kappa}',\omega')>$, from Equation (E-5) we must have $\overline{\kappa}' = \overline{\kappa}, \omega' = \omega$; this means we can replace κ_y by κ_y', and hence we obtain

$$c_{-2}^t(\omega) = 4 \iint d^2\overline{\kappa} \iiint d^2\overline{\kappa}' d\omega' <-i\kappa_x H(\overline{\kappa},\omega) i\kappa_y H*(\overline{\kappa}',\omega')> \kappa^{-2} e^{-i(\overline{\kappa}-\overline{\kappa}')\cdot\overline{r}}$$

$$= 4 \iint d^2\overline{\kappa} \iiint d^2\overline{\kappa}' d\omega' <H_x(\overline{\kappa},\omega) H_y^*(\overline{\kappa},\omega)> \kappa^{-2} e^{-i(\overline{\kappa}-\overline{\kappa}')\cdot\overline{r}} \quad .$$

$$(E-10)$$

We can now average to obtain the cross spectrum of surface slopes, $S_{\zeta_x\zeta_y}(\overline{\kappa},\omega)$. Also, since κ and ω are uniquely related by the dispersion relation ($\kappa = \omega^2/g$), the κ^{-2} can be removed and treated as a constant. Thus we obtain

$$\frac{\omega^4}{g^2} c_{-2}^t(\omega) = 4 \iint d^2\overline{\kappa} \, S_{\zeta_x\zeta_y}(\overline{\kappa},\omega) = \frac{2\omega^5}{g^3} c_{-2}(\frac{\omega^2}{g}) \quad , \quad (E-11)$$

where the rightmost expression is the spatial Fourier coefficient. But the integral can be shown to be the average temporal cross spectrum of the slopes; i.e., if we take the temporal Fourier transform of $\zeta_x(x,y,t)$ and multiply it by the conjugate of the temporal Fourier transform of $\zeta_y(x,y,t)$ (call them $H_x(\omega)$ and $H_y^*(\omega')$), we see that

$$\iint d^2\overline{\kappa} \, S_{\zeta_x\zeta_y}(\overline{\kappa},\omega) = \int <H_x(\omega)H_y^*(\omega')> d\omega' \quad . \quad (E-12)$$

From finite-length time records of length T, appropriately windowed and Fourier transformed to give $_TH_i(\omega)$, a spectral estimate of the right side is $\frac{<_TH_x(\omega)\,_TH_y^*(\omega)>}{T}$. Thus we have proved that

$$\frac{2\omega^5}{g^3} c_{-2}(\omega) = 4\frac{<_TH_x(\omega)\,_TH_y^*(\omega)>}{T} \quad ; \text{ likewise we can show that}$$

$$\frac{2\omega^5}{g^3} c_2(\omega) = \frac{2<_TH_x(\omega)\,_TH_x^*(\omega)> - 2<_TH_y\,H_y^*(\omega)>}{T} \quad ; \quad (E-13)$$

$$\frac{2\omega}{g} c_0(\omega) = \frac{<_TH(\omega)\,_TH^*(\omega)>}{T} \quad ; \quad \frac{2\omega^3}{g^2} c_1(\omega) = \frac{2i<_TH(\omega)\,_TH_x^*(\omega)>}{T} \quad ;$$

$$\frac{2\omega^3}{g^2} c_{-1}(\omega) = \frac{2i<{}_T H(\omega){}_T H_y^*(\omega)>}{T} \ .$$

From comparison with Equation (E-3), one readily sees that the averages on the right side of Equation (E-13) are identically the "moments" employed by *Tyler et al.* (1974).

Finally, we can form N-sample averages for c_{-2}, obtaining

$$<{}_N c_{-2} \ {}_N c_{-2}> =$$

$$4 \iint d^2\overline{\kappa}_1 \iint d^2\overline{\kappa}_3 \iiint d^2\overline{\kappa}_2 d\omega_2 \iiint d^2\overline{\kappa}_4 d\omega_4$$

$$\left[\frac{1}{N^2} \sum_{i,j}^{N\ N} <(-i\kappa_1\cos\theta_1)(-i\kappa_1\sin\theta_1) \right.$$

$$\times (-i\kappa_3\cos\theta_3)(-i\kappa_3\sin\theta_3) H^i(\overline{\kappa}_1,\omega_1)H^i(\overline{\kappa}_2,\omega_2)H^j(\overline{\kappa}_3,\omega_3)H^j(\overline{\kappa}_4,\omega_4)>$$

$$\left. \times e^{i(\overline{\kappa}_1+\overline{\kappa}_2+\overline{\kappa}_3+\overline{\kappa}_4)\cdot\overline{r}} \right] \ . \tag{E-14}$$

By employing the same procedures as above, using Equation (C-9) for Gaussian variable quadruple products, and noting that $<H^iH^{*j}> = 0$ for $i \neq j$ (i.e., different samples of an ensemble are uncorrelated), we obtain two terms from Equation (E-14). The third is zero, because we end up with averages of nonconjugated H-factors. Thus we obtain

$$< {}_N c_{-2} \ {}_N c_{-2}> = c_{-2}^2 + c_{-2}^2/N, \text{ and hence}$$

$$\text{Var}[{}_N c_{-2}] = <{}_N c_{-2} \ {}_N c_{-2}> - <{}_N c_{-2}><{}_N c_{-2}> = c_{-2}^2/N \ . \tag{E-15}$$

Proofs for the variances of the other c_n proceed identically.

APPENDIX F

This appendix derives relationships (21) and (22), based on the use of the model given on the left side of Equation (20). First we establish the fact that C_1 and C_2 can be expressed as

$$C_1 \equiv \frac{c_1}{c_o} = f_1(s)\cos\theta_o \quad ; \quad C_2 \equiv \frac{c_2}{c_o} = f_2(s)\cos 2\theta_o. \tag{F-1}$$

(Proofs for C_{-1} and C_{-2} proceed analogously and will not be given here.)

Multiplying Equation (14) by $tf_1(\theta)$ and $tf_2(\theta)$ respectively and integrating, we have

$$\pi c_1 = A(s) \int_{-\pi}^{\pi} \cos\theta \cos^s\left(\frac{\theta-\theta_o}{2}\right)d\theta \; ;$$

$$\pi c_2 = A(s) \int_{-\pi}^{\pi} \cos 2\theta \cos^s\left(\frac{\theta-\theta_o}{2}\right)d\theta \; ,$$

or

$$C_1 = 2I_o^{-1}(s) \int_{-\pi}^{\pi} \cos\theta \cos^s\left(\frac{\theta-\theta_o}{2}\right)d\theta \; ;$$

$$C_2 = 2I_o^{-1}(s) \int_{-\pi}^{\pi} \cos 2\theta \cos^s\left(\frac{\theta-\theta_o}{2}\right)d\theta \; , \tag{F-2}$$

where we define

$$I_o(s) \equiv \int_{-\pi}^{\pi} \cos^s\left(\frac{\theta}{2}\right)d\theta; \quad I_1(s) \equiv \int_{-\pi}^{\pi} \cos\theta \cos^s\left(\frac{\theta}{2}\right)d\theta;$$

$$I_2(s) \equiv \int_{-\pi}^{\pi} \cos 2\theta \cos^s\left(\frac{\theta}{2}\right)d\theta \; . \tag{F-3}$$

Hence we want to prove that

$$f_1(s)\cos\theta_o = 2I_o^{-1}(s) \int_{-\pi}^{\pi} \cos\theta \cos^s\left(\frac{\theta-\theta_o}{2}\right)d\theta \tag{F-4}$$

in order to establish the validity of the first part of Equation (F-1). If Equation (F-4) is valid, then by setting $\theta_o = 0$ we have

$$f_1(s) = 2I_1(s)/I_o(s) \quad \text{and} \quad f_2(s) = 2I_2(s)/I_o(s) \ . \qquad \text{(F-5)}$$

Now, multiply the left side of Equation (F-4) by $\cos\theta_o$ and integrate:

$$\int_{-\pi}^{\pi} f_1(s)\cos^2\theta_o d\theta_o = \pi f_1(s) = 2\pi I_1(s)/I_o(s) \ ,$$

from Equation (F-5). Doing the same to the right side of Equation (F-4) we obtain

$$2I_o^{-1}(s) \int_{-\pi}^{\pi} d\theta_o \int_{-\pi}^{\pi} d\theta \cos\theta_o \cos\theta \cos^s\left(\frac{\theta-\theta_o}{2}\right) \ .$$

Hence we must establish that

$$\int_{-\pi}^{\pi} \cos\theta \cos^s\left(\frac{\theta}{2}\right) d\theta = \frac{1}{\pi} \int_{-\pi}^{\pi} d\theta_o \int_{-\pi}^{\pi} d\theta \cos\theta_o \cos\theta \cos^s\left(\frac{\theta-\theta_o}{2}\right) \ .$$

$$\text{(F-6)}$$

This is easily done by using the following identity on the right side of Equation (F-6): $\cos\theta_o\cos\theta = \frac{1}{2}\cos(\theta-\theta_o) + \frac{1}{2}\cos(\theta+\theta_o)$, and then changing variables to $\theta' = \theta-\theta_o$, eliminating θ_o. We then obtain

$$\frac{1}{2\pi} \int_{-\pi}^{\pi} d\theta' \int_{-\pi}^{\pi} d\theta \cos\theta' \cos^s\left(\frac{\theta'}{2}\right) + \frac{1}{2\pi} \int_{-\pi}^{\pi} d\theta' \int_{-\pi}^{\pi} d\theta \cos(2\theta-\theta')\cos^s\left(\frac{\theta'}{2}\right) \ .$$

The second term above is zero, as can be seen by expanding $\cos(2\theta-\theta') = \cos2\theta\cos\theta' + \sin2\theta\sin\theta'$, and performing the integration over θ. The first term is identically the left side of Equation (F-6) after the integration over θ is done. Hence we have proved Equation (F-6) and established the validity of the form of Equation (F-1) for C_1; the proof for C_2 is identical.

Now we derive explicit expressions for $f_1(s)$ and $f_2(s)$ by establishing recurrence-like relations among the $I_n(s)$. First we integrate $I_1(s)$ by parts obtaining

$$I_1(s) = \int_{-\pi}^{\pi} \cos\theta \cos^s(\tfrac{\theta}{2}) d\theta$$

$$= \frac{s}{2} \int_{-\pi}^{\pi} \sin\theta \sin(\tfrac{\theta}{2}) \cos(\tfrac{\theta}{2}) \cos^{s-2}(\tfrac{\theta}{2}) d\theta = \frac{s}{4} \int_{-\pi}^{\pi} \sin^2\theta \cos^{s-2}(\tfrac{\theta}{2}) d\theta$$

$$= \frac{s}{8} \int_{-\pi}^{\pi} \cos^{s-2}(\tfrac{\theta}{2}) d\theta - \frac{s}{8} \int_{-\pi}^{\pi} \cos 2\theta \cos^{s-2}(\tfrac{\theta}{2}) d\theta \ .$$

Thus we have

$$I_1(s) = \frac{s}{8} I_o(s-2) - \frac{s}{8} I_2(s-2) \quad . \tag{F-7}$$

Using integration by parts on $I_0(s)$, we find

$$I_o(s) = \int_{-\pi}^{\pi} \cos^s(\tfrac{\theta}{2}) d\theta = (s-1) \int_{-\pi}^{\pi} \sin^2(\tfrac{\theta}{2}) \cos^{s-2}(\tfrac{\theta}{2}) d\theta$$

$$= (s-1) \int_{-\pi}^{\pi} \cos^{s-2}(\tfrac{\theta}{2}) d\theta - (s-1) \int_{-\pi}^{\pi} \cos^s(\tfrac{\theta}{2}) d\theta$$

$$= (s-1)I_o(s-2) - (s-1)I_o(s) \ ,$$

from which we obtain

$$sI_o(s) = (s-1)I_o(s-2) \ . \tag{F-8}$$

Finally, we use the trigonometric identity $\cos\theta = 2\cos^2\frac{\theta}{2} - 1$ on $I_1(s)$

$$I_1(s) = \int_{-\pi}^{\pi} \cos\theta \cos^s(\tfrac{\theta}{2}) d\theta$$

$$= 2 \int_{-\pi}^{\pi} \cos^{s-2}(\tfrac{\theta}{2}) d\theta - \int_{-\pi}^{\pi} \cos^s(\tfrac{\theta}{2}) d\theta = 2I_o(s+2) - I_o(s) \ .$$

By replacing s by s-2 here we obtain

$$I_1(s-2) = 2I_o(s) - I_o(s-2) \quad . \tag{F-9}$$

If we use Equation (F-8) in Equation (F-9) and replace s-2 by s, we obtain

$$I_1(s) = \frac{s}{s+2} I_o(s) \; , \tag{F-10}$$

whence using the first of Equation (F-5) we see that

$$f_1(s) = \frac{2s}{s+2} \quad . \tag{F-11}$$

Then solving for $I_2(s)$ in terms of $I_o(s)$ using Equations (F-7) through (F-10) we likewise obtain

$$I_2(s) = \frac{s(s-2)}{(s+2)(s+4)} I_o(s) \; , \tag{F-12}$$

whence we have

$$f_2(s) = \frac{2s(s-2)}{(s+2)(s+4)} \quad . \tag{F-13}$$

If expansions employing coefficients c_n for $|n| > 2$ are used, the general pattern is seen to be

$$f_n(s) = \frac{2s(s-2) \; \dots \; (s+2-2|n|)}{(s+2)(s+4) \; \dots \; (s+2|n|)} \cdot \tag{F-14}$$

ACKNOWLEDGEMENT

Part of this work was performed under Office of Naval Research contract N0014-77C-0356 while Dr. Lipa was at Stanford University.

REFERENCES

Abramowitz, M. and I. A. Stegun. 1964. *Handbook of Mathematical Functions*, Chapter 6, U.S. Government Printing Office, Washington, D.C.

Barrick, D. E., M. W. Evans, and B. L. Weber. 1977. Ocean surface currents mapped by radar, *Science*, *198*, 138-144.

Barrick, D. E. and J. B. Snider. 1977. The statistics of HF sea-echo Doppler spectra, *IEEE Trans. on Antennas and Propagation*, *AP-25*, 19-28.

Barrick, D. E. and B. L. Weber. 1977. On the nonlinear theory of gravity waves on the ocean's surface. Part II: interpretation and applications, *J. Phys. Oceanogr.*, *7*, 11-21.

Brandt, S. 1970. *Statistical and Computational Methods in Data Analysis*, North Holland Publishing Co., Amsterdam, 414 pp.

Kraus, J. D. 1950. *Antennas*, McGraw-Hill, New York, 155-172.

Lipa, B. J. 1977. Derivation of directional ocean-wave spectra by integral inversion of second order radar echoes, *Radio Science*, *12*, 425-434.

Longuet-Higgins, M. S., D. E. Cartwright and N. D. Smith. 1963. Observations of the directional spectrum of sea waves using the motions of a floating buoy, In: *Ocean Wave Spectra*, Prentice-Hall, Englewood Cliffs, N.J. 111-136.

Phillips, D. L. 1962. A technique for the numerical solution of certain integral equations of the first kind, *J. Assoc. Comp. Mach.*, *9*, 84-97.

Stewart, R. H. 1977. A discus-hulled wave measuring buoy, *Ocean Engineering*, *4*, 101-107.

Stewart, R. H. and J. R. Barnum. 1975. Radio measurements of oceanic winds at long ranges: an evaluation, *Radio Science*, *10*, 853-857.

Twomey, S. 1963. On the numerical solution of Fredholm integral equations of the first kind by inversion of the linear system produced by quadrature, *J. Assoc. Comp. Mach.*, *10*, 97-101.

Tyler, G. L., C. C. Teague, R. H. Stewart, A. M. Peterson, W. H. Munk, and J. W. Joy. 1974. Wave directional spectra from synthetic aperture observations of radio scatter, *Deep Sea Res.*, *21*, 989-1016.

Weber, B. L. and D. E. Barrick. 1977. On the nonlinear theory of gravity waves on the ocean's surface. Part I: derivations, *J. Phys. Oceanogr.*, *7*, 3-10.

Synthetic Aperture HF Radar Wave Measurement Experiments

Calvin C. Teague

Center for Radar Astronomy

Stanford University

ABSTRACT

The directional distribution of ocean waves at one or several discrete wave lengths can be measured to 5 or 10 degree resolution by a high-frequency surface-wave radar system using a synthetic aperture antenna. For infrequent observations, the synthetic aperture antenna may be simpler than a large antenna installation having the same directional resolution. The application of the synthetic aperture to oceanographic observations is illustrated by experiments which measured the directional distribution of waves in the open ocean from a small island and from a ship in the Pacific Ocean, and the fetch-limited growth of waves following a polar frontal passage at Galveston, Texas.

INTRODUCTION

The measurement of the directional ocean wave spectrum in the open ocean is at best a difficult task with conventional oceanographic techniques. Several remote sensing techniques utilizing high-frequency (HF) or microwave radar systems are being developed to supplement the oceanographers' tools. This paper discusses one such technique, that of a synthetic aperture HF radar system using surface-wave signals. By using this technique, the detailed directional distribution of ocean waves at one or more wave lengths can be measured with 5 or 10 degree resolution and the total energy of these wave lengths can be determined. The observations can be made from land, and can cover a large area, usually a circle of radius 100 km or more, from a single site. The directional resolution can

be up to an order of magnitude better than that of a tilt buoy.
The measurements also can be made in the open ocean where a fixed
array of surface-height or pressure sensors would be difficult
or impossible. The use of a synthetic aperture antenna may be
considerably less costly than that of a large array, particularly
for experiments which are performed infrequently. However, the
technique is not applicable to all experimental situations, and
the requirements for its successful application are indicated. At
present, the synthetic aperture technique is best suited to basic
measurements which cannot be obtained with sufficient directional
resolution by more conventional methods, but it could be mechanized
to allow routine observations to be made.

A synthetic aperture antenna is formed by carrying a simple,
relatively broad-beamed antenna along a straight path at a uniform
velocity. When properly processed, the data provide the directional
resolution of a very large physical antenna. Although neither the
uniform velocity nor the straight path is a requirement so long as
the path and velocity at every instant are known, their use simplifies the theory considerably and is assumed throughout this paper.
As is shown below, the directional resolution of such an antenna
is equivalent to that of a large physical antenna with the same
length as the synthetic aperture. After a summary of the theory of
the synthetic aperture antenna, its application to oceanographic
measurements is illustrated by a review of experiments designed to
measure the directional distribution of 7-second waves in the open
ocean under steady-state conditions and the fetch-limited growth
of waves from a long, straight shore after a polar frontal passage.

BACKGROUND

Since few investigators are working on the theoretical aspects
of the synthetic aperture antenna as applied at HF to the measurement
of the ocean environment, a review of the theory is appropriate. The
synthetic aperture antenna (*Harger*, 1970) is often applied at microwave frequencies to airborne or satellite-based radar systems for
mapping the surface of the earth. Its application in the HF (2-30
MHz) region is less common, however. *Clarke* (1970) describes a
method for using incoherent processing of signals from two antennas
to synthesize a large aperture for observing ionospheric HF radar
echoes; *Shearman et al.* (1973) describe its implementation. *Lynch*
(1970), on the other hand, describes an aperture formed by coherent
processing of ionospherically-propagated HF signals from a single
antenna carried by an airplane and finds that a synthetic aperture
up to 70 km in length is possible. The coherent processing technique is employed here, and because of the spectral characteristics
of the ocean echoes, the maximum platform velocity is constrained
to only a few m/s, for which surface-based platforms are feasible.

Also, only surface-wave signals are considered, so that the frequency resolution of the measurements is not limited by the
ionospheric path. A typical coherent integration time is 5 minutes,
corresponding to a frequency resolution of 0.003 Hz. A very simple
theory then suffices for the problem of measuring ocean wave directional spectra.

     Assume that it is desired to measure the azimuthal direction
of arrival θ of a signal of known frequency f_s and wave length λ,
with plane wave fronts arriving at the receiver. If a receiver is
moved at a constant velocity V_r as shown in Figure 1, there is a
Doppler shift of $(V_r/\lambda) \cos \theta$ induced by this motion and the
observed frequency is

$$f = f_s + \frac{V_r}{\lambda} \cos \theta \qquad\qquad (1)$$

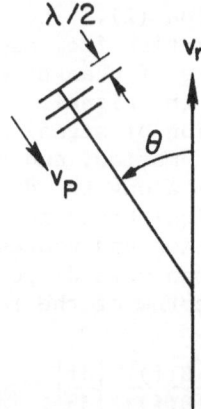

Fig. 1. Geometry for synthetic aperture antenna. An antenna
element is carried at velocity V_r along a straight line in the horizontal plane. A signal arrives from an angle θ with respect to the
velocity vector. For a radar system operating at wave length λ,
this signal is produced by first-order Bragg scattering by ocean
waves of $\lambda/2$ with phase velocity V_p. (from *Teague*, 1975).

Since every other quantity is known, θ can be found simply from

$$\theta = \cos^{-1} [(f-f_s)\lambda/V_r] \tag{2}$$

If two signals both with frequency f_s but from different directions θ_1 and θ_2 are present, two Doppler shifted frequencies f_1 and f_2 can be found by spectral analysis. This can be extended to a continuum of signals with respect to θ as would be observed when a radar transmitter illuminates the ocean surface as next described.

Assume an azimuthal energy distribution $E(\theta)$ for the signals all of frequency f_s, and for the moment assume that $E(\theta)$ is nonzero only for $0 \leq \theta \leq \pi$ (this restriction will be removed later). Also, assume that the moving antenna element has a directional power pattern $A(\theta)$. Then the frequency spectrum of the wave form received by the moving antenna is

$$S(f) = E[\theta(f)] \, A[\theta(f)] \, \left|\frac{d\theta}{df}\right| \quad \text{for} \quad |f-f_s| \leq \frac{V_r}{\lambda} \tag{3}$$

where $\theta(f)$ is given by Equation (2). The factor $d\theta/df$, which is the Jacobian of the transformation from angle to frequency, accounts for the difference in the rate of change of Doppler shift as a function of arrival angle of the signal. The Doppler shift varies most rapidly when the direction of signal arrival is nearly perpendicular to the direction of motion, and approaches a constant value for signals coming from along the direction of antenna motion. Consequently, as with a physical broadside antenna, resolution is best in the broadside dierection and poorest along the axis. Inverting Equation (3), the ocean wave directional spectrum can be found from the frequency spectrum of the received signals from

$$E[\theta(f)] = \frac{S(f)}{A[\theta(f)]} \left|\frac{df}{d\theta}\right|$$

$$= \frac{S(f)}{A[\theta(f)]} \left| \left(\frac{V_r}{\lambda} \sin \theta(f)\right) \right| \tag{4}$$

where $\theta(f)$ is given by Equation (2).

The antenna power pattern $A(\theta)$ is usually quite broad, e.g., $A(\theta) = 1$ independent of direction for a vertical whip, and $A(\theta) = \sin^2\theta$ for a loop antenna with the null aligned with the direction of motion. In some cases, the left-right ambiguity (that is, the inability to distinguish signals arriving from $+\theta$ from those arriving from $-\theta$) with this simple system is not serious, for example, when measurements are made along a straight coast where the path is parallel to the coastline. The ambiguity can easily be removed, however, by making simultaneous observations with two antennas having different, known patterns as indicated in Figure 2. These patterns need not be highly directive. For each frequency f, the received energy is the result of a signal E_1 from $+\theta$, and E_2 from $-\theta$. Call the patterns from the two antennas $A_1(\theta)$ and $A_2(\theta)$. Then

$$S_1 = (A_{11}E_1 + A_{12}E_2) \left| \frac{d\theta}{df} \right|$$

$$S_2 = (A_{21}E_1 + A_{22}E_2) \left| \frac{d\theta}{df} \right| \tag{5}$$

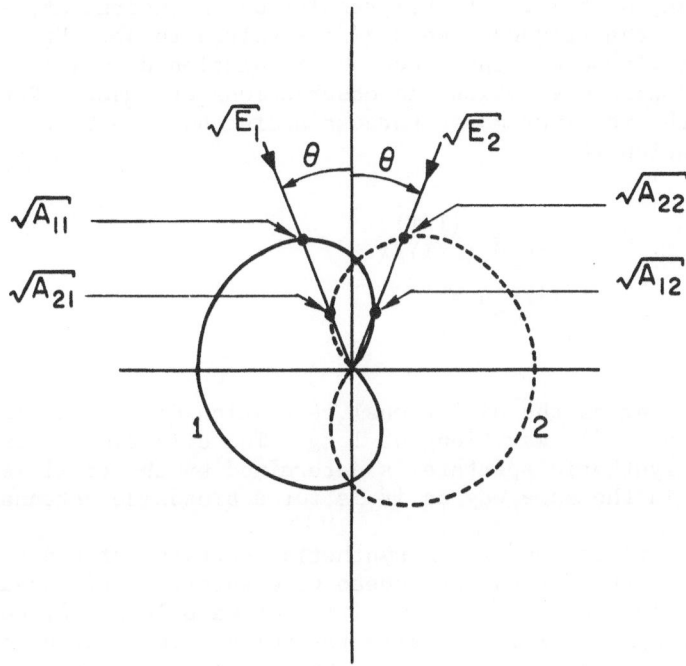

Fig. 2. Left-right ambiguity resolution using two different antenna patterns. The left-pointing main lobe is indicated by 1 and the right-pointing main lobe is indicated by 2. Echo powers E_1 and E_2 come from the left and right, respectively, at an angle θ to V_r. Antenna voltage coefficients are $\sqrt{A_{ij}}$, where i = 1 or 2 for the two antennas, and j = 1 or 2 for the two echo directions. (from *Teague*, 1975).

where S_i is the spectral density observed using antenna i having a power response A_{ij} to signal j. This pair of equations can be solved for E_1 and E_2 by using standard matrix methods. A convenient antenna system for this procedure is a loop and whip antenna combined to give a cardioid pattern, with the main lobe switched from one side of the path to the other for successive data samples.

The directional resolution of the synthetic aperture antenna can be obtained from Equation (2). Differentiating, and letting $f_{max} = V_r/\lambda$,

$$d\theta = \frac{1}{\sin \theta} \frac{df}{f_{max}} \qquad (6)$$

The maximum resolution is obtained in the broadside direction (corresponding to $\theta = 90°$), and if the vehicle Doppler shift is measured to 10%, the angular resolution in the broadside direction is 0.1 radian, or 5.7°. If this resolution is desired at, say $\theta = 30°$, then the frequency should be resolved to 5%. Examined from another viewpoint, the frequency resolution df = 1/T, where T is the total time over which the observations are made. During this time, the receiver moves through a distance $L = V_rT$. Rewriting Equation (6),

$$d\theta = \frac{1}{\sin \theta} \frac{(1/T)}{(L/T)/\lambda}$$
$$= \frac{(1/\sin \theta)}{L/\lambda} \qquad (7)$$

This is familiar as the directional resolution of a broadside physical array (L/λ) wave lengths long. The directional resolution of the synthetic aperture is determined by the total length of the path in the same way as it is for a broadside antenna array.

In the application of the synthetic aperture antenna to the measurement of the directional ocean wave spectrum, the ocean surface is illuminated by a transmitter of wave length λ, pulsed to provide range resolution, where the pulses are coherent in order to allow spectral analysis of the echo signals. The receiver and transmitter are located close together compared to the distance to the scattering patch, so that the incident and reflected rays can be considered parallel. From the theory developed by *Crombie* (1955), *Barrick* (1972) and others, the frequency spectrum of the echo consists of two first-order lines having frequency $\pm 2V_p/\lambda$ generated by those ocean waves of wave length $\lambda/2$ traveling radially toward or away from the radar with phase velocity V_p, and a second-order, or higher-order, continuum produced by wave-wave interactions involving ocean waves of all wave lengths. In order to apply the synthetic aperture equations above, the signal returned to the receiver, in the absence of any antenna motion, must have a single, well-defined frequency. That is, the first-order component must dominate the higher-order terms. This implies that the measurements should be made at the low end of the HF spectrum, usually below 10 MHz, for typical wind and sea conditions subject to the constraint of noninterference to and from other users of the spectrum. Furthermore, the Doppler shift produced by the wave motion must be predictable. For measurements in the open ocean, the phase velocity is given by the deep-water dispersion relation $V_p = \sqrt{g/k}$, where g is the gravitational acceleration and k is the ocean wave number. But in coastal areas, the presence of the bottom may affect the phase velocity of the waves, and care must be exercised to insure that the phase velocity is accurately known. Genrally this effect is negligible if the water depth is at least one-half the deep ocean wave length. Finally, since the effect of a current is to add an additional Doppler shift to that produced by the wave phase velocity (*Stewart and Joy*, 1974), the currents in the region to be mapped should be small compared with both the wave phase velocity and the antenna velocity, or they should be known so that their effects can be removed from the data.

When these conditions are satisfied, the frequency spectrum obtained with a stationary antenna and after translating the signals to a final intermediate frequency of approximately 0.5 Hz appears as in Figure 3a. There are two distinct lines: one near 0.4 Hz which arises from waves of wave length $\lambda/2$ radially approaching the radar, and one near 0.7 Hz arising from waves of the same wave length receding from the radar. In the case of observations along a coast, there will be a third line at zero Doppler shift relative to the carrier due to echoes from stationary targets. The synthetic aperture processing can be applied separately to each of these three regions. In order to separate approaching wave, receding wave and land echoes, these three regions of the spectrum must not overlap. The maximum frequency from the stationary targets is V_r/λ, and the minimum frequency from the approaching waves is $(2V_p-V_r)/\lambda$. If $V_r < V_p$, the difference $(2V_p-V_r)/\lambda-V_r/\lambda = (2V_p-2V_r)/\lambda > 0$ and there is no overlap. An example is shown in Figure 3b, where only ocean wave targets are present. If it is known that there are no stationary targets visible (as might be the case for measurements made from a ship or very small island in the open ocean), this requirement can be relaxed to $V_r < 2V_p$. In some cases, it may be desirable to move both the transmitting and receiving antennas on the same platform (perhaps a ship). Then all the above relations apply with V_r replaced by $2V_r$.

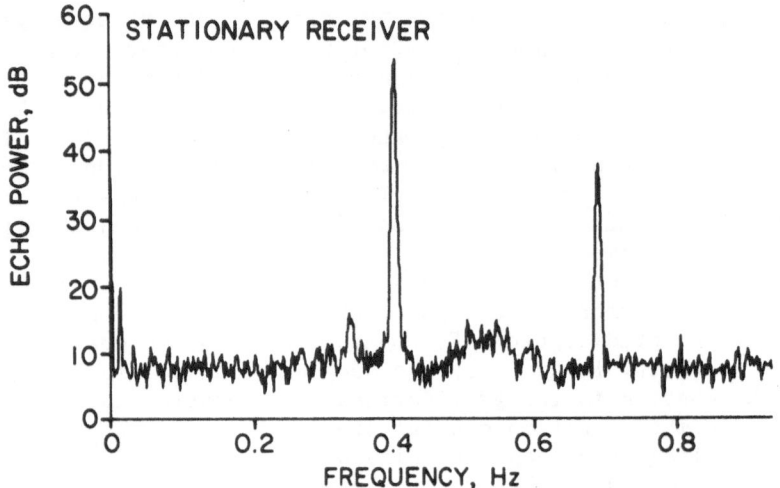

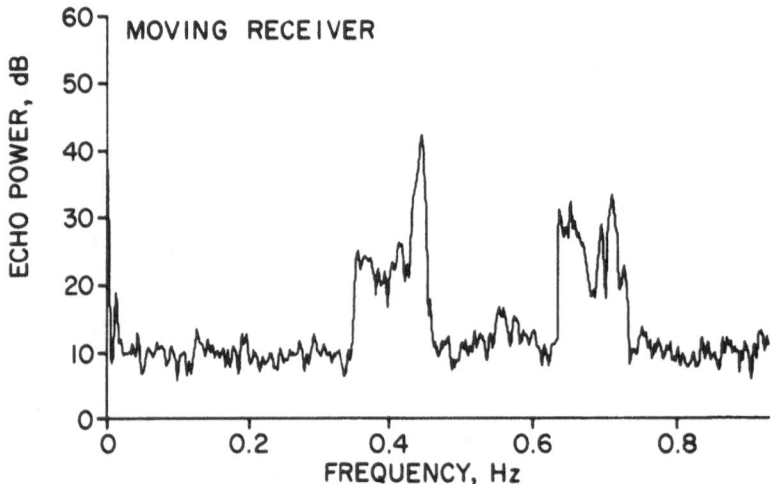

Fig. 3.   First-order ocean wave backscatter frequency spectra.
The spectra were obtained at Wake Island at a frequency of 1.95 MHz.
The receiver local oscillator was approximately 0.55 Hz above the
transmitter frequency.   Signals near 0.4 Hz are due to 77 m (7 s)
waves approaching the island, and those near 0.7 Hz are due to
receding waves.   Since the island is well removed from other land,
there is no line at 0.55 Hz due to stationary targets.   (a - top)
Signals obtained with a stationary receiver.   The second-order
peak near the approaching line is 40 dB down from the first-order
peak and can be ignored.   (b - bottom) Signals obtained with the
receiver moving at approximately 7 m/s.   Signals were arriving from
all directions, with the peak in the approaching spectrum coming
from a direction near $-V_r$ in Figure 1.   (after *Tyler et al.*, 1974).

APPLICATIONS

During the past five years in joint experiments performed by Stanford University and Scripps Institution of Oceanography the synthetic aperture radar has been used for several different oceano- graphic experiments: the measurement of the directional spectrum of a steady-state equilibrium sea in the trade winds region from Wake Island to confirm the technique and to compare the radar measurements to those provided by a tilt buoy; the measurement of a steady-state directional spectrum from a ship in the open ocean during a 1974 NORPAX experiment to determine if the synthetic aperture radar system would be usable from a ship; and, the meas- urement of fetch-limited wave growth off the long, straight coast- line near Galveston, Texas where the synthetic aperture radar provided the only practical way of making the measurement. Each of these experiments is now summarized.

During November 1972, a week-long series of observations was conducted at Wake Island in the mid-Pacific Ocean under conditions of steady trade winds which slowly increased in response to a nearby typhoon (*Teague et al.*, 1973, *Tyler et al.*, 1974). The purpose of the experiment was to compare the synthetic aperture measurements with those of a tilt buoy, and to make a detailed measurement of the directional wave spectrum at one wave length under conditions typical of the open ocean. The island is very small, about 6.5 km in maximum extent, and the ocean bottom is many wave lengths deep just a few kilometers from the island. A LORAN A transmitter at 1.95 MHz ($\lambda = 154$ m) on the island was used as the illumination source. The transmitter emits high-power coherent pulses of approximately 50 μs width with a repetition period of approximately 30 ms. The synthetic aperture paths were along the taxiway (about 3 km long) and parking ramp of the air- port, and along the main road of the island. The LORAN transmitter resonantly selects 7 s (77 m) ocean waves having a phase velocity of 11 m/s. A step van driven at approximately 7 m/s was used to transport a loop and sense antenna operating as a cardioid with the main lobe switched on a pulse-to-pulse basis to resolve the left-right ambiguity. The wind slowly increased from 5 m/s to 13 m/s during the course of the experiment. Comparative data were available from a tilt buoy deployed on two days.

Radar measurements were made each day of the experiment and typically consisted of 10-30 runs of 3-7 minutes duration each. The runs for each day were separately processed and the results were then averaged together to form one or two directional spectra for that day. For each averaged spectrum a curve of the form $S(\theta) = \alpha + (1-\alpha) \cos^s[(\theta-\theta_0)/2]$ was fit to the measurements, with the values of α, s, and the mean wave direction θ_0 adjusted to give a best least-squares fit. Values of s ranged from 2.8 (a very

broad spectrum) to 12.8 (a sharp spectrum). The lower values of
s were generally associated with higher wind speeds, although the
wind direction did change slightly during the course of the experiment. The value of α was on the order of 0.01. That is, the
measurements indicated that approximately 1% of the wave energy
was traveling against the wind. This may represent the noise level
of the measurement and should be considered an upper limit. An
example of a measured average spectrum and curve fit is shown in
Figure 4. This spectrum is typical of those measured during the
course of the experiment, and has a directional half-power width
of approximately 135°.

In order to compare the radar and buoy directional measurements, the first five directional moments N_{pq}, where

$$N_{pq} = \int_{0}^{2\pi} \cos^p \theta \, \sin^q \theta \, S(\theta) \, d\theta \, , \qquad (8)$$

and $S(\theta)$ is the ocean wave directional distribution for the 77 m
waves, were computed for both the radar and buoy measurements.
The results, normalized by N_{oo}, are indicated in Table 1. The
agreement is good and indicates that the radar and buoy are
measuring comparable parameters of the directional distribution.
Tyler et al. (1974) provide data processing details.

The absolute value of the radar cross-section* (and hence
ocean wave heights) is ordinarily a difficult measurement to make.
In this experiment, we were able to make this measurement by comparing the received ocean echo to the direct signal from the transmitter. The receiver was located about 3 km from the transmitter,
and ocean echoes were obtained from a range of 20-100 km. An
attenuator was switched into the receiver at the time of each
transmitter pulse, so that the transmitter signal did not overload
the receiver. The attenuator was then removed during the time that
the echo was present. The range to the transmitter was known from

---

*The radar cross-section is the cross-sectional area of a hypothet-
ical sphere which scatters the incident radio energy isotropically,
and which is located at the position of the target. Its value is
such that the energy that the sphere would scatter in the direction
of the radar receiver is equal to that scattered by the actual
target. It thus represents the apparent radar "size" of the target.
For the case of extended targets such as the ocean, it is usually
expressed as the ratio of the actual radar cross-section to the
physical area of the target and is called σ_o. The value of σ_o at
each wave length is linearly related to the ocean wave energy
present at the corresponding Bragg wave length (*Barrick*, 1972).

TABLE 1.   Comparison of radio scatter with buoy observations. The wave directional moments and associated factors are for approaching 0.15 Hz ocean waves. Bandwidth for buoy analysis is 0.098 Hz, for radio analysis 0.001 Hz. (from *Tyler et al.*, 1974).

	13 Nov. 1972 wind 8.1 m/s from 60°T		15 Nov. 1972 wind 9.5 m/s from 62°T	
	Radio	Buoy	Radio	Buoy
Local time	1103–1307	1040–1325	1108–1308	1121–1344
Relative power	1.0	1.0	3.5	2.9
N_{10}/N_{00}	−0.47	−0.60	−0.59	−0.58
N_{01}/N_{00}	−0.38	−0.28	−0.21	−0.32
N_{20}/N_{00}	0.51	0.64	0.59	0.57
N_{02}/N_{00}	0.49	0.39	0.41	0.39
N_{11}/N_{00}	0.03	−0.01	0.00	0.04
θ_0 (° true)	57°,48°[1]	63°	75°, 57°[1]	67°
s	2.8,4.4[1]	3.9, 5.3, 1.9[2]	4.0, 4.9[1]	3.9,4.3,2.8[2]

[1] The two radio values correspond to linear and logarithmic fitting.

[2] The three buoy values correspond to three different expressions for relating the moments to s and are described more fully in *Tyler et al.* (1975).

a map, and the range to the target was known from the time delay of the echo. The size of the target was computed from the pulse width and range. The radar cross-section could then be computed from the ratio of the echo power to the power of the direct pulse, scaled by a geometrical factor to account for the propagation path. By making the measurement as a *ratio,* all quantities (such as absolute antenna gain, transmitter output power, receiver gain, etc.) which are common to both paths affecting the direct and echo signals cancel out. The radar-inferred wave height was approximately a factor of 2 (2.3 − 3.4 dB) below that measured by the buoy (*Teague et al.*, 1975). Subsequently, we have discovered a difference between the definition of the directional ocean wave spectrum and the interpretation required by the scattering theory of *Barrick* (1972) which accounts for a factor of 2, and now believe our results agree with the buoy measurements to less than 1 dB.

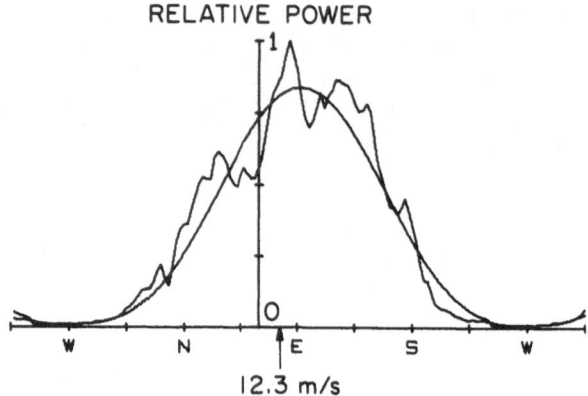

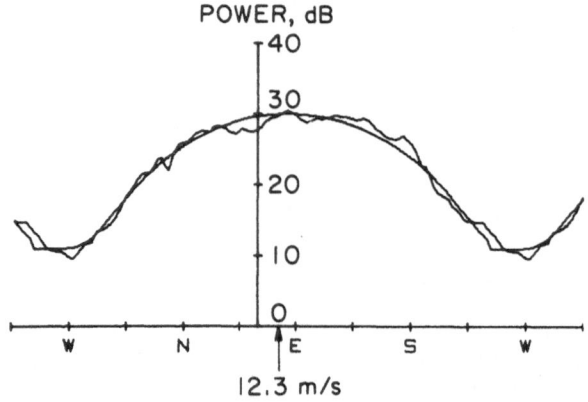

Fig. 4. Radar-inferred ocean wave directional spectrum. (a -
top) Linear scale. (b - bottom) Logarithmic scale. The data were
obtained at Wake Island on November 18, 1972, and the mean wind
direction and speed over the preceding 24 hours are indicated by
the arrow. The rough curves represent the measured data, and the
smooth curves represent a function of the form $\alpha + (1-\alpha)$
$\cos^s[(\theta-\theta_0)/2]$, with α, s, and θ_0 adjusted for a least-squares
fit to the linear and long data. Values of s are (a) 2.8 and
(b) 3.4.

A similar experiment (*Teague*, 1975, and *Teague et al.*, 1975)
was performed as part of the NORPAX experiment in January and
February 1974 in the open ocean about 1500 km northwest of Hawaii.
The purpose of this experiment was to see if a synthetic aperture
HF radar system would be feasible from a ship. Both the trans-
mitter and receiver were carried on board the R/V THOMAS WASHINGTON

along straight courses of approximately 350 m in length in several
different directions. Four frequencies, 4.80, 6.78, 13.38, and
21.77 MHz, were generated by a portable radar developed by Stanford
University. A peak power of 1 W and a pulse width of 50 s were
used. A set of quarter-wave vertical wires was used as the transmit antenna, and four loops and sense antennas having cardioid
patterns, alternately switched to port and starboard, were used as
receiving antennas. A transponder on the spar buoy FLIP, at a
range of about 2 km, was used to measure the resulting element
patterns, which were far from ideal because of numerous structural
components and communications and navigation satellite antennas
also on the ship.

An example of one of the directional spectrum measurements is
shown in Figure 5. The directional spectra were computed separately
for the approaching wave field and the receding wave field. Since
the measurement was made in the open ocean with nothing to block
the receding waves, the two results should be identical. They
differ for several reasons: the signal-to-noise ratio was limited
because of low transmitter power; second, ionospherically-propagated
signals from shortwave broadcast stations and communications users
were almost always present; and third, even though the weather was
moderate, the ship tended to roll quite heavily when it was oriented
in the crosswind direction in order to fill in gaps in coverage for
measurements made with the ship headed into the wind. The roll
modified the antenna element responses and introduced sidebands in
the received signal which raised the noise level of the measurement.
The shipboard synthetic aperture system was moderately successful,
but it probably would work considerably better from a ship somewhat
larger than the 50 m vessel used here (perhaps a large oil tanker)
and with higher transmitter power than the 1 W used in this
experiment.

An experiment under conditions complementary to those of the
open ocean of Wake Island and NORPAX was performed at Galveston,
Texas during the first four months of 1976. The purpose of this
experiment was to study the directional growth of the receding
waves under fetch-limited conditions. For these measurements,
the synthetic aperture antenna was the primary sensor. Several
observations were made immediately after polar frontal passages
when the wind abruptly shifted to the offshore direction along the
long, straight coastline near Galveston. As at Wake Island, a
LORAN A transmitter at 1.90 MHz (157.9 m wave length) was used for
illumination, and a loop and sense antenna were mounted on a van
which was driven along the flat beach and various straight roads
on Galveston Island. The path lengths ranged from 1 to 4 km, with
a resulting directional resolution as fine as 5°. The wind speeds
during the experiments ranged from 5 to 15 m/s. Data processing
is still in progress, and some preliminary results are presented

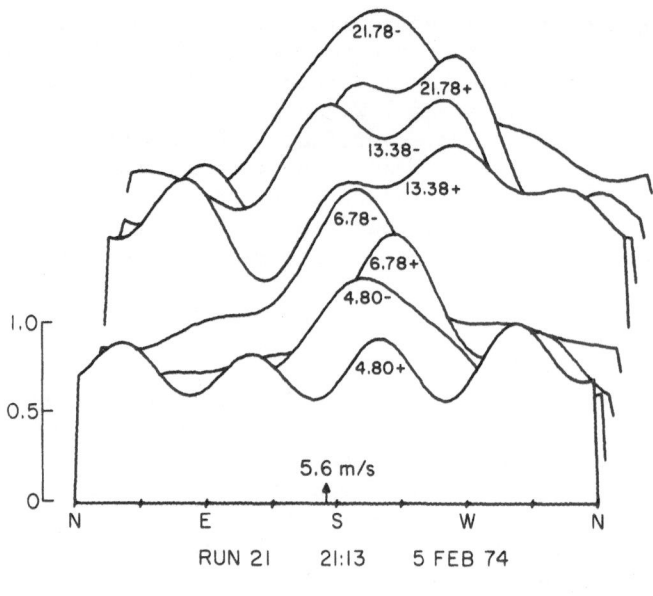

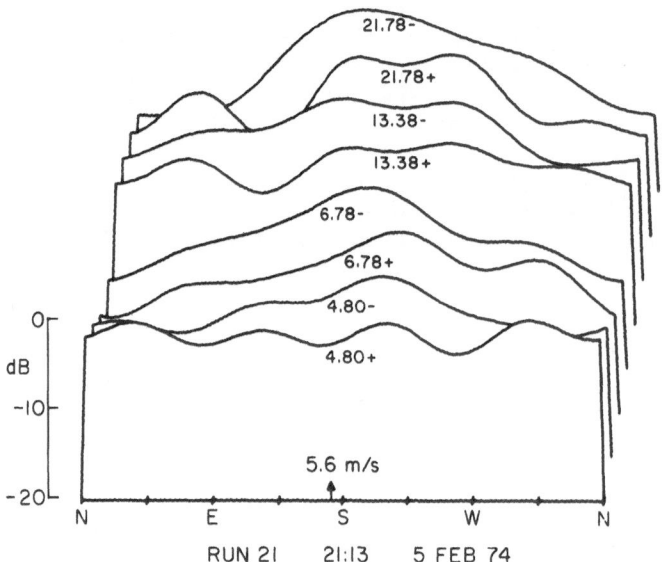

Fig. 5. Radar-inferred ocean wave directional spectra from shipborne observations. (a – top) Linear scale. (b – bottom) Logarithmic scale. The data were obtained during the 1974 NORPAX experiment approximately 1500 km NW of Hawaii, from a radar on board the R/V THOMAS WASHINGTON. Four radar frequencies, 4.80, 6.78, 13.38, and 21.78 MHz were used. The directional spectra were computed separately from the approaching waves (+) and the receding waves (−). The data have been highly smoothed. (from *Teague*, 1975).

here. A paper describing the experiment in more detail is in
preparation.

     An example of the receding wave growth data from one day of
the experiment is shown in the form of a shaded contour map in
Figure 6. The plot indicates the relative energy of 7s waves
traveling radially away from the radar as a function of position
on the ocean surface. Each contour represents an increase in
ocean energy of 5 dB, or $\sqrt{10}$ in power. It is evident that the
contours tend to run perpendicular to the mean wind direction.
The wave growth along 5 different directions, averaged over 18°
segments, is plotted in Figure 7. In this figure, the abscissa

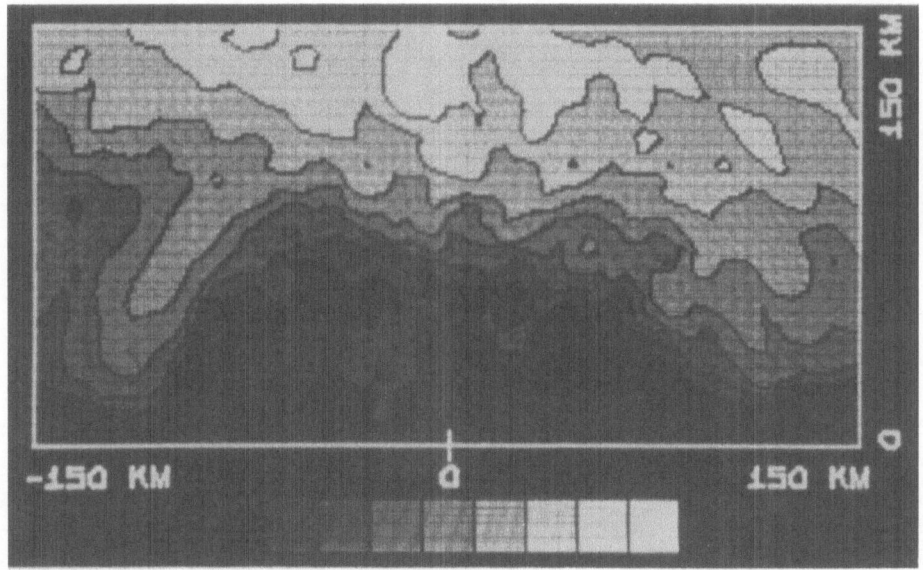

     Fig. 6. Shaded contour plot of 7 s fetch-limited directional
wave growth. The data were obtained on February 22, 1976, at
Galveston, Texas. The contours indicate the energy and location
of radially-traveling receding waves after an offshore wind of
approximately 10 m/s has been blowing for 30 hours in a direction
perpendicular to the predominant contour direction. The contour
interval is 5 dB and each contour level corresponds to an increase
in ocean energy of $\sqrt{10}$. An 8-step grey scale is included for
calibration. The positive x-axis is oriented at 234°T, parallel
to the coastline, and the y-axis points offshore toward the lighter
shading from the radar located at the origin. The feature near
x = -100 km, y = 50 km is probably due to vertical incidence E-layer
ionospheric reflection rather than ocean energy.

is R cos θ, where R is the radar (radial) range, and θ is the
angle between a particular wave component and the mean wind direction.   This quantity is the distance that the waves have traveled
in the direction of the wind.   When this measure is used, it appears
that the growth rate is independent of the angle to the wind, out
to at least ±45° from the wind.   The growth is exponential over
2.5 orders of magnitude, with a rate of 1.3%/ocean wave length or
0.73 dB/km.

The finite depth of the water in the first 75 km from the
shore has a noticeable effect on the phase velocity of the 77 m
waves to which the radar is sensitive.   While this perturbation

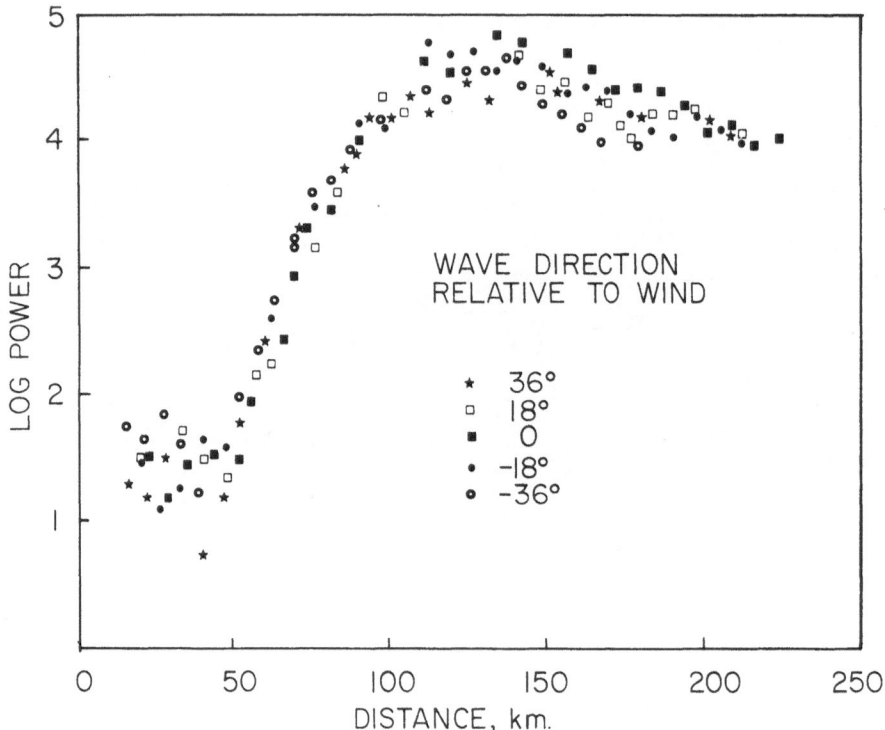

Fig. 7. Directional growth of 7 s waves vs. distance. The
data of Figure 6 have been averaged over 18° segments and plotted
as a function of R cos θ, where R is the radial range and θ is
the direction of the waves relative to the wind direction, 336°T.
The growth is exponential over 2.5 orders of magnitude with a rate
of 1.3%/ocean wave length, and the rate is independent of θ out to
±45° from the wind. R cos θ is the distance the waves have traveled
in the direction of the wind.

is minimal for the study of the receding wave growth, since most of the growth occurs at a considerable distance from the shore, it should be considered in the study of approaching wave fields.

SUMMARY

A synthetic aperture antenna, formed by carrying a simple antenna along straight lines at speeds of a few m/s for distances up to 4 km in length, has been used in several experiments to measure the directional distribution of ocean wave energy in the open ocean. The experiments made use of surface-wave propagation of HF signals in the range of 2-20 MHz. The technique requires that first-order scattering dominate higher-order scattering, that the wave phase velocity, and hence the Bragg frequency, is predictable from the deep-water dispersion relation, and that the currents in the scattering region are negligible or are known from other measurements. The direction of arrival of the ocean echoes is determined by a spectral analysis of the received signals. The simplest system can resolve signals covering a range of approximately $150°$; a simple extension using a pair of antennas can resolve the left-right ambiguity and distinguish signals over a range of $300°$. The remaining directions can be resolved by combining data taken on paths in different directions. Although the technique has been used on board a ship, it is much easier to apply to land-based measurements. The directional spectrum of 7 s waves under equilibrium conditions can be approximated by a function of the form $\cos^s(\theta/2)$ with s in the range of 3 to 12. The fetch-limited growth of 7 s waves was observed to be exponential with respect to $R \cos \theta$, where R is the range and θ is the angle between wind and waves, and the rate was independent of θ out to $\pm45°$.

For each of these measurements, the synthetic aperture antenna was an essential element, providing a directional resolution an order of magnitude better than previously existing ocean wave directional measurements. Considerable care was required for the experiments and data processing, and the measurements were in the form of basic research rather than routine operational observations. The technique is best applied at the low end of the HF spectrum, and frequencies must be chosen so as to avoid interference to other users of the spectrum. The synthetic aperture radar system is most attractive for observing the directional distribution of ocean waves at a few discrete wave lengths where a large surface height or pressure sensor array is impractical of where a tilt buoy provides insufficient directional resolution, and for making an initial survey at a new site, before committing the funds for a large permanent antenna installation. It could be automated, however, perhaps by the use of a vehicle on a track, for a more permanent installation.

ACKNOWLEDGEMENT

This work was supported by the Office of Naval Research under contract N00014-75C-0356 and represents a joint research program performed by Stanford University and Scripps Institution of Oceanography.

REFERENCES

Barrick, D. E. 1972. First-order theory and analysis of MF/HF/VHF scatter from the sea, *IEEE Trans. on Antennas and Propagation*, *AP-20*, 2-10.

Clarke, J. 1970. Aperture-synthesis technique for h.f. ionospheric radar, *Proc. IEEE*, *117*, 1633-1638

Crombie, D. D. 1955. Doppler spectrum of sea echo at 13.56 Mc/s, *Nature*, *175*, 681-682.

Harger, R. O. 1970. *Synthetic Aperture Radar Systems: Theory and Design*, Academic Press, New York and London.

Lynch, J. T. 1970. Aperture synthesis for HF radio signals propagated via the F layer of the ionosphere, Stanford Electronics Laboratories, Stanford, California, Technical Report, SU-SEL-70-066.

Shearman, E. D. R., P. Bickerstaff, and L. Fotiades. 1973. Synthetic-aperture skywave radar: techniques and first results. *IEEE Conference Proceedings: Radar-Present and Future*, 105.

Stewart, R. H. and J. W. Joy. 1974. HF radio measurement of surface currents, *Deep Sea Res.*, *21*, 1039-1049.

Teague, C. C., G. L. Tyler, J. W. Joy, and R. H. Stewart. 1973. Synthetic aperture observations of directional height spectra for 7 s ocean waves, *Nature, Physical Science*, *244*, 98-100.

Teague, C. C. 1975. In situ decametric radar observations of ocean wave directional spectra during the 1974 Norpax "pole" experiment: Final report, Radioscience Lab., Stanford University, Technical Report, SEL-75-003.

Teague, C. C., G. L. Tyler, and R. H. Stewart. 1975. The radar cross-section of the sea at 1.95 MHz: comparison of in situ and radar determinations, *Radio Science*, *10*, 847-852.

Teague, C. C., G. L. Tyler, and R. H. Stewart. 1977. Studies of the sea using HF radio scatter, *IEEE Trans. on Antennas and Propagation*, *AP-25*, 12-19.

Tyler, G. L., C. C. Teague, R. H. Stewart, A. M. Peterson, W. H. Munk, and J. W. Joy. 1974. Directional spectra from synthetic aperture observations of radio scatter, *Deep Sea Res.*, *12*, 989-1016.

HIGH FREQUENCY SKYWAVE MEASUREMENTS OF WAVES AND CURRENTS ASSOCIATED WITH TROPICAL AND EXTRA-TROPICAL STORMS

Joseph W. Maresca, Jr.

SRI International

ABSTRACT

The capability of HF skywave radar to measure surface winds, waves, and currents at distances up to 3000 km for tropical and extra-tropical storm conditions is summarized. Significant wave height and wave spectral estimates made using the Wide Aperture Research Facility skywave radar were compared to similar measurements obtained by NOAA Data Buoy Office buoys EB20 (41°N, 138°W) and EB71 (26°N, 93.5°W). Agreement to within 10% was found for wave heights under varying conditions, ranging from less than 1 m to hurricane wave conditions greater than 5 m.

INTRODUCTION

High frequency (HF) skywave radar estimates of the surface wind speed and direction, surface current, significant wave height, and wave spectrum can be made over several million square kilometers of ocean by remotely measuring the Doppler spectrum of the sea-echo signal. Measurements made with the SRI-operated Wide Aperture Research Facility (WARF) HF skywave radar demonstrate that the skywave radar can track large extra-tropical storms, and small intense tropical storms and hurricanes; the radar can also provide detailed surface maps of the winds, waves, and currents within a storm. The large coverage area of the WARF skywave radar is possible because the HF radio waves are transmitted to, and returned from, ocean areas by means of one or more ionospheric "reflections." For one ionospheric reflection, ocean areas up to 3000 km away from the radar can be monitored. The purpose of this paper is to summarize the capabilities of HF skywave radar that utilizes ionospheric reflection, with special emphasis on the measurement of significant wave height and the one-dimensional wave frequency spectrum.

The skywave radar estimates of the surface wave parameters are obtained from the HF sea-echo Doppler spectrum. An example of the sea-echo spectrum recorded at WARF for 51.2 s of coherent integration is shown in Figure 1. The spectrum consists of two strong first-order echoes produced by the resonant interaction of ocean waves of wave number k_o. For near grazing angles,

$$k \cong 2k_o.$$

In addition, a second-order continuum that is sensitive to changes in the directional ocean wave spectrum surrounds the first-order echoes. *Barrick* (1972a,b) has derived theoretical expressions that accurately describe the HF scattering process. These expressions have been mathematically inverted to recover parts of the directional wave spectrum directly from the Doppler spectrum.

Barrick and Lipa (1978) summarize the theory of HF scattering from ocean waves; the methods for computing the wave height, the wave spectrum, and the surface current; and the results of several experiments conducted to validate these methods through the use of HF ground-wave radar. In this paper, experiments conducted at the WARF skywave radar to determine the accuracy of these methods, including a simple method of mapping wind direction, are described. The WARF radar measurements have been compared to *in situ* point measurements made from NOAA Data Buoy Office (NDBO) data buoys and research vessels. The relative agreement between the radar and *in situ* measurements have been found to be within the measurement accuracy of the *in situ* measurement.

WARF SKYWAVE RADAR

WARF is a bistatic HF skywave radar located in central California. The radar is operated in the HF band between 6 and 30 MHz. A 20-kW swept-frequency continuous-wave signal is transmitted from the transmitter site at Lost Hills, California; after round-trip ionospheric propagation, the signal is received at the receiving site 185 km to the north at Los Banos, California. The receiving array is 2.5 km long and forms a beam of 1/2° at 15 MHz.

The WARF radar coverage area is shown in Figure 2. The radar can be electronically steered in 1/4° increments anywhere within the coverage area. The minimum range or skip distance of the radar is typically 800 km, and the maximum range for one ionospheric

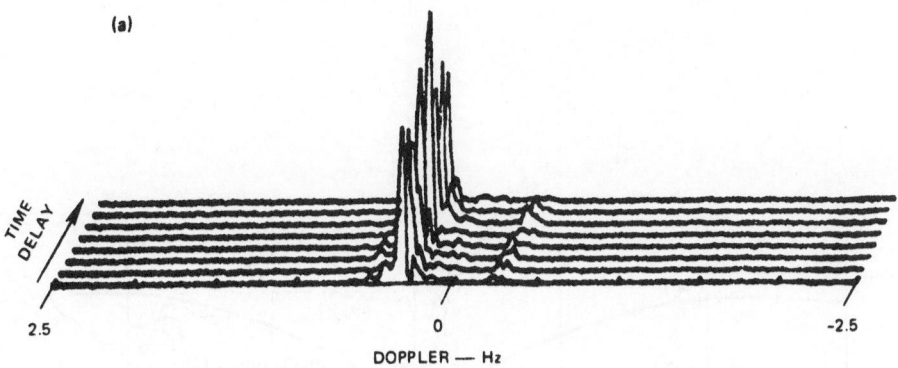

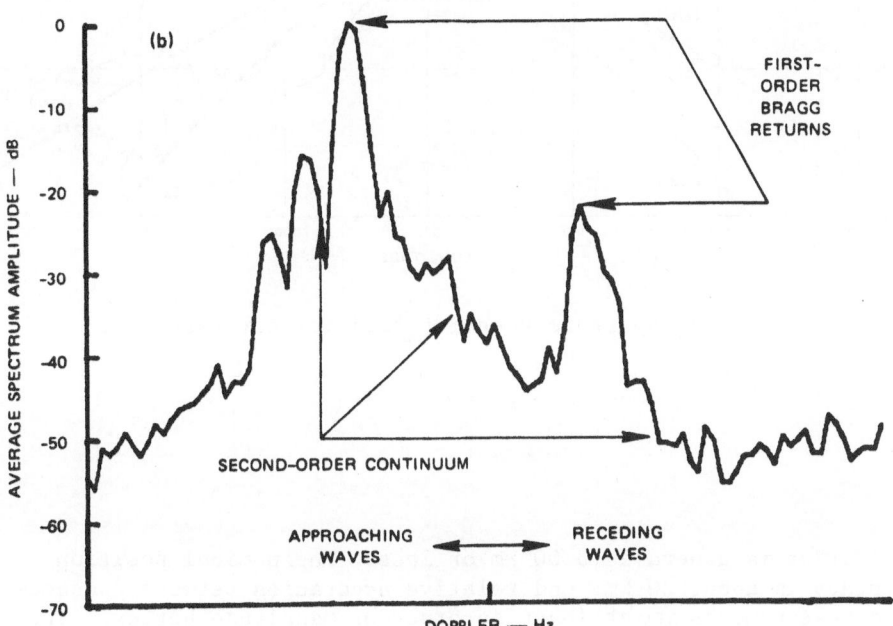

Fig. 1.   Range-doppler-processed sea-echo Doppler spectrum.
The mean Doppler spectrum is an average of Doppler spectra recorded at different range lines separated by 3 km.   The first-order
echoes produced by a resonant interaction between the radio waves
and the ocean waves is sensitive to changes in the wind-direction
field.   The second-order sideband structure surrounding the stronger
Bragg line is sensitive to changes in the directional ocean wave
spectrum.

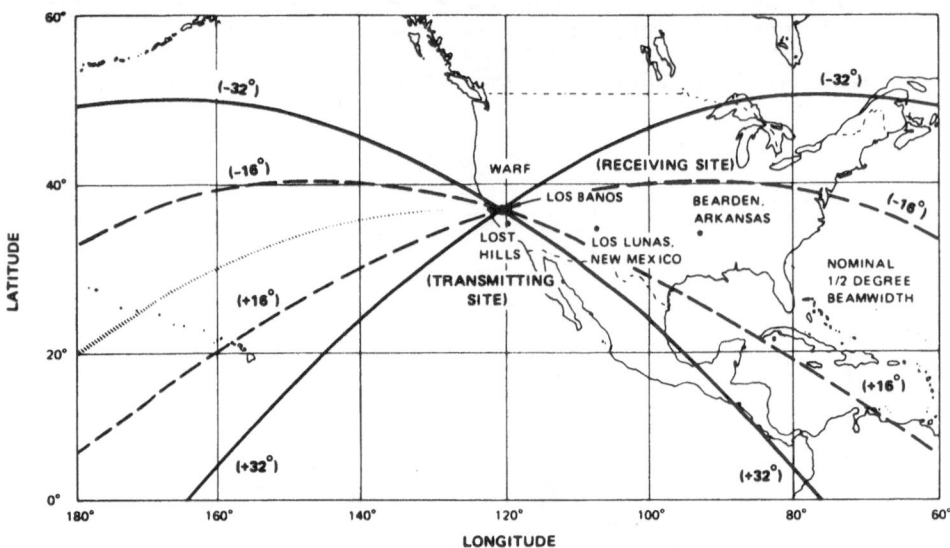

Fig. 2. Wide Aperture Research Facility coverage map.

reflection is generally 3000 km or less. The nominal position
accuracy is about 20 km, and relative accuracies between consecutive measurements are at least an order of magnitude better. The
range to the ocean patch illuminated is a function of time and radar
frequency. In general, ranges out to 2200 km can be monitored 95%
of the time; coverage beyond 2200 km drops to 50%.

The size of the ocean patch monitored by the WARF radar is a
function of the sampling parameters and can be specified by the
radar operator. The characteristic length scale of the ocean wave
conditions is considered in selecting the size of the ocean patch
to be illuminated by the radar. For example, the ocean patch monitored for a hurricane would be smaller than the ocean patch

monitored for a Pacific storm. The minimum resolution cell routinely achievable is 3 km in range by 15 km in cross-range. The cross-range resolution is a function of the radar range and beamwidth. For a maximum one-hop range of 3000 km, a 1/2° beam would result in a cross-range distance of about 35 km. Generally, 21 resolution cells or independent Doppler spectrum measurements spaced at 3-km increments make up the illuminated ocean patch. If the Doppler spectra at each range interval is averaged, the size of ocean patch will be 63 x 25 km. For Pacific Ocean measurements, the 63 x 25 km scattering patch is used for wave measurements; for hurricanes, a smaller ocean patch, 15 x 25 km, is used.

SURFACE WIND DIRECTION MAPS

The capability and technique for mapping the surface wind direction field for large weather systems over the ocean has previously been demonstrated (*Long and Trizna*, 1973; *Barnum et al.*, 1977). This mapping technique has also been applied directly to tropical storm and hurricane wind fields (*Maresca and Barnum*, 1978; *Maresca and Carlson*, 1977b).

Wind direction is estimated from the power ratio of the first-order echo returns (*Stewart and Barnum*, 1975; *Barnum et al.*, 1977). Agreement between the WARF radar and the anemometer measurements of wind direction over the Pacific Ocean is ±16° (*Stewart and Barnum*, 1975). For hurricane winds, agreement between the WARF radar and NDBO buoy wind direction measurements is better than 10° (*Maresca and Carlson*, 1977b). Figure 3 shows an example of the WARF-inferred surface wind map produced as Hurricane Eloise passed between NDBO buoys EB04 and EB10 in the Gulf of Mexico. The WARF-inferred wind direction maps have been used to locate the center of hurricanes. The earliest measurements were made on Eloise, and they agreed to within 35 km of the smooth track produced by the National Hurricane Center (NHC). Recently, skywave radar track was compiled for Hurricane Anita from 17 wind direction maps recorded over a 4-day period. The mean difference between the WARF position estimates and the NHC smooth track was 19 km (*Maresca and Carlson*, 1978b).

SURFACE CURRENT CAPABILITY

The radial component of the surface current can be inferred from the measured phase velocity, or Doppler shift, of the ocean waves producing the first-order echoes (*Barrick et al.*, 1974; *Stewart and Joy*, 1975; *Barrick et al.*, 1977). Unlike the skywave radar wind direction measurement, the effects of the ionospheric motion must be known to make a skywave measurement of current. The entire Doppler spectrum can be shifted by ionospheric motion.

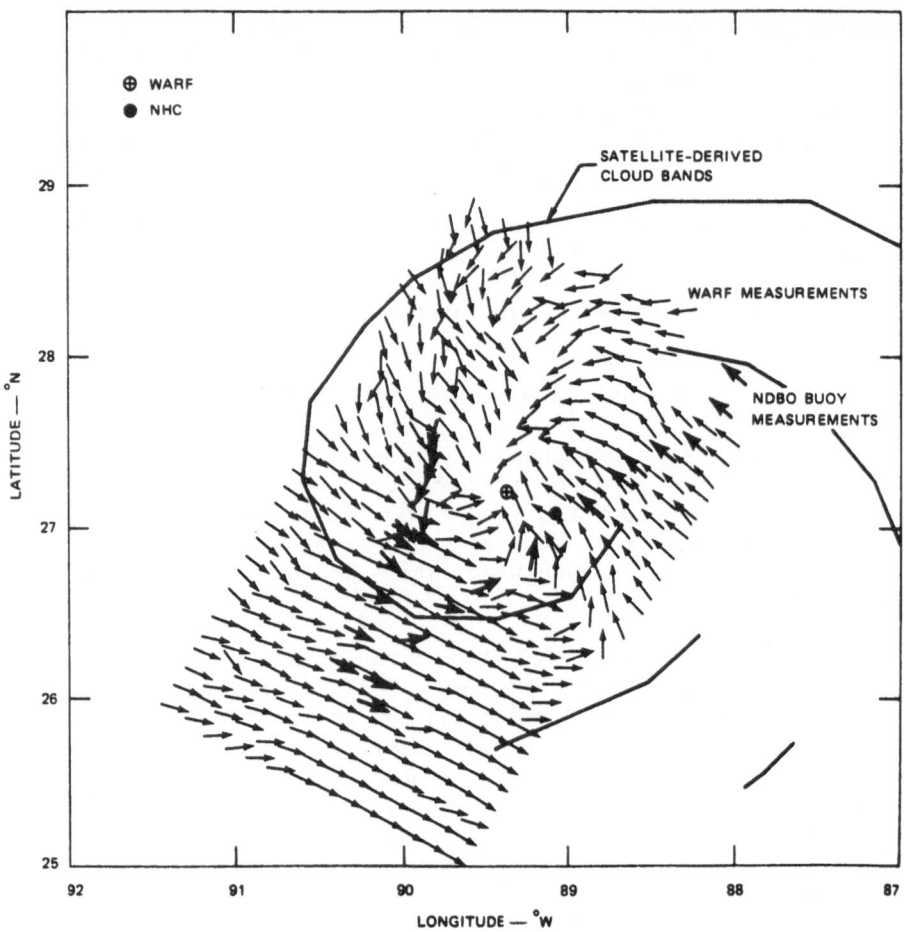

Fig. 3.   Comparison of WARF-inferred surface wind directions
with NDBO data buoy directions for Hurricane Eloise. Agreement is
within 10°. The hurricane eye position (⊕) determined from the
WARF wind direction measurements is near the hurricane eye position (●) provided by the National Hurricane Center.

It has been shown by *Maresca et al.* (1976) and *Maresca and Carlson*
(1977b) that land and offshore oil platform echoes received during
coastal scans along the Gulf of Mexico are sufficient to remove
the effects of the ionosphere from the data. The analysis of the
ocean surface would then be identical to the ground-wave radar
analysis. The requirement of a reference would generally limit
the measurement to coastal regions. The expected accuracy of the
surface current estimates is about 10 cm/s.

SIGNIFICANT WAVE HEIGHT AND WAVE SPECTRUM ESTIMATES

The measurement of the significant wave height and wave spectrum is made from the weak second-order region of the sea-echo Doppler spectrum. The accuracy of the wave estimates is dependent not only upon the method used to compute the wave parameters, the number of radar samples averaged, and the natural variability of the waves, but also on the contamination effects of the ionosphere. The quality of the skywave radar data depends on the ionospheric propagation path. Ionospheric smearing effects and multiple-path propagation can destroy the second-order spectral contributions (*Maresca and Barnum*, 1977; *Maresca and Carlson*, 1977b). In general, the potential contamination caused by the ionosphere would be the largest source of error in the measurement. Recent work by SRI and NOAA (*Georges and Maresca*, 1978; *Maresca and Georges*, 1978a) has resulted in improved methods for collecting uncontaminated data and for recovering the wave spectra even when the data is contaminated by the ionosphere.

Each of the HF inversion formulas summarized from the measured Doppler spectrum by *Barrick and Lipa* (1978), *Barrick* (1977a,b), and *Lipa* (1977, 1978) to compute the ocean wave spectrum must be tested for skywave data. Presented below are results of three wave-height verification experiments conducted at the WARF radar. In one of these experiments, the one-dimensional wave frequency spectrum is also computed from the radar data. Significant wave heights during the experiments ranged from less than 1 m to more than 5 m. The radar estimates were compared to NDBO data buoy measurements in the Gulf of Mexico (EB71) and Pacific Ocean (EB20). Agreement between the NDBO buoy measurements and the WARF radar measurements for significant wave heights were within the accuracy, approximately ±3 m, of the NDBO buoy measurements.

WAVE SPECTRAL ESTIMATES FROM THE HF DOPPLER SPECTRUM

Barrick (1977a,b) derived the following approximate closed-form expression to compute the one-dimensional wave frequency spectrum, $S(\omega)$, and rms wave height, h_*, from the Doppler spectrum:

$$S(\omega_B|\nu-1|) = \frac{4\sigma_2(\omega_B\nu)/W(\nu)}{k_o^2 \int_0^\infty \sigma_1(\omega_B\nu)d(\omega_B\nu)} \tag{1}$$

and

$$h_*^2 = \frac{2 \displaystyle\int_{-\infty}^{\infty} [\sigma_2(\omega_B \nu)/W(\nu)]d\,\omega_B \nu}{k_o^2 \displaystyle\int_{-\infty}^{\infty} \sigma_1(\omega_B \nu)d(\omega_B \nu)} \tag{2}$$

where $\acute{\omega} = (\omega_B|\nu-1|)$ is the radian ocean wave frequency; ω_B is the radian Bragg frequency; ω_D is the radian Doppler frequency; $\nu = \omega_D/\omega_B$; $\sigma_1(\omega_B \nu)$ and $\sigma_2(\omega_B \nu)$ are the first- and second-order power contribution to the Doppler spectrum expressed as radar cross section per mean surface area per radian per second of bandwidth; k_o is the radian radio wave number; and $W(\nu)$ is the weighting function derived by *Barrick* (1977a).

The weighted second-order power contribution used in Equation (1) is obtained by averaging the return from both sides of the stronger first-order echo. The rms wave height is obtained by dividing the total weighted second-order power surrounding both sides of the stronger first-order echo by the total first-order power. If one assumes that the process is Gaussian, the significant wave height, H_s, is $4h_*$.

The accuracy of these expressions can be evaluated by inverting theoretical Doppler spectra produced from known input directional wave spectra. The results indicate that Equations (1) and (2) tend to overestimate wave height by a constant amount. *Barrick* (1977a) has shown that when the ratio of the actual wave height to the radar-measured wave height h/h_*, is plotted as a function of the parameter $k_o h_*$, h/h_* is a constant for $k_o h > 0.20$. *Maresca and Carlson* (1977b) also showed the dependence of h/h_* on radar frequency and wave directionality.

Barrick (1977a) has tested Equation (2) by using surface wave HF radar data and *in situ* buoy measurements, and the agreement was found to be within 22%. The rms error is primarily dependent on the accuracy of the method used to compute wave height and the number of independent Doppler spectra averaged before computing the wave height. Each Doppler spectrum of the sea-echo signal in *Barrick's* test was an average of 9 samples. If the number of samples averaged had been increased, then better estimates of the sea echo would have resulted and the 22% error might have been reduced.

WARF MEASUREMENTS OF SIGNIFICANT WAVE HEIGHT AND THE WAVE SPECTRUM

The results of three WARF wave height verification experiments conducted on 2 June 1977, 23 August 1977 and 1 September 1977 are reported here to demonstrate the accuracy of the skywave radar measurement technique over a range of seas and ionospheric conditions. The radar measurements were recorded in the vicinity of NDBO buoys EB71 (26°N, 93.5°W) in the Gulf of Mexico and EB20 (41°N, 138°W) in the Pacific Ocean. EB71 is located approximately 2900 km from WARF; EB20 is located about 1700 km from WARF. The Doppler spectra were produced from 102.4 s of coherent integration. Equation (1) was used to compute the ocean wave spectrum, and Equation (2) was used to compute the rms wave height. The significant wave height was computed by multiplying the rms wave height by 4.0. At least 40 independent samples of the Doppler spectra were averaged. The errors in the spectral estimates are proportional to $1\sqrt{N}$, where N is the number of samples (*Barrick and Snider*, 1977). The sampling error based on 40 or more averages for all three experiments is 16% or less. The nominal accuracy of NDBO moored buoys measurement of significant wave height reported by the Data Quality Division of NDBO is ±0.3 m. The WARF measurements of significant wave height made for all three experiments agree to within ±0.3 m of the buoy measurements.

June 2, 1977 Experiment

The placement of NDBO data buoy EB20 on station in the Pacific Ocean in 1977 provided the first opportunity to compare the WARF-estimates of significant wave height, H_s with an independent *in situ* measurement. The Doppler spectra were taken on 2 June 1977 to the west of an atmospheric front. The front passed through EB20, travelling to the east. During a period of less than 6 hours, H_s increased from 1.4 m to 3.0 m and remained constant at approximately 3 m for more than 6 hours. The significant wave height measured over a 12-hour period is given in Table 1.

The WARF radar measurement reported here was centered 125 km west of EB20 at 1819Z. Since the wave height and wave spectral measurements at EB20 remained constant (±0.1 m) from 1800 to 2400Z, we expect that the WARF measurement of the significant wave height should be about 3.0 m. WARF radar estimates of significant height computed from Equation (2) and the one-dimensional wave frequency spectrum computed from Equation (1) were compared to the measurements made at EB20. The WARF estimate of H_s is 3.0 m; the WARF measured wave spectrum is shown in Figure 4. We compared these estimates to those made at EB20 at 2100Z because the WARF measurement was located west of the buoy where the seas were probably

TABLE 1.   Significant wave height measured
at EB20 June 2, 1977.

Time (GMT)	Significant Wave Height (m)
1200	1.4
1500	1.8
1800	3.0
2100	3.0
2400	2.9

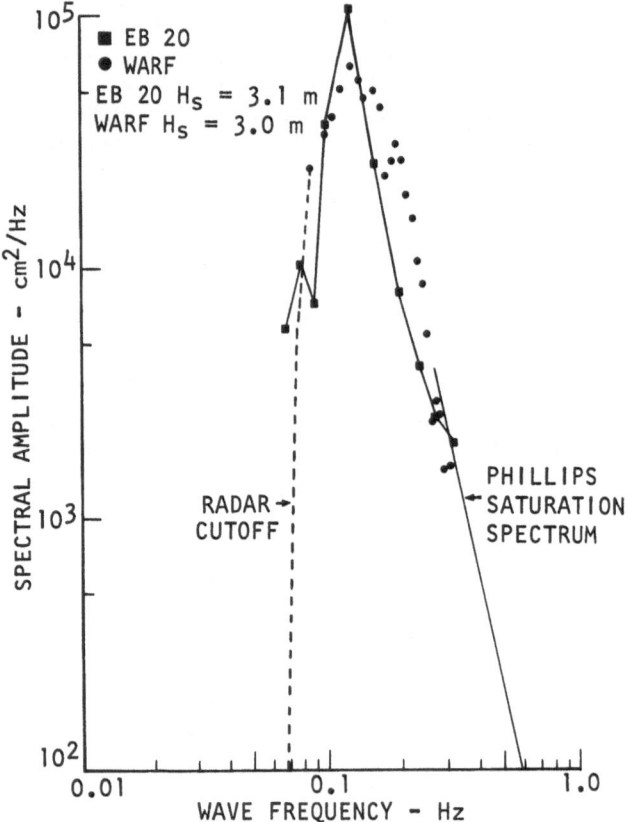

Fig. 4.   Comparison of the one-dimensional wave frequency
spectrum derived from WARF measurements of the Doppler spectrum
with that of NDBO data buoy EB20.

slightly larger. However, the difference in the EB20 wave spectra measurements at 1800Z and 2100Z are negligible, and comparison of the EB20 and WARF measurements any time between 1800Z and 2100Z produce essentially identical results. The agreement between the significant wave heights measured at EB20 and WARF is within the ±0.3 m accuracy of the NDBO moored buoys. Further details of this experiment are available in *Maresca and Carlson* (1977a) and *Maresca and Georges* (1978b).

August 23, 1977 Experiment

The placement of NDBO data buoy EB71 on station in 1977 provided the first opportunity to compare WARF estimates of significant wave height in the Gulf of Mexico with an *in situ* measurement. WARF measurements recorded at 2140Z on 23 August 1977 were centered on EB71. Wave conditions at EB71 remained fairly constant before, during, and after the experiment, as shown in Table 2. Wave conditions were generally 1 m or less. The WARF estimate of H_s is 1.0 m, which compares reasonably well with the 0.9 m measurement made at EB71. Theoretical simulations suggest that the accuracy of the method used to calculate h_* decreases significantly for $H_s < 1$ m, and this experiment probably represents a lower limit on the measurement capability.

TABLE 2.   Significant wave height measured
at EB71 August 23, 1977.

Time (GMT)	Significant Wave Height (m)
1200	0.9
1500	0.9
1800	1.0
2100	0.9
2400	0.8

September 1, 1977 Experiment

Hurricane Anita was tracked by the WARF radar from 29 August through 2 September 1977. WARF measurements of H_s were made at 2324Z on 31 August 1977 in the right rear quadrant of Anita as the storm passed near EB71. The WARF measurement was centered about 80 km from the radar-derived storm center. Comparison of the WARF

estimates of H_s with EB71 estimates of H_s was difficult because
small differences in location of the measurement with respect to
the center of the hurricane, where conditions change rapidly over
short distances, can result in different values of H_s. The areal
distribution of the significant wave-height field was generated
from EB71 measurements of H_s as the storm passed the buoy. It was
assumed that the spatial distribution of the hurricane wave conditions remained constant for ±12 hours. The WARF estimate of
significant wave height was then compared to an interpolated value
of H_s computed from the EB71 significant wave height field (Figure
5). The WARF estimate of H_s was 5.4 m, and the interpolated value
of H_s from the EB71 estimates was 5.5 m. This is within the 0.3 m
error associated with the NDBO buoy measurement. Estimates of the
WARF significant wave heights computed in all four quadrants of the
hurricane will be described in more detail in *Maresca and Carlson*
(1978a).

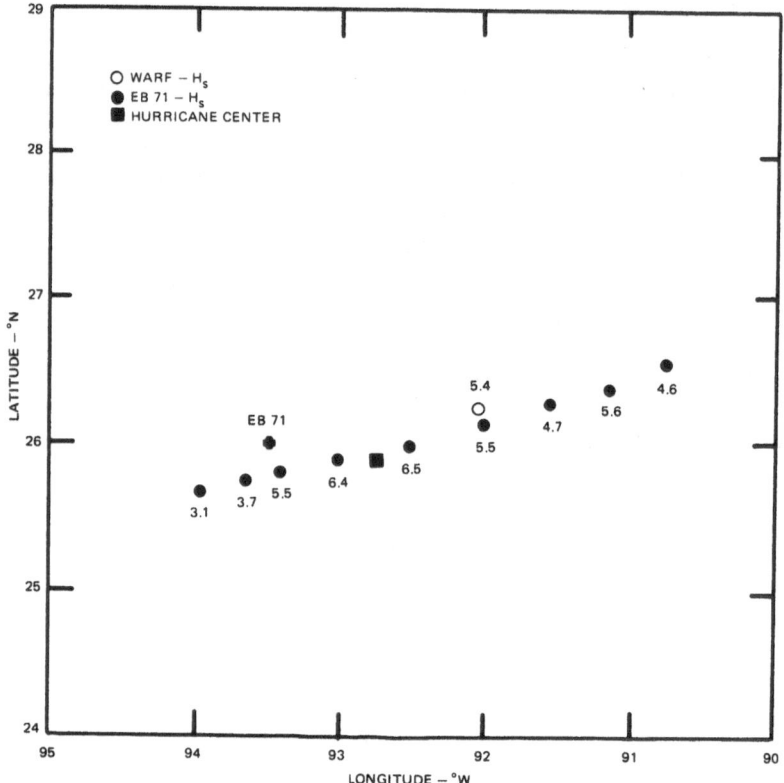

Fig. 5. Comparison of the significant wave height measured
by the WARF skywave radar and NDBO EB71. The EB71 significant
wave height estimates were compiled from 24 hours of data and
were plotted with respect to the hurricane center.

SUMMARY AND CONCLUSION

This paper summarizes the capability of measuring significant wave height and the wave spectrum by HF skywave radar. Agreement between the radar estimates of significant wave height and NDBO data buoy measurements was within 10% for a wide range of wave conditions. Similar accuracy was found for the wave spectrum estimate. The demonstrated capability by skywave radar for continously tracking and monitoring the surface winds and waves throughout all regions of a hurricane is unique. One skywave radar similar in design to WARF could routinely monitor the incident wave conditions along entire continental shelf regions off the east or west coast of the United States.

ACKNOWLEDGMENTS

This work was funded in part by the Air Force Office of Scientific Research (AFOSR) and in part by the NOAA Wave Propagation Laboratory. The wave spectral data was obtained from the NOAA Data Buoy Office.

REFERENCES

Barnum, J. R., J. W. Maresca, Jr., and S. M. Serebreny. 1977. High-resolution mapping of oceanic wind fields with skywave radar, *IEEE Trans. on Anetnnas and Propagation*, AP-25, 128-132.

Barrick, D. E. 1972a. First-order theory and analysis of MF/HF/VHF scatter from the sea, *IEEE Trans. on Antennas and Propagation*, AP-20, 2-10.

Barrick, D. E. 1972b. Remote sensing of sea state by radar, In: *Remote Sensing of the Troposphere*, (ed. V. E. Derr), Chapter 12, U.S. Government Printing Office.

Barrick, D. E., J. M. Headrick, R. W. Bogle, and D. D. Crombie. 1974. Sea backscatter at HF: interpretation and utilization of the echo, *Proceedings of the IEEE: 62*, 673-680.

Barrick, D. E. and J. B. Snider. 1977. The statistics of HF sea-echo Doppler spectra, *IEEE Trans. Antennas and Propagation*, AP-25, 19-28.

Barrick, D. E., M. W. Evans, and B. L. Weber. 1977. Ocean surface currents mapped by radar, *Science, 198*: 138-144.

Barrick, D. E. 1977a. Extraction of wave parameters from measured HF sea-echo Doppler spectra, *Radio Science, 12*: 415-424.

Barrick, D. E. 1977b. The ocean wave height nondirectional spectrum from inversion of the HF sea-echo Doppler spectrum, *Remote Sensing of Environment, 6*: 201-227.

Barrick, D. E. and B. J. Lipa. 1978. Ocean surface features observed by HF coastal ground-wave radars: a progress review,

In: *Ocean Wave Climate*, M. D. Earle and A. Malahoff (Eds.), Plenum, New York (This volume.)

Georges, T. M. and J. W. Maresca, Jr. 1978. The effect of radar beamwidth on the quality of sea-echo Doppler spectra measured with HF skywave radar, in preparation.

Lipa, B. J. 1977. Derivation of directional ocean-wave spectra by integral inversion of second-order radar echoes, *Radio Science*, *12*: 425-434.

Lipa, B. J. 1978. Inversion of second order radar echoes from the sea, *J. Geophys. Research*, *83*: 959-962.

Long, A. E. and D. B. Tizna. 1973. Mapping of North Atlantic winds by HF radar sea backscatter interpretation, *IEEE Trans. on Antennas and Propagation*, AP-21, 680-685.

Maresca, J. W. Jr., J. R. Barnum, and K. L. Ford. 1976. HF skywave radar measurements of coastal and open ocean surface currents (Abstract), 1976 Annual Meeting of USNC/URSI University of Massachusetts, Amherst, Massachusetts, October 11-15, 1976.

Maresca, J. W. Jr., and J. R. Barnum. 1977. Measurement of oceanic wind speed from HF sea scatter by skywave radar, *IEEE Trans. on Antennas and Propagation*, AP-25, 132-136.

Maresca, J. W. Jr. and C. T. Carlson. 1977a. Remote measurement of the ocean wave frequency spectrum by HF skywave radar (Abstract), 1977 Fall Meeting American Geophysical Union, San Francisco, California, December 5-9, 1977.

Maresca, J. W. Jr. and C. T. Carlson. 1977b. Tracking and monitoring of hurricanes by HF skywave radar over the Gulf of Mexico, Technical Report 1, SRI International, Menlo Park, California.

Maresca, J. W. Jr. and J. R. Barnum. 1978. Remote measurements of the position and surface circulation of hurricane Eloise by skywave radar, *Monthly Weather Review*, in press.

Maresca, J. W. Jr. and C. T. Carlson. 1978a. HF skywave radar measurements of wave height for hurricane Anita, in preparation.

Maresca, J. W. Jr. and C. T. Carlson. 1978b. HF skywave radar track of hurricane Anita, in preparation.

Maresca, J. W. Jr. and T. M. Georges. 1978a. Estimating significant wave height from the HF Doppler spectrum of the sea echo received from two or more ionospheric paths, in preparation.

Maresca, J. W. Jr. and T. M. Georges. 1978b. HF skywave radar measurement of the ocean wave spectrum, in preparation.

Maresca, J. W. Jr. and T. M. Georges. 1978c. Ionospheric diagnostics and sea scatter propagation management, in preparation.

Stewart, R. H. and J. R. Barnum. 1975. Radio measurements of oceanic winds at long ranges: an evaluation, *Radio Science*, *10*: 853-857.

Stewart, R. H. and J. W. Joy. 1975. HR radio measurements of surface currents, *Deep-Sea Research*, *21*: 1039-1049.

AN ASSESSMENT OF GEOS-3 WAVE HEIGHT MEASUREMENTS

Chester L. Parsons

NASA Wallops Flight Center

Wallops Island, Virginia

ABSTRACT

An iterative technique has been developed for the fitting of an averaged narrow-pulse radar altimeter return waveform from the Geodynamics Experimental Ocean Satellite (GEOS-3) with a four-parameter function derived from fundamental microwave rough scattering theory. For a signal reflected from the earth's oceans, the parameter associated with the slope of the leading edge of this waveform is directly relatable to the significant wave height of the sea. The technique was used during February 1976 when an intensive effort was made to map the sea state of the North Atlantic Ocean for comparison with several forms of truth information. Underflights of GEOS-3 orbits were made by a NASA C-54 aircraft with other narrow pulse radar systems, and shipboard observations of the significant wave height were received from four European-staffed Ocean Weather Stations. More recently, comparisons of GEOS-3 measurements with NOAA Data Buoy Office data in the Gulf of Alaska have also been made. Excellent agreement exists between the aircraft remote sensor data, the buoy data, and the GEOS-3 measurements if the effect of tracking jitter is included in the GEOS-3 data processing. The average difference between the GEOS-3 measurements and the corresponding comparison data set values is shown to be .34 m with a standard deviation of the differences of .61 m. The agreement between satellite and Ocean Weather Ship data is not as good. The presence of systematic biases in the shipboard observations is suggested.

INTRODUCTION

On March 20, 1975, a Delta launch vehicle was fired from the
Air Force Western Test Range, Vandenberg AFB, California, and
placed into orbit the Geodynamics Experimental Ocean Satellite
(GEOS-3). Equipped with a 13.9 GHz radar altimeter, this satellite
has as one of its major objectives the demonstration that radar
altimeter data when suitably processed are capable of yielding
measurements of sea state along the GEOS-3 groundtrack.

The value of such sea state information is obvious when
considering the advantages inherent in the use of a satellite
platform. A measurement device located at satellite altitudes
is free of the wave motion interference problem found with *in situ*
measurements and the atmospheric turbulence interference of airplane flights. With a stable sensor design, then, a satellite sea
state measurement device is capable of generating consistent data
over long time periods. Additionally, in contrast to the large
logistical problem of deploying aircraft and *in situ* sensors to
map the sea state of a large geographical area, the GEOS-3 radar
altimeter is able to profile arcs across an ocean in near real-
time for the first time and to perform this activity routinely.
This capability may have significant impact on the various activities to which knowledge of sea state is important. Ship routing
can be greatly improved economically, and the safety of search and
rescue operations and coastal recreation areas can be enhanced by
satellite sea state measurements. In terms of research and development, ocean-scale measurements of sea state may significantly affect
man's knowledge of the air-sea interface, a field of study that has
a direct bearing on the prediction of future weather using numerical
weather prediction techniques.

In this paper, it is demonstrated that the GEOS-3 radar
altimeter is capable of consistently producing measurements of
sea state that are of comparable quality with the best aircraft
and *in situ* measurements available. Thus, the satellite altimeter
is a legitimate data source of great value for the determination
of sea state on a global scale.

THE GEOS-3 SIGNIFICANT WAVE HEIGHT MEASUREMENT METHOD

The feasibility of determining sea state from an active nadir-
looking microwave device has been discussed in the literature by
Miller and Hammond (1972), *McGoogan* (1975), and *Walsh* (1974).
Figure 1 is an illustration from *Walsh et al.* (1977) that aptly
illustrates the means by which significant wave height can be
measured by a pulse-limited altimeter. Two extreme cases are
shown. At the top, a spherically expanding pulse is shown impinging

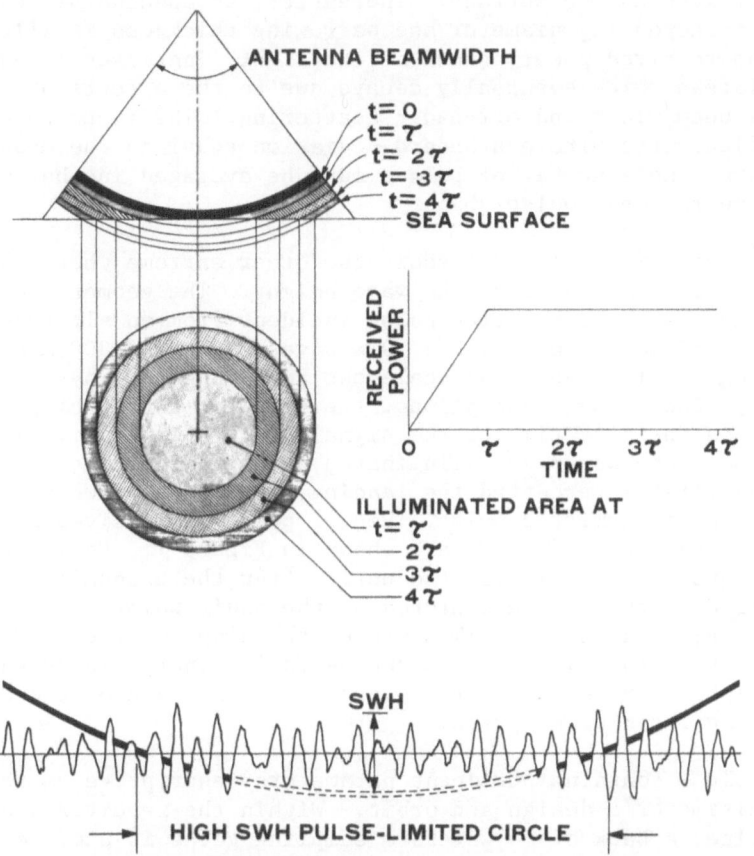

Fig. 1. Two examples of pulse-limited altimetry. The top and middle show a time sequence for a rectangular pulse incident on a sea surface whose wave height is small compared to the pulse width. The situation for wave heights large in comparison to the pulse width is depicted at the bottom (after *Walsh et al.*, 1977)

on a sea surface whose wave heights are much smaller than the
pulse width. The center of the figure shows the sea surface area
illuminated at time increments equal to the duration of the transmitted pulse. For a rectangular pulse the geometry is such that
the area increases linearly with the time until the back of the
pulse arrives at the surface. Thereafter, an annulus of constant
area but increasing diameter and narrowing thickness is illuminated.
The mean received power, shown at the right, increases linearly
to a plateau which eventually decays due to the effects of the
antenna beam width and off-nadir scattering. The instantaneous
power fluctuates with a standard deviation equal to the mean at
each point and a number of pulses must be averaged incoherently to
determine the mean pulse shape.

The bottom of Figure 1 shows the other extreme where the pulse
width is much smaller than the wave height. The geometry corresponds to a 3 ns transmitted pulse incident from an altitude of
843 km on a sea surface with a 10 m wave height and 300 m dominant
wave length. (The vertical scale has been magnified by a factor
of 100.) Under these conditions, the vertical extent of the sea
determines the rise time of the signal's return waveform. The
crests of these waves are illuminated prior to the time when the
calm sea first intercepted the leading edge of the downwardly
propagating transmitted pulse. Hence, power is received by the
altimeter before the $t = 0$ mark shown in Figure 1. Equivalently,
the troughs are not illuminated until after the transmitted pulse
has filled in the circle centered at the nadir point at time $t = \tau$
for the calm sea example. Therefore, the slope of the leading
edge of the return waveform is decreased for increasing significant
wave height. By measuring the slope, the sea state at the altimeter's nadir can be computed.

Table 1 contains pertinent parameters descriptive of the GEOS-3
radar altimeter's design and orbit. Within the receiver system of
the device, a bank of 16 waveform sampling gates is used to construct a sampled picture of the return waveform. The sampled waveforms from 200 radar pulses are averaged onboard the satellite to
produce a 2 second mean waveform, the subject of the mathematical
computation described below.

The mean return waveform for a short-pulse altimeter operated
at nadir can be conceptually viewed as a convolution of the altimeter system point target impulse response, the flat-sea impulse
response, the wave height probability density function, and the
tracking loop jitter (*Brown*, 1977). The latter factor is the error
in the placement of the waveform sampling gates in time caused by
the changing altitude of the satellite above the ocean's surface
and the noisiness of the timing circuitry. By assuming that a
Gaussian function is a reasonable approximation for the system point

Table 1. Pertinent GEOS-3 radar altimeter parameters

Satellite Orbit
Altitude	843 km
Inclination	115°
Eccentricity	0
Period	101.8 min.
Groundspeed	7 km s^{-1}

Radar Altimeter Transmitter
Frequency	13.9 GHz
Transmitter Power	2.5 kw
Pulse Duration	12.5 nsec
Pulse Repetition Frequency	100 sec^{-1}

Radar Altimeter Sample-and-Hold Gate Array
Number of Gates	16
Gate Spacing	6.25 nsec
Gate Width	12.5 nsec

target response and the wave height probability density function, that tracking loop jitter is unimportant in a first-order computation of significant wave height using GEOS-3 data, and that pointing angle errors have negligible effect on the leading edge of the return waveform, the generalized expression for the return power as a function of time (*Miller and Brown*, 1974) can be reduced to the following simplified form

$$y(t) = aP\left[\frac{t-b}{c}\right] + d \qquad (1)$$

where a   = return signal amplitude
       b   = time origin
       c   = return waveform risetime
       d   = return waveform baseline amplitude

$$P(z) = \int_{-\infty}^{z} Z(q)dq$$

and $Z(q) = \frac{1}{\sqrt{2\pi}} \exp\,(-q^2/2)$

The task to be performed consists of optimally choosing, in some sense, the four parameters, a, b, c, and d which best characterize the mean waveform produced by the 16 GEOS-3 sample-and-hold gates. With the average waveform mathematically modelled, the parameter c can be used to compute the GEOS-3 significant wave height estimate.

The determination of the four parameters is based on an iterative, least squares approach developed by *Hayne* (1977). Let the four-parameter function, $y(t)$, which is a function of a, b, c, d, and t, be approximated at the point a_0, b_0, c_0, and d_0 in parameter space by the Taylor series expansion

$$\tilde{y}_i = y_0 + y_a\alpha + y_b\beta + y_c\gamma + y_d\delta \doteq y_i \tag{2}$$

where

$$y_0 = y(a_0, b_0, c_0, d_0, t) = a_0 P\left(\frac{t-b_0}{c_0}\right) + d_0$$

$$y_a = \frac{\partial y}{\partial a}\bigg|_{a_0, b_0, c_0, d_0, t} = P\left(\frac{t-b_0}{c_0}\right)$$

$$y_b = \frac{\partial y}{\partial b}\bigg|_{a_0, b_0, c_0, d_0, t} = \frac{-a_0}{c_0} Z\left(\frac{t-b_0}{c_0}\right)$$

$$y_c = \frac{\partial y}{\partial c}\bigg|_{a_0, b_0, c_0, d_0, t} = \frac{-a_0}{c_0}\left(\frac{t-b_0}{c_0}\right) Z\left(\frac{t-b_0}{c_0}\right)$$

$$y_d = \frac{\partial y}{\partial d}\bigg|_{a_0, b_0, c_0, d_0, t} = 1$$

and $\alpha = a-a_0$; $\beta = b-b_0$; $\gamma = c-c_0$; and $\delta = d-d_0$. Then form a sum over all N values of the errors $(\tilde{y}_i-y_i)$ squared.

$$E = \sum_{i=1}^{N} (\tilde{y}_i-y_i)^2 = \sum_{i=1}^{N} (y_0+y_a\alpha+y_b\beta+y_c\gamma+y_d\delta-y_i)^2 \tag{3}$$

Minimizing E with respect to α, β, γ, and δ, the following system of four equations is obtained.

$$\begin{bmatrix} \Sigma y_a^2 & \Sigma y_a y_b & \Sigma y_a y_c & \Sigma y_a y_d \\ \Sigma y_a y_b & \Sigma y_b^2 & \Sigma y_b y_c & \Sigma y_b y_d \\ \Sigma y_a y_c & \Sigma y_b y_c & \Sigma y_c^2 & \Sigma y_c y_d \\ \Sigma y_a y_d & \Sigma y_b y_d & \Sigma y_c y_d & \Sigma y_d^2 \end{bmatrix} \cdot \begin{bmatrix} \alpha \\ \beta \\ \gamma \\ \delta \end{bmatrix} = \begin{bmatrix} \Sigma y_a(y_i-y_0) \\ \Sigma y_b(y_i-y_0) \\ \Sigma y_c(y_i-y_0) \\ \Sigma y_d(y_i-y_0) \end{bmatrix} \tag{4}$$

Direct substitution of the expressions for y_0, y_a, y_b, y_c, and y_d from (2) into (4) completes the mathematical formulation of the problem. The iterative inversion of the 4x4 matrix will produce the desired solution for the values α, β, γ, and δ.

The initial guesses a_0, b_0, c_0, and d_0 are corrected to produce revised estimates a, b, c, d since $a = a_0 + \alpha$, $b = b_0 + \beta$, $c = c_0 + \gamma$, and $d = d_0 + \delta$. Only the first order terms are retained in the original Taylor series in equation (2), and thus a is an approximate estimate. Similarly, b, c, and d are estimates.

In this iterative procedure, the next step then is to use a, b, c, d in place of the original a_0, b_0, c_0, d_0, and repeat the entire cycle to produce improved estimates of a, b, c, d. The elements of the matrixes in (4) are, of course, re-evaluated each time through the cycle; the indicated sums are for the N=16 separate values from the 16 GEOS-3 waveform samplers. The interative procedure is repeated until a defined relative error criterion is satisfied; a typical loop exit criterion is to require that the total sum of errors squared decreases by less than 0.1% of the sum of errors squared in the last previous time through the loop. For a reasonable first guess a_0, b_0, c_0, d_0, the problem will converge in typically two to four iterations to reasonable final answers for a, b, c, and d. Given the risetime parameter, c, determined by this iterative least-squares procedure, the estimation of significant wave height is as follows.

The convolution model for the mean sea-scattered pulse return assumes that the risetime parameter c is the composite result from the assumed Gaussian calm sea equivalent radar pulse width σ_c and the rough-sea rms height σ_s. Because of the convolution of two Gaussian functions of widths σ_s and σ_c,

$$\text{or} \quad \begin{aligned} c^2 &= \sigma_s^2 + \sigma_c^2 \\[6pt] \sigma_s^2 &= c^2 - \sigma_c^2 \end{aligned} \quad (5)$$

Here σ_s is the two-way ranging time in units of nanoseconds. Conversion of ranging times in nanoseconds to surface height in meters is accomplished by multiplying by the two-way speed of light. Because significant wave height (SWH) is four times the rms surface height relative to mean sea level, σ_s must also be multiplied by four. Thus,

$$\text{SWH} = 0.6\sqrt{c^2 - \sigma_c^2} \text{ , SWH in m, c and } \sigma_c \text{ in ns.} \quad (6)$$

The value of σ_c used in this study was 7.49 ns and was based upon the observation of GEOS-3 waveforms over known regions of calm seas early in the lifetime of the satellite.

The above calculations are based on the assumption that tracking jitter is of negligible importance in the procedure. Recent work by *Walsh* (1978) indicates that such is not the case.

Biases introduced by the non-uniform spacing of the individual
gates in the sample-and-hold gate bank and by non-uniform output
voltage amplitudes have been quantitatively determined and the
effect of tracking jitter is also examined. Using these results,
an empirical correction to the significant wave height determined
using equation (6) can be made according to the following expression:

$$SWH^* = \begin{cases} SWH & 0 \leq SWH < 2.0 \text{ m} \\ .77 \text{ SWH} + .47 & 2.0 \leq SWH < 5.0 \text{ m} \\ .7 \text{ SWH} + .8 & 5.0 \leq SWH \end{cases} \quad (7)$$

where SWH* is the corrected significant wave height.

As an example of GEOS-3 significant wave height measurements,
data produced during orbit 4281 across the North Atlantic on February 6, 1976, is shown in Figure 2. The groundtrack is pictured in
Figure 3 as an overlay on the 1200 GMT surface pressure field from
the National Weather Service. The record begins at point A near
40°N latitude along the groundtrack. The wave height builds from
about 4 m at A to near 15 m near point B, due south of the center
of the low pressure system. The intensity of the storm is indicated by the 944 mb pressure reading at the core. A smaller amplitude
band of high seas is encountered partway between points B and C.

The data points in Figure 2 are the results of the processing
of onboard-averaged 2 second waveforms by the technique culminating
in equation (7). The dotted lines indicate portions of the data
record in which the altimeter had difficulty in maintaining its
"lock" on the position of the sea surface, probably the result of
insufficient power in the signal backscattered by the rough seas.

A statistical study of all GEOS-3 data collected over the
North Atlantic during February 1976 has been performed to characterize the fluctuations of the data as a function of the computed,
or observed, significant wave height. The results are shown in
Figure 4. The standard deviation σ of the GEOS-3 measurements is
seen to vary around 2.0 m over the entire range from 0 to 14 m. A
minimum does exist in seas of about 6 m and σ does begin to increase slightly in high seas. If the values computed using the
onboard-averaged (per frame) waveform are averaged over multiple
frames, then σ can be reduced by the inverse of the square-root of
the number of frames averaged in accord with random process theory.
The wave height standard deviation for an average of nine frames,
or nine single measurements, is also shown in Figure 4 over the
full range of sea state. If a fractional accuracy is desired, the
standard deviation can be divided by the average observed significant

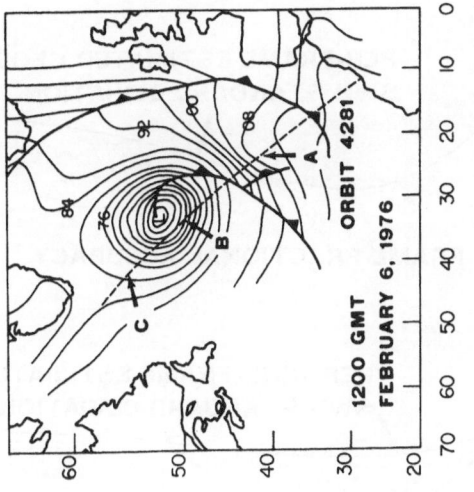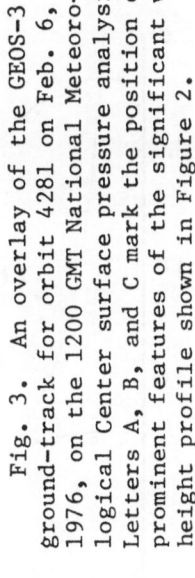

Fig. 3. An overlay of the GEOS-3 ground-track for orbit 4281 on Feb. 6, 1976, on the 1200 GMT National Meteorological Center surface pressure analysis. Letters A, B, and C mark the position of prominent features of the significant wave height profile shown in Figure 2.

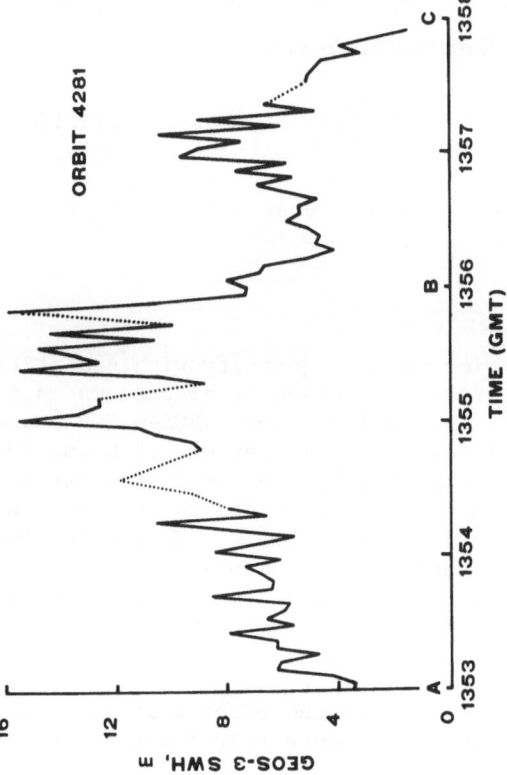

Fig. 2. The variation of significant wave height (SWH) as computed by the GEOS-3 radar altimeter during orbit 4281, Feb. 6, 1976.

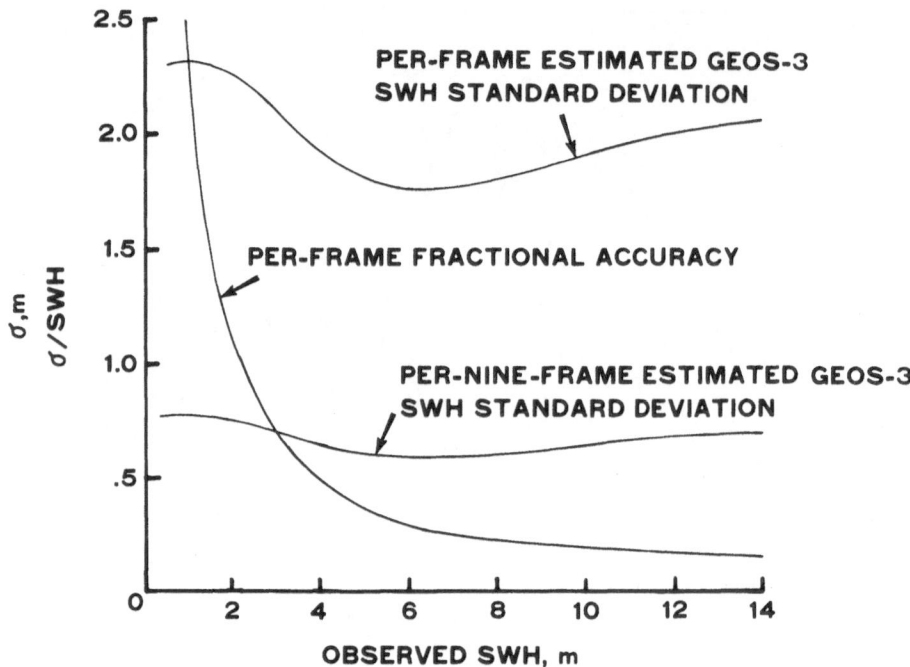

Fig. 4.   Experimental variations of GEOS-3 significant wave height (SWH) standard deviations both without averaging and with nine point averaging. The fractional accuracy, the ratio of σ to the computed wave height, with no averaging is also presented.

wave height. The variation of this quantity with sea state for the case of no multiple frame averaging is also shown in the illustration. The error in any individual GEOS-3 SWH measurement is seen to be about 100% for a wave height of 2.5 m and 50% at 4 m. If a nine-frame average is acceptable in terms of a degradation in spatial resolution, then 100 % error occurs at a wave height of approximately .75 m and 50% error at 1.5 m. Therefore, in the process of incorporating GEOS-3 measurements in some application activity, the trade-off between the accuracy of the computed values and the resolution with which such measurements are needed along a groundtrack must be addressed.

This discussion has concerned the error that may occur in a GEOS-3 measurement of significant wave height due to the inherent

noisiness of the measurement process. In contrast to this evaluation of system precision, an assessment must also be made of the
accuracy of the GEOS-3 measurement. This is treated in the following section.

COMPARISON WITH OTHER MEASUREMENTS OF SIGNIFICANT WAVE HEIGHT

Newfoundland Expedition

     To test the performance of the GEOS-3 measurement capability,
it was desirable to monitor a wide range of sea states for which
surface truth information is available. Because of the inaccessi-
bility of high sea state regions to normal measurement techniques
and the degradation of GEOS-3 measurements near calm seas discussed
above, an expedition was conducted during February 1976 to collect
aircraft measurements of significant wave height to test the performance of the GEOS-3 algorithm. The Wallops Flight Center (WFS)
C-54 research vehicle, based in St. Johns, Newfoundland for the
entire month, conducted eleven overflights of GEOS-3 groundtracks.
These are pictured in Figure 5. Equipped with a pulse compression
radar altimeter (*Hughes Aircraft Company*, 1976), the Naval Research
Laboratory nanosecond radar (*Walsh*, 1974), and various meteorological
instruments, the C-54 traversed approximately 150 km of each track.
With a satellite groundspeed of 7 km s^{-1}, the entire aircraft instrumentation record must be compared with a single nine-frame average
measurement from the satellite. Therefore, the comparison of the
GEOS-3 SWH data with aircraft data is reduced to a comparison of
eleven elements.

     During the month of data collection, the pulse compression
radar altimeter performed well. The nanosecond radar, however,
suffered from equipment failure during six of the eleven C-54
flights. Table 2 is a tabulation of the pulse compression radar
measurements, the valid nanosecond radar measurements, and the
appropriate GEOS-3 nine-frame averaged significant wave height
values accompanied by the C-54 flight dates. The latitudes and
longitudes of the starting and stopping points are those of the
pulse compression altimeter's data collection period but are
approximately correct for the nanosecond radar's operation also.

Gulf of Alaska Buoy Data Comparisons

     NOAA Data Buoy Office (NDBO) data from buoys EB03, EB16, EB17,
EB19, and EB35 located in the Gulf of Alaska at 56.0N, 147.9W;
42.5N, 130.0W; 52.0N, 156.0W; 51.0N, 136.0W; and 55.3N, 157.0W,
respectively, have also been obtained. Collected from February
through April 1977, these data have been scanned and those values
that were measured by a buoy lying within 200 km of a GEOS-3

TABLE 2.   Newfoundland expedition aircraft data summary. Significant wave heights (SWH) measured with a pulse compression radar (AAFE), a Naval Research Laboratory (NRL) nanosecond radar, and GEOS-3 are compared in the last three columns.

DATE	GEOS-3 ORBIT	START			STOP			AAFE SWH (M)	NRL SWH (M)	GEOS-3 SWH (M)
		LATITUDE (°N)	LONGITUDE (°W)	TIME (GMT)	LATITUDE (°N)	LONGITUDE (°W)	TIME (GMT)			
2/8/76	4310	49°12.4'	48°3.3'	143521	50°6'	49°16.8'	145856	1.9	–	2.7
2/9/76	4324	50°53.2'	45°0'	144800	51°45.8'	46°23.6'	150016	3.7	–	4.7
2/11/76	4353	45°8.9'	51°47.3'	153810	44°12.3'	50°40.2'	154942	4.2	–	4.4
2/12/76	4367	47°54'	49°37.3'	155300	47°13.4'	48°40.3'	154942	4.7	–	5.4
2/13/78	4381	47°54.7'	44°9.8'	150200	46°58.9'	43°4.6'	152611	4.5	–	4.9
2/17/76	4438	46°31.8'	45°55.3'	160504	45°59.8'	45°16.3'	161533	7.1	–	9.3
2/20/76	4481	43°2.8'	51°20.2'	164600	42°14.4'	50°42.7'	170353	5.4	5.0	6.0
2/23/76	4523	48°43'	40°56.7'	160100	49°43.6'	41°43.6'	162101	–	8.0	7.6
2/24/76	4538	42°32.1'	53°46.8'	174817	41°45.7'	52°58.9'	180917	5.3	4.7	5.2
2/28/76	4594	51°53.8'	43°50.2'	162200	51°23.3'	42°58.8'	164405	4.3	4.2	3.9
3/3/76	4651	48°54.3'	42°42.4'	170700	48°4.1'	41°40.9'	172905	5.5	6.1	6.1

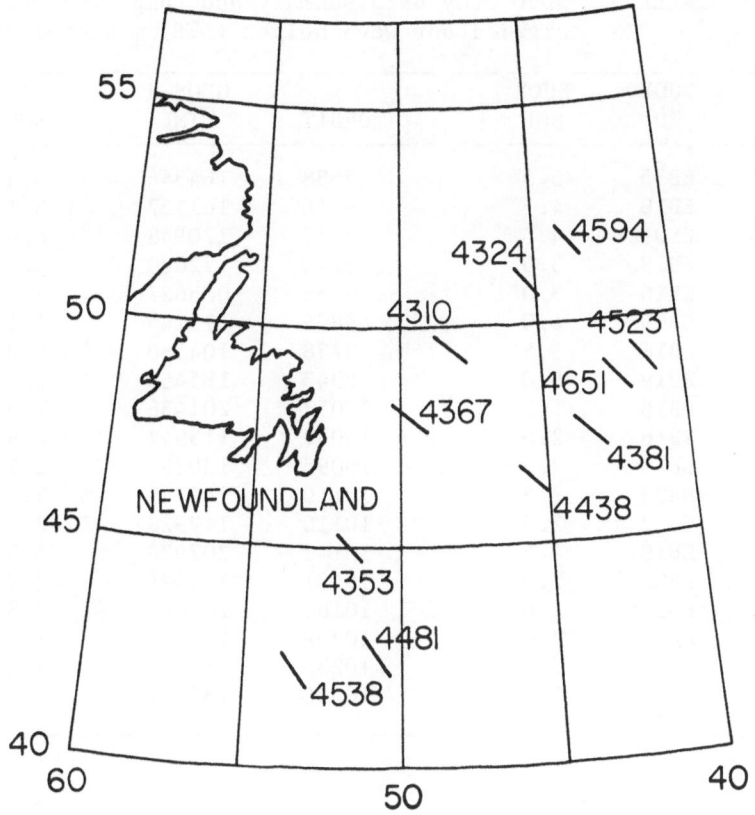

Fig. 5. Groundtracks of the eleven aircraft underflights
during the February 1976 Newfoundland Expedition.

groundtrack have been selected as another comparison data set. The
NDBO data and the corresponding GEOS-3 measurements are tabulated
in Table 3 and are shown in Figure 6 with the aircraft comparison
data sets. Good agreement is seen to exist with the 1976 data.
The average difference between the GEOS-3 measurements and all of
the corresponding corroborative data set values shown in Figure 6
is .34 m and the standard deviation of the differences is .61 with
no apparent dependency upon the magnitude of the sea state. Table 4
summarizes the good agreement found between GEOS-3 significant
wave height measurements and the supporting data sets. Certainly,
these deviations are well within the error bounds that may be
quoted for any of the data sources involved.

TABLE 3.   NDBO buoy data summary and comparison to GEOS-3
significant wave height (SWH) measurements.

DATE	BUOY ID	BUOY SWH	ORBIT	GEOS-3 TIME	SWH
2/15	EB35	5.5	9588	164346	5.4
2/17	EB19	4.5	9616	161157	5.1
2/20	EB03	4.5	9659	170948	4.6
2/26	EB19	3.5	9744	172000	4.8
2/27	EB16	3.0	9753	085637	3.3
3/9	EB19	4.0	9896	112745	4.1
3/12	EB16	5.5	9738	104550	5.9
3/12	EB19	8.0	9943	185455	7.2
3/20	EB16	1.5	10057	201436	1.3
3/22	EB16	2.0	10080	113949	2.9
3/23	EB19	2.5	10095	130253	2.7
3/24	EB03	2.5	10110	142745	3.4
3/24	EB17	2.5	10110	142928	2.6
3/26	EB19	4.5	10142	202930	4.3
3/29	EB03	2.5	10181	145421	2.9
3/29	EB17	3.0	10181	145607	3.3
3/31	EB16	2.0	10208	124746	1.2
4/2	EB03	3.5	10238	153543	4.1
4/2	EB17	3.5	10238	153471	4.0

Ocean Weather Station Reports

Additional comparison data in the form of Ocean Weather
Station reports from stations C, L, M, and R were received from
the British Meteorological Office, Brackness, Berkshire for the
entire month of February. The approximate locations of these
four fixed vessels were 52.7N, 35.5W; 57.1W, 20.7W; 66.5N, 3.1E;
and 47.0N, 16.9W, respectively. The GEOS-3 data were scanned and
values found at locations within 200 km of an Ocean Weather Station
were tabulated for comparison with the appropriate shipboard
observations of significant wave height. Twenty-five data points
were found in this manner. These are displayed in Figure 7. There
appear to be systematic differences between the four weather ship
data sets. When compared with GEOS-3, Ship L consistently reports
higher wave heights than GEOS-3 while Ship C sees lower wave height.
No trends involving data from ships R and M can be distinguished
because of the small sample sets. More data is required before
the presence of bias can be substantiated and quantified.

TABLE 4.   Summary of agreement between GEOS-3 and corroborative SWH data.

Data Set	Mean Difference	σ	No. Pts.
NDBO Buoys	.24	.53	19
Nanosecond Radar	.16	.59	5
Pulse Compression Altimeter	.62	.71	10
Total	.34	.61	34

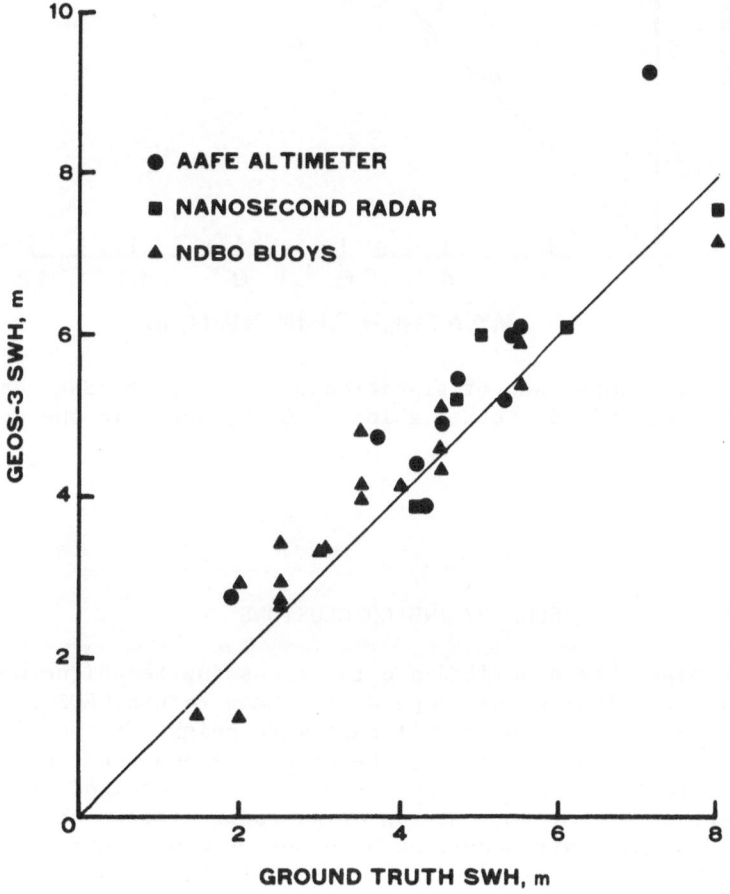

Fig. 6.   The comparison of GEOS-3 significant wave height (SWH) measurements with the corresponding truth information from the pulse compression altimeter (AAFE), the NRL nanosecond radar, and selected NDBO data buoys.

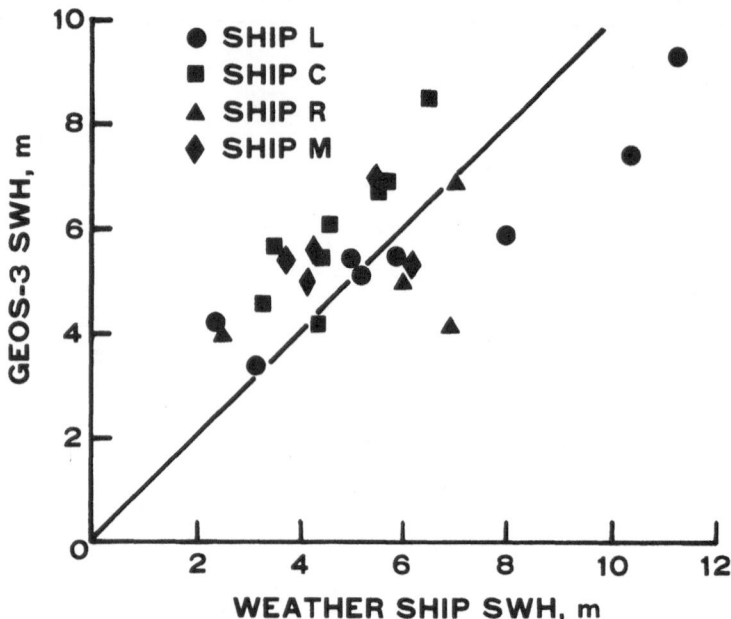

Fig. 7. Comparison of significant wave height (SWH) data from GEOS-3 and fixed weather ships L, C, R, and M in the North Atlantic.

SUMMARY AND CONCLUSIONS

This paper has described a data processing technique which is capable of relating the slope of the mean return GEOS-3 radar altimeter waveform to the significant wave height of the sea. Mathematically straightforward, the method is nonetheless difficult to accredit because of the absence of a recognized standard measurement of sea state. Narrow pulse radar systems aboard the NASA/WFC C-54 aircraft underflew a number of GEOS-3 orbits to gather data for comparison with the satellite's product. Shipboard observations from four fixed weather station vessels in the North Atlantic were also collected. In a separate comparison, NDBO buoy data from the Gulf of Alaska were used. The agreement between GEOS-3, the aircraft instrumentation, and the buoys was good with the average difference between the supporting data and

GEOS-3 being .34 m and the standard deviation .61 m.   In the weather ship comparison, systematic biases in the shipboard observations may be responsible for an increase in scatter.

Based upon these sources of truth data, it is concluded that the GEOS-3 return waveform when suitably processed is capable of producing measurements of significant wave height in high sea states of quality comparable to that of any other sources available.   The consistency of operation and the potentially global coverage of the satellite radar altimeter dictate that the device be considered henceforth as a primary sea state data source.

ACKNOWLEDGMENTS

The author wishes to express his indebtedness to B. Yaplee, D. Hammond, and R. Mannella of the Naval Research Laboratory for the nanosecond radar data, and to W. Townsend of NASA for the invaluable AAFE Pulse Compression Radar Altimeter measurements of significant wave height.   Special thanks are due to R. Dwyer of the Computer Science Corporation for the implementation of the operational version of the algorithm and for its utilization during February 1976.

REFERENCES

Brown, G. S.   1977. The average impulse response of a rough surface and its applications, *IEEE Trans. Antennas and Propagation,* AP-25, 67-74.

Hayne, G. S.   1977. Initial development of a method of significant waveheight estimation for GEOS-III, NASA CR-141425, pp. 11.

Hughes Aircraft Company.   1976.   Final report of the advanced application flight experiment breadboard pulse compression radar altimeter program.   (FR-76-14-183, Hughes Aircraft Company; NASA Contract NAS6-2558) NASA CR-141411.

McCoogan, J. T.   1975.   Satellite altimetry applications, *IEEE Trans. Microwave Theory and Techniques,* MIT-23, 970-978.

Miller, L. S. and G. S. Brown.   1974.   Engineering studies related to the GEOS-C radar altimeter, NASA CR-137462.

Miller, L. S. and D. L. Hammond.   1972.   Objectives and capabilities of the Skylab S-193 Altimeter Experiment, *IEEE Trans. Geoscience Electronics,* GE-10, 9, 711-722.

Walsh, E. J.   1974.   Analysis of experimental NRL radar altimeter data, *Radio Science, 9,* 711-722.

Walsh, E. J., E. A. Uliana, and B. S. Yaplee.   1977.   Ocean wave heights measured by a high resolution pulse-limited radar altimeter, *Boundary Layer Meteorology, 13,* 187-200.

Walsh, E. J.   1978.   Extraction of ocean wave height from GEOS-3 altimeter data, submitted for publication to *J. Geophys. Res.*

On the Use of Aircraft in the Observation of One- and Two-Dimensional Ocean Wave Spectra

Duncan B. Ross

Atlantic Oceanographic and Meteorological Laboratories

National Oceanic and Atmospheric Administration

ABSTRACT

Aircraft have been utilized for a number of years to observe various characteristics of wind-generated ocean waves. The techniques involved are both active and passive and incorporate sensors which operate in the microwave and optical regions of the electromagnetic spectrum. This paper reviews a number of such sensors and presents data in support of their utility. It is found that aircraft can play an important role in a wave monitoring program.

INTRODUCTION

The problems at hand in the area of wave research range from the study of capillary waves and their interactions with long gravity waves to the development and verification of wave prediction models. Unfortunately, no single sensor exists which can be used to provide the wave measurements required for all of these problems. Definition of a wave monitoring program, therefore, first requires the objectives to be firmly established; and only then can the measurement systems be developed which will yield the required data.

Aircraft platforms as a means of obtaining measurements possess certain unique advantages. For example, severe transient phenomena such as tropical and extra-tropical cyclones can be probed in a manner which will yield virtually synoptic wind and wave measurements in all regions of the storm. Fixed platforms, of course, are ideal for long-term monitoring but are dependent upon the vicissitudes of nature with regard to where data is taken within a

storm.   Fixed platform data, however, when combined with specialized data sets such as from aircraft, spacecraft, or over-the-
horizon radar, can produce a complete picture of the response of
the ocean to the storm.

   This paper reviews some of the techniques which can be used
from aircraft for observing both one- and two-dimensional wave
information. *Pierson* (1976) provides additional details on some
of the techniques discussed herein.

                            WAVE PROFILES

   The earliest attempt to observe waves directly from low-flying
aircraft involved the use of sensitive radar altimeters (*Morrow*,
1964, and *Barnett and Wilkerson*, 1967) to obtain a profile of the
surface waves. A principal requirement for this approach is the
removal of aircraft motion. *Barnett and Wilkerson* (1967) removed
aircraft heave displacements by doubly integrating and subtracting
the output of a vertical accelerometer from the altimeter output.
Low-frequency energy remained, however, due to pitch and roll
errors and was removed by digital high-pass filtering prior to
Fourier analysis of the wave time series. The spectrum thus obtained must be mapped from a moving to a fixed coordinate system
since the apparent frequency obtained from the moving platform is
related to the true frequency.

$$\omega_e = \omega + (\omega^2/g)V \cos \theta \tag{1}$$

where ω_e is the apparent frequency, ω is the true wave frequency,
g is the acceleration due to gravity, V is the aircraft velocity,
and θ is the heading of the waves relative to the aircraft track.
The quantity ω^2/g is the wave number.

   Laser altimeters may also be used as profilometers (*Ross*, 1967)
and both radar and laser systems have been used for fetch-limited
wave growth studies. For example, data obtained by the author in
the North Sea (*Ross et al.*, 1970) and along the east coast of the
United States (*Ross and Cardone*, 1974) have been combined with data
obtained by *Barnett and Wilkerson* (1967) and are shown plotted as
a function of fetch (distance from shore) in Figure 1 in the dimensionless form suggested by *Kitaigorodskii* (1961). Aircraft motion
in the case of the laser data was removed by simply high-pass
filtering the laser time series. This approach is based upon the
assumption of little aircraft motion in the pass band of the waves
(cf. *Ross et al.*, 1970) and is substantiated by the fact that a
characteristic dip normally occurs in the spectrum in the region

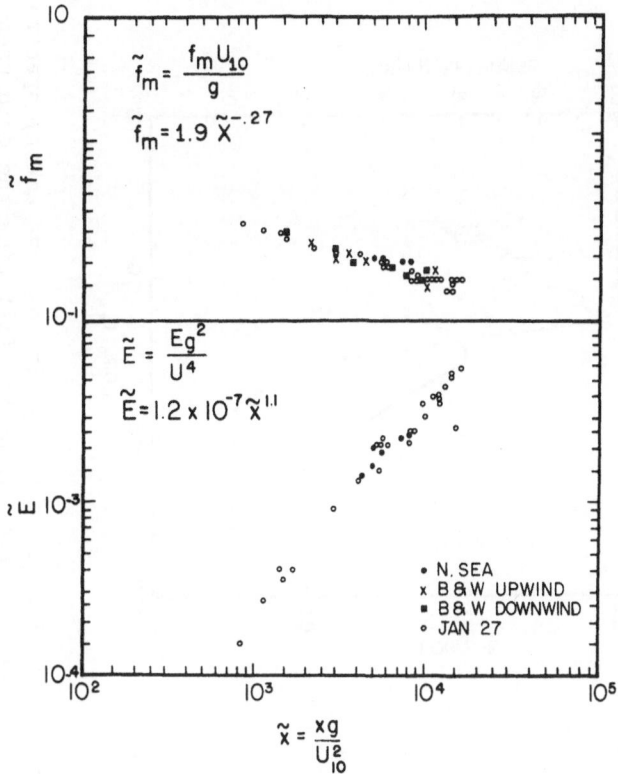

Fig. 1.   Observations of the fetch ($\tilde{X}$) dependence of the total energy ($\tilde{E}$) (twice the variance) and spectral peak frequency ($\tilde{f}_m$) nondimensionalized by the surface wind as measured at 10 meters, and gravity.   The closed and open circles are the data of *Ross et al.* (1970) and *Ross and Cardone* (1974); the crosses and boxes are the data of *Barnett and Wilkerson* (1967).

separating wave energy from aircraft motion. Figure 2, for example, shows the spectrum from an unfiltered time series obtained in Pacific hurricane Ava (*Ross et al.*, 1974) in June, 1973. Figure 3 shows the spectrum of the same time series following high-pass filtering. From the filter response also plotted on Figure 3, it can be seen that the spectral estimate which is 60% passed has a numerical value which is only 13% of the value at the spectral peak. The low frequency side of the peak frequency is, therefore, real.

A limitation of the profilometer approach manifests itself in the presence of a wind sea which contains a swell. Since the aircraft tracks are normally flown approximately up or downwind (parallel to the local sea), swell travelling at an angle to the track

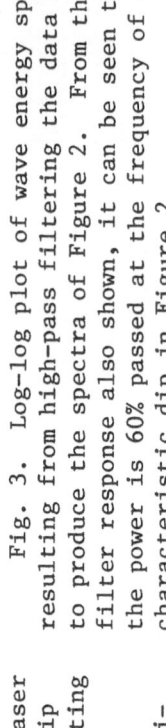

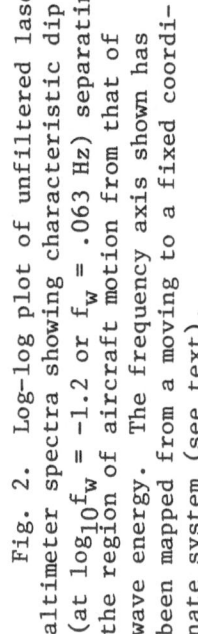

Fig. 3. Log-log plot of wave energy spectra resulting from high-pass filtering the data used to produce the spectra of Figure 2. From the filter response also shown, it can be seen that the power is 60% passed at the frequency of the characteristic dip in Figure 2.

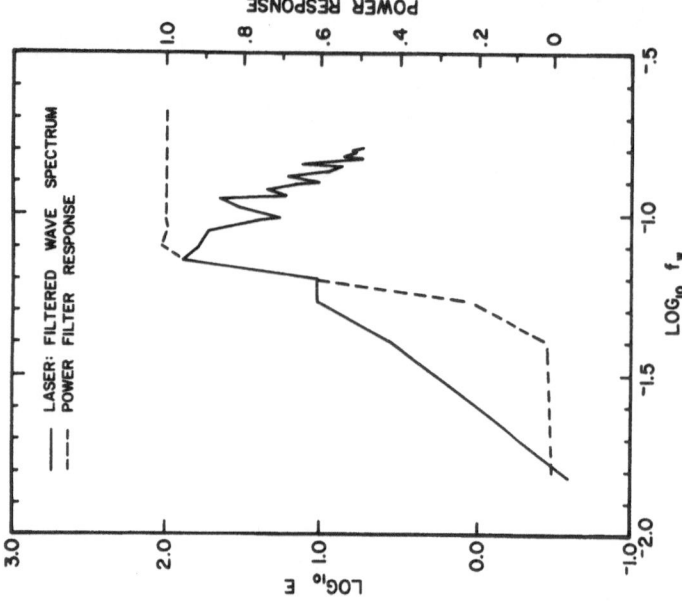

Fig. 2. Log-log plot of unfiltered laser altimeter spectra showing characteristic dip (at $\log_{10} f_w = -1.2$ or $f_w = .063$ Hz) separating the region of aircraft motion from that of wave energy. The frequency axis shown has been mapped from a moving to a fixed coordinate system (see text).

will appear at a lower frequency which may be irretrievably mixed
with the aircraft motion, or compromise the filtering procedure.
Should the swell energy not be lost, the mapped frequency spectrum
is still incorrect unless the swell direction is known and mapped
independently of the wind wave frequencies. Figure 4 is an example
of an aircraft-obtained spectrum which contains multiple swell frequencies compared to a nearby buoy spectrum. In this case the
local sea was correctly mapped while the swell frequencies are
somewhat in error. It can be seen that the significant wave heights
are in good agreement. The utility of this spectrum, therefore,
depends upon the importance of knowing exactly the frequency of
the swell energy.

The aircraft technique is particularly useful in stormy
situations such as fetch-limited conditions where the local sea
strongly dominates the spectrum. Surprisingly, hurricane spectra
appear to be largely a local sea-dominated situation in the region

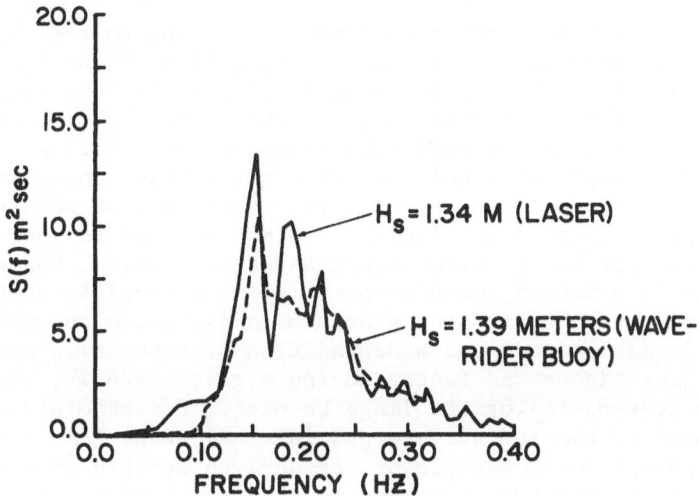

Fig. 4. Comparison of laser- and buoy-derived wave spectra.
Surface winds were 9 m/sec. Significant wave heights (H_s) compare
well. The spectra match more closely for the higher frequency
level sea than for the lower frequency swell.

of higher winds. Figure 5 is an example of three spectra obtained
in hurricane Ava (*Ross et al.*, 1974) compared to fetch-limited
seas observed in the North Sea (*Ross et al.*, 1970). The similarity
of these spectra support the validity of parameterization of the
growth process as proposed by *Hasselmann et al.* (1976).

IMAGING RADAR

Side-Looking Airborne Radar (SLAR) imaging was first applied
to the wave problem by *Bondarenko et al.* (1972). Using optical
Fourier transforms, they compared the radar spectra to similar
transforms of vertical photography and conducted a detailed study
of the angular distribution of wave energy. They found that their
data was well approximated by a function of the type $\cos^m(\theta)$ with
m varying from about 10 for a swell system to 2.5 for wind-driven
waves. In a follow-up study *Byelousov et al.* (1975) found values
of m = 30 for swell and m = 10 for a locally generated sea. These
values for swell systems are somewhat larger than those reported
by *Mitsuyasu et al.* (1975) and *Ewing* (personal communication), sug-
gesting further studies exploiting the high resolution inherent to
the radar technique are appropriate.

The term SLAR is usually used when referring to real aperture
radar systems where the resolution is a function of the footprint
size which is determined by the antenna pattern and the length of
the pulse. Synthetic Aperture Radar (SAR), however, can usually
achieve significantly improved resolution by transmitting a wide-
beam coherent radar pulse and recording the Doppler phase information present in the return signal. The resolution of SAR systems
is determined mainly by the length of time required to generate
the synthetic aperture. Since ocean waves are moving, however,
the Doppler information normally peculiar to a specific spot on the
earth relative to the airborne platform is distorted somewhat; the
footprint is mislocated; and a degradation of resolution occurs.
Obviously, the higher and faster moving a given wave is, the worse
the motion degradation of the image becomes. Nevertheless, SAR
has been used to observe surface and internal waves (*Brown et al.*,
1976). Figure 6 is an example of SAR-derived spectra obtained
while near hurricane Ava, compared to spectra obtained from simultaneous vertical photography and a composite spectrum consisting
of a laser profilometer spectrum obtained nearby superimposed upon
the spectrum one expects for the 11 m/s surface wind present assuming fully-developed conditions. Good agreement is seen in the high
frequency (short wavelength) peak observed by all instruments. The
SAR spectra, however, was not mapped from moving to fixed coordinates, and this discrepancy easily accounts for the apparently
shorter wavelength for the swell as observed by the SAR which was
traveling in a direction opposite to the aircraft track.

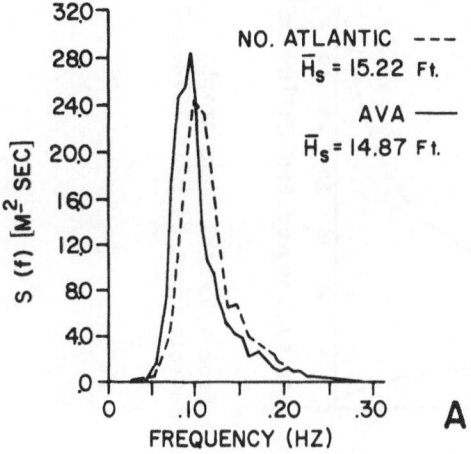

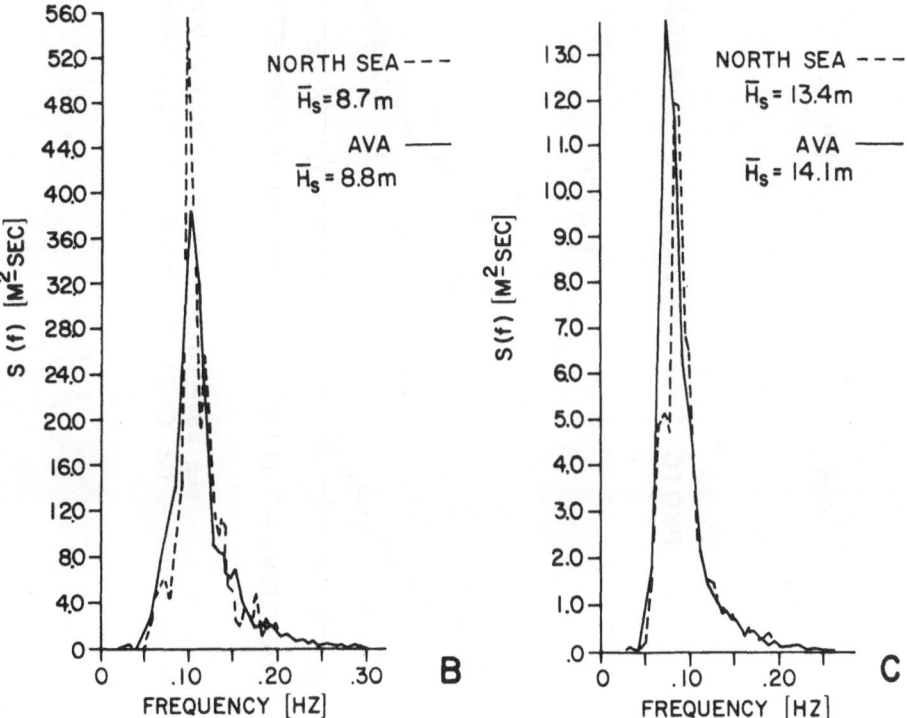

Fig. 5. Comparison of laser-derived wave spectra obtained in hurricane Ava (*Ross et al.*, 1974) and in the North Sea under fetch-limited conditions (*Ross et al.*, 1970) showing similar spectral shapes.

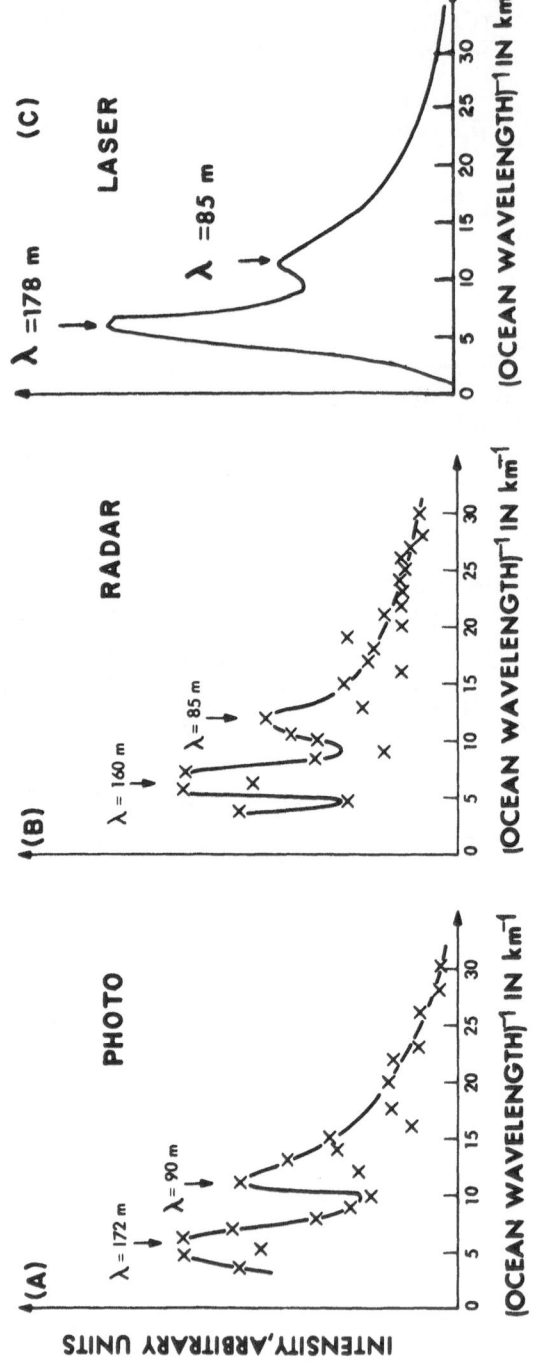

Fig. 6. Comparison of ocean wave spectra near hurricane Ava derived from vertical photography, A, L-Band Synthetic Aperture Radar (SAR), B, and a composite laser spectrum (cf. *Ross et al.*, 1974), C.

For aircraft application, SLAR is particularly useful since optimum altitudes are low (150-500 m) allowing determination of accompanying surface winds and simultaneous profilometer data. Furthermore, SLAR is not degraded due to surface wave motion although the resulting spectra must also be mapped from moving to fixed coordinates as in the case of the profilometer approach. The principal limitation to SLAR is resolution, typically 30 to 100 meters in azimuth with the exception of one system, developed by Westinghouse, which is capable of 8 meter resolution. SAR on the other hand can achieve excellent resolution (≈3 meters) even from satellite altitudes and is therefore more appropriate for space applications.

Presently, an exact transfer function relating a radar image transform spectrum to the two-dimensional surface wave spectrum is not known. This function is dependent upon the local tilt of the waves with respect to the incoming radar signal and the amplitude of the resonant short waves responsible for the backscatter (*Wright*, 1968). Since the short waves are modulated by longer waves, in a poorly known manner, present use of remote radar imagery is restricted to wave length and direction information.

SHORT-PULSE RADAR RETURN

When an approximately square wave radar pulse impacts the surface of the ocean, the leading edge of the reflected radar energy as a function of time has a characteristic slope which is determined by the height of the surface roughness elements. Furthermore, the integrated return of the reflected pulse is a function of the rms slope of the water surface for narrow beam radars (*Barrick*, 1968) while the slope of the trailing edge is a function of the amplitude of the resonant Bragg waves for wide-beam radar. Figure 7 illustrates the relationship between the transmitted and reflected pulse shapes for a flat surface and a rough surface.

Hasselmann (1972), *Brown* (1977), and *Hammond et al.* (1977) review the geometry and discuss details of this approach and the latter authors present preliminary results of an aircraft system designed to simultaneously exploit both the leading and trailing edge return shapes to infer significant wave height and windspeed respectively. Figure 8 from *Hammond et al.* (1977) shows calculated and observed return pulse shapes for a range of windspeeds. Preliminary experimental results that they have obtained for low-wave and variable wind conditions show promise for the determination of surface winds. One major drawback to their approach is the requirement for a 60° beam width which necessitates high transmitted power levels and currently limits the operating altitude to less than 1500 meters. Narrow beam systems operating at nadir, however,

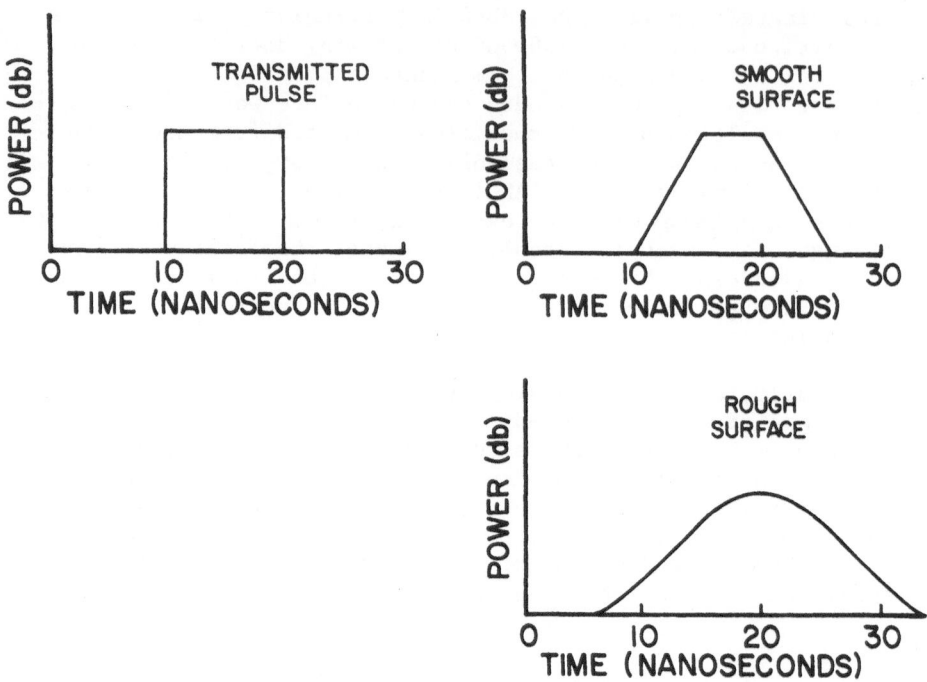

Fig. 7.   Behavior of short-pulse radar vertically incident to a smooth and rough ocean surface.

suffer no power limitations, and studies of the leading edge return pulse shape of the GEOS-3 radar altimeter suggest that significant wave height can be measured to an accuracy of about 1 meter above a threshold of about 2-3 meters (*Alpers and Rufenach,* 1977, *Ross et al.*, 1978). *Parsons* (1978, this volume) explains methods for processing GEOS-3 radar altimeter data to obtain significant wave heights and compares satellite-observed significant wave heights to observations from weather ships, aircraft underflights, and NOAA data buoys. Imposition of parametric techniques such as that of *Hasselmann et al.* (1976), utilizing the significant height and wind-speed information present in this type of data present the possibility of specification of the one-dimensional spectrum in simple situations.

DUAL-FREQUENCY TECHNIQUES

The possibility of remotely observing the complete two-dimensional wave spectrum exists by the use of a two-frequency

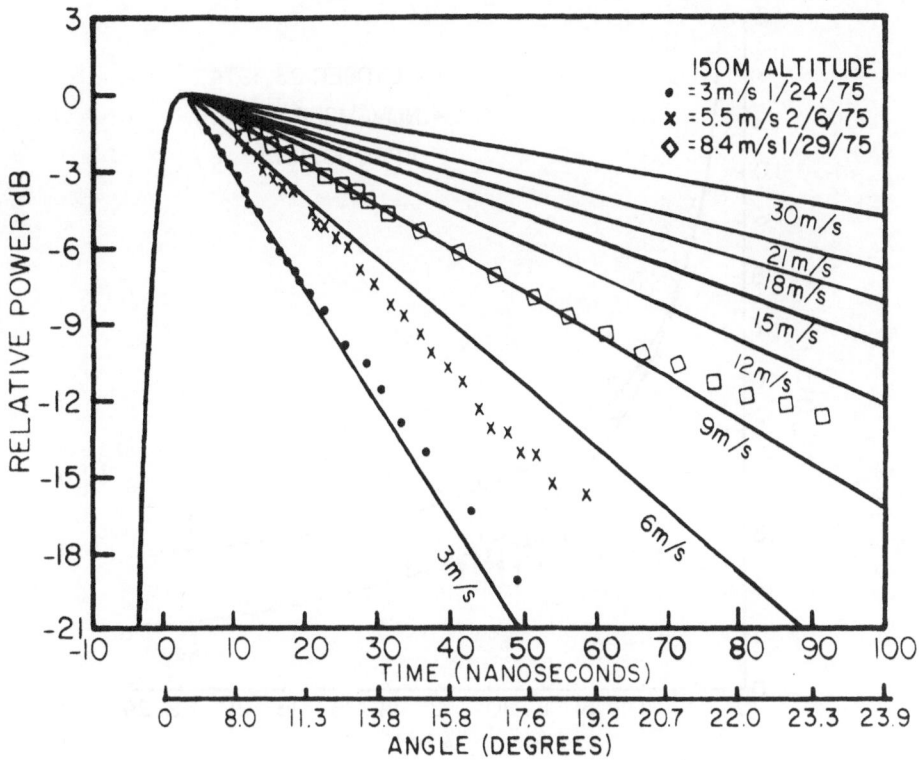

Fig. 8. Comparison of calculated and observed behavior of
trailing edge of pulsed vertical radar showing the effect of
varying windspeeds (after *Hammond et al.*, 1977).

system. In this approach, microwave energy is beamed to the surface
simultaneously at two nearby frequencies. If the reflecting surface
elements are modulated by longer gravity waves, a low-frequency
modulation is present on the reflected microwave energy which can
be processed to yield information on a wave component which is proportional to the difference between the two radiated microwave frequencies. By varying the frequency, incidence angle, and azimuth
angle of the radar, details of the complete two-dimensional wave
spectrum can be obtained if a transfer function properly accounting
for the hydrodynamic interactions between the long waves and the
reflecting Bragg waves is applied.

Alpers and Hasselmann (1977) discuss this technique in detail
and propose a transfer equation which accounts for the hydrodynamic
interactions. Figure 9 from *Plant* (1977) convincingly demonstrates
the presence of wave number information in the two-frequency approach
by comparing the calculated dispersion relationship of the measured

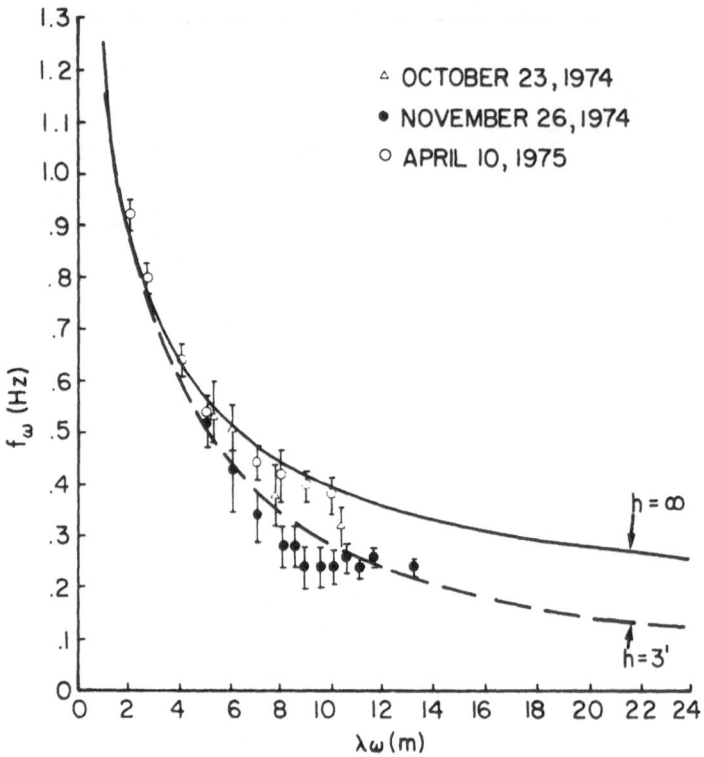

Fig. 9. The gravity wave dispersion relationship calculated theoretically and from dual-frequency measurements. The gravity wave dispersion relationship is $(2\pi f_\omega)^2 = gk \tanh\left(\frac{2\pi h}{\lambda_\omega}\right)$, where h is the water depth (after *Plant*, 1977).

waves to theoretical curves for waves in both deep and shallow water. *Weissman et al.* (1977) presents experimental results from an airborne system which suggests that the correlation function of the two received signals for off-nadir incidence angles is proportional to rms wave height.

The attractive aspect of the two-frequency approach from aircraft would be in an azimuth-scanning version which would allow calculations of the two-dimensional spectrum without need for large data storage and processing capabilities such as in the case of imaging radar. The difficulty of properly specifying a transfer function, however, would seem to suggest considerable research remains before this technique is exploited routinely.

PHOTOGRAPHIC TECHNIQUES

Analysis of vertical photography has been used since the Second World War to study diffraction and refraction patterns of ocean waves in shallow water. The presence of wave information in vertical and near vertical photography is based upon geometric optics and exploits the angular reflection of diffuse skylight in those portions of the photograph not affected by reflections due mainly to the sun (i.e., sun glitter). The wave information may be extracted by processing stereo pairs or by Fourier transforms of single photographs. Results using the former technique are reviewed by *Kinsman* (1965). An optical Fourier analysis technique of off-nadir photography has been proposed by *Stilwell* (1969) and carried to second order by *Kasevich* (1975). Similar analysis techniques applied to vertical photography obtained in the open ocean cited earlier have been compared to imaging radar and measured one-dimensional spectra in Figure 6.

The principal limitation to the photographic techniques is the requirement of uniform skylight from either very clear or continuous overcast skies. The data in Figure 6, for example, was obtained at an altitude of 6000 meters beneath the cirrus outflow of hurricane Ava. This limitation has restricted the application of the photographic approach to specific experiments.

CONCLUSIONS

Airborne and spaceborne platforms have been successfully utilized to obtain valuable one- and two-dimensional wave information in support of a variety of research applications. They provide a means of quickly obtaining specialized data in support of predictive models to generate climatological conditions based upon historical weather patterns. They are, therefore, an important part of a wave monitoring program which must utilize a variety of techniques to obtain wave data required for proper resource management and engineering development.

REFERENCES

Alpers, W. R., and K. Hasselmann 1977. The two-frequency microwave technique for measuring ocean wave spectra from an airplane or satellite, In: *Proc. of a URSI Symposium on Radio Oceanography*, 29 September to 6 October, 1976, Hamburg, FRG, Reidel Publishing Co., Dordrecht, Holland. In press.

Alpers, W. R. and C. L. Rufenach. 1978. Measurements of ocean wave heights using the GEOS-3 altimeter. To appear, *J. Geophys. Res.*

Barnett, T. P. and J. C. Wilkerson. 1967. On the generation of ocean wind waves as inferred from airborne radar measurements of fetch-limited spectra, *J. Mar. Res.*, *25*, 292-328.

Barrick, D. E. 1968. Rough surface scattering based on the specular point theory, *IEEE Trans.*, AP-16, 449.

Bondarenko, I. M., A. A. Zagorodnikov, V. S. Loschchilov, and K. B. Tchelyshev. 1972. The relationship between wave parameters and the spatial spectrum of aerial- and radar-pictures of the sea surface, *Okeanologiya*, Vol. XII, No. 6.

Brown, W. E., Jr., C. Elachi, and T. W. Thompson. 1976. Radar imaging of ocean surface patterns, *J. Geophys. Res.*, *81*, 2657-2667.

Brown, G. S. 1977. The average impulse response of a rough surface and its applications, *IEEE Trans. on Antennas and Propagation*, AP-25, No. 1, 67-74.

Byelousov, P. S., Y. O. Zhilko, A. A. Zagorodnikov, V. I. Korniyenko, B. S. Loschchilov, and K. B. Tchelyshev. 1975. Investigations of wind-driven sea parameters by means of side-looking airborne radar, In: *Proc. of the Final Symposium on the Results of the Joint USSR/USA Bering Sea Experiment*, Leningrad, 12-17 May 1974, K. Ya. Kondratyev, Yu. I. Rabinovich, and Dr. W. Nordberg, editors, Gidrometeoizdat, Leningrad, 68-79.

Hammond, D. L., R. A. Minella, and E. J. Walsh. 1977. Short-pulse radar used to measure sea surface windspeed and significant wave height, *IEEE Trans. on Antennas and Propagation*, AP-25, No. 1, 61-66.

Hasselmann, K. 1972. The energy balance of wind waves and the remote sensing problem, NOAA Tech. Rept. ERL 228-AOML 7-2, 25-1 - 25-55.

Hasselmann, K., D. B. Ross, P. Muller, and W. Sell. 1976. A parametric wave prediction model, *J. Phys. Oceanogr.*, *6*, No. 2, 200-228.

Kasevich, R. S. 1975. Directional wave spectra from daylight scattering, *J. Geophys. Res.*, *80*, 4535-4541.

Kinsman, B. 1965. *Wind Waves*, Prentice Hall, Inc., Englewood, Cliffs, N.J., 675 pp.

Kitaigorodskii, S. A. 1961. Applications of the theory of similarity to the analysis of wind-generated wave motion as a stochastic process, *Isv. Akad. Nauk.*, USSR, Ser. Geofiz., No. 1, 73-80.

Mitsuyasu, H., F. Tasai, T. Suhara, S. Mizuno, M. Ohkusu, T. Honda, and K. Rikiishi. 1975. Observations of the directional spectrum of ocean waves using a cloverleaf buoy, *J. Phys. Oceanogr.*, *5*, No. 4, 750-760.

Morrow, C. M. 1964. Ocean wave profiling systems, Naval Research Laboratory, Report No. 6052.

Parsons, C. 1978. An assessment of GEOS-3 wave height measurements, In: *Ocean Wave Climate*, M. D. Earle and A. Malahoff (Eds.) Plenum, N.Y., (this volume).

Pierson, W. J. 1976. The theory and applications of ocean wave measuring systems at and below the sea surface on the land, from aircraft, and from spacecraft, NASA Contractor Report NASA CR-2646, National Aeronautics and Space Administration, Washington, D.C. 388 pp.

Plant, W. J. 1977. Studies of backscattered radar return with a CW, dual-frequency, X-band radar, *IEEE Trans. on Antennas and Propagation*, AP-25, No. 1, 28-35.

Ross, D. B. 1967. Recent development in remote sensing of deep ocean waves, *Proc. of the Marine Technical Society*, May 1967, 371-393.

Ross, D. B., V. J. Cardone, and J. W. Conaway, Jr. 1970. Laser and microwave observations of sea-surface conditions for fetch-limited 17- to 25-m/s winds, *IEEE Trans.*, Vol. GE-8, No. 4, 326-336.

Ross, D. B. and V. J. Cardone. 1974. Observations of oceanic whitecaps and their relation to remote measurements of surface wind speed, *J. Geophys. Res.*, *79*, No. 3, 444-452.

Ross, D. B., B. Au, W. Brown, and J. McFadden. 1974. A remote sensing study of Pacific hurricane Ava, In: *Proc. of the Ninth International Symposium on Remote Sensing*, Environmental Research Institute of Michigan, Ann Arbor, Michigan, 163-180.

Ross, D. B., V. J. Cardone, P. Black, W. Alpers, and C. Rufenach. 1978. GEOS-3 measurements of significant wave height and wind-speed in hurricane Caroline. In preparation.

Stilwell, D., Jr. 1969. Directional energy spectra of the sea from photographs, *J. Geophys. Res.*, *74*, No. 8, 1974-1986.

Weissman, D. E. and J. W. Johnson. 1977. Dual-frequency correlation radar measurement of the height statistics of ocean waves, *IEEE Trans. on Antennas and Propagation*, AP-25, No. 1, 74-83.

Wright, J. W. 1968. A new model for sea clutter, *IEEE Trans. on Antennas and Propagation*, AP-16, 217-223.

Directional Wave Spectra from Wave Sensors

Leon E. Borgman

University of Wyoming

ABSTRACT

The measurement of directional ocean wave spectra with in situ instruments in the ocean environment requires arrays consisting of relatively few sensors. More elaborate systems are difficult and costly to maintain.

A unified mathematical procedure to treat sparse arrays of sensors is presented. The basic starting point for the analysis is a system of integral equations which must be inverted to estimate the directional spectral density at each frequency. The system is reduced to a set of linear equations if a Fourier series for the spreading function on all or a part of the circle is introduced. Various parameterized models may be fitted using the Fourier coefficients of the series. A sequential estimate enhancement procedure involving the higher harmonics of the fitted model and the lower order harmonics included in the linear system of equations is outlined.

The formulas for a variety of particular sensor arrays are summarized in the Appendix. In addition, some sample computations showing realistic, although artificial, data are presented.

INTRODUCTION

The constraints of cost and operational difficulty in oceanic measurements severely limit most data arrays. A problem which would be very simple and straightforward if 100 sensors were available,

becomes much more difficult if only five or fewer sensors are
provided. Hence, the basic problem is to overcome these limitations
and to obtain directional estimates with reasonable resolution from
arrays requiring relatively few measuring devices.

It is probably unrealistic to expect more than a mean
direction and a measure of dispersion about that direction from ·
most arrays. If the directional distribution of wave energy is
unimodal, various two-parameter models such as the generalized
cosine-squared, the von Mises, or the wrapped-normal formulas can
be used. The parameters for these models can be estimated from
the lower order Fourier coefficients of the directional distribution at each frequency.

Skewed distributions, bimodal distributions or wave trains
with discrete directions cause more difficulty. These can usually
be handled if it is known they are needed, provided enough sensors
are available.

A good review of the procedures that have been used to make
directional spectra estimates is given by *Panicker* (1974) and more
recently by *Rikiishi* (1977). Some typical arrays that have been
suggested are (1) a floating buoy which measures water level elevations and the x- and y-components of the water slope (*Longuet-
Higgins, et al.*, 1963), (2) a clover-leaf of buoys (*Cartwright and
Smith*, 1964; *Ewing*, 1969; *Mitsuyasu, et al.*, 1974), (3) a line of
water level elevations gages (*Cummins*, 1959; *Stevens*, 1965;
Hasselman, et al., 1973; *Lowe, et al.*, 1972), (4) polygonal and
star arrays of wave staffs (*Mobarek*, 1965; *Chakrabarti*, 1971;
Chakabarti and Snider, 1973, 1974; *Panicker and Borgman*, 1970;
Borgman, 1969; *Ploeg*, 1972), (5) x- and y-components of the force
on a sphere together with bottom pressure fluctuations (*Suzuki*,
1969), (6) a wave staff and x- and y-components of the horizontal
water particle velocities at some depth near the staff (*Nagata*,
1964; *Bowden and White*, 1966; *Simpson*, 1969).

All of the different methods can be set into a single unified
computational pattern. This pattern is established in the following discussion. Some methods are presented for sharpening the
resolution of the estimates. The basic equations are developed in
the text. Several specific examples are outlined in more detail
in Appendix A. These include (a) a triangular space array of water
level elevation measurements, (b) a line array of water level
elevation measurements, (c) measurements of wave height and horizontal water particle velocity components, and (d) measurements of
water level elevation and components of surface slope with a
floating buoy. The formulas for several of these sensor configurations are accompanied in the Appendix with computational examples
taken from data and from computer simulations.

The starting point or foundation for the unified formulation are the spectra and cross-spectra for and between the various sensors. The estimation of these functions is not discussed here. However, it is worthwhile to mention the significant new developments in data-adaptive spectral estimation whereby substantially better estimates appear to be produced (*Capon*, 1969; *Burg*, 1972; *Lacoss*, 1971; *Ulrych*, 1972). These procedures deserve continuing attention and study.

The data-adaptive procedures represent one way to enhance spectral estimates. They seem to be particularly appropriate for conditions where a spectral line is present (waves at a single frequency from a single direction). Another enhancement procedure which appears to be suitable for a continuum of wave directions spread over a small angular sector is developed in the following discussion. The procedure is sequential in nature. (1) A first estimate is made of the full-circle Fourier coefficients for the spreading function. (2) A two-parameter model is fitted to the lower order coefficients. (3) On the basis of the model parameters, the angular sector where the spreading function is essentially zero is established. (4) A new Fourier series expansion on the sector where the spreading function is nonzero is made. The cross-spectra and spectra are used to estimate the lower order coefficients in the sub-circle expansion. (5) A low parameter model is fitted to the new Fourier coefficients. (6) The tail of the Fourier series (higher-order coefficients) are established from the fitted model. (7) The lower order Fourier coefficients are re-estimated and again fitted with the model and (8) steps 6 and 7 repeated again and again until the coefficient estimates stabilize.

An example of one procedure for establishing the higher order Fourier coefficients using the characteristic equation is given in Appendix B. The procedure is illustrated with the fit of a normal probability function.

A STANDARD FORMULATION

Some of the linear wave properties that have been used in the estimation of directional spectra are (1) the water level elevation, η, (2) the x- and y-components of the water surface slope, η_x and η_y respectively, (3) bottom pressure fluctuations, p, about static pressure, and (4) the x- and y-components of the horizontal water particle velocities, v_x and v_y. The basic linear formulas for these properties are given below. The waves are listed as traveling in a direction θ measured counterclockwise from the positive x-axis.

$$\eta = \alpha \cos(kx \cos\theta + ky \sin\theta - 2\pi ft + \phi)$$

$$\eta_x = -\alpha k \cos\theta \sin(kx \cos\theta + ky \sin\theta - 2\pi ft + \phi)$$

$$\eta_y = -\alpha k \sin\theta \sin(kx \cos\theta + ky \sin\theta - 2\pi ft + \phi)$$

$$p = \alpha pg \frac{\cosh k(d-z)}{\sinh kd} \cos(kx \cos\theta + ky \sin\theta - 2\pi ft + \phi)$$

$$v_x = \alpha 2\pi f \frac{\cosh k(d-z)}{\sinh kd} \cos\theta \cos(kx \cos\theta + ky \sin\theta - 2\pi ft + \phi)$$

$$v_y = \alpha 2\pi f \frac{\cosh k(d-z)}{\sinh kd} \sin\theta \cos(kx \cos\theta + ky \sin\theta - 2\pi ft + \phi)$$

$$\tag{1}$$

In the above, a right-hand coordinate system, with z positive upward and measured from mean water level, is used. The wave number k = 2π/wave length and frequency (cycles per second) are usually taken as being interrelated by the formula

$$(2\pi f)^2 = gk \tanh kd \tag{2}$$

where d is mean water depth and g is the acceleration due to gravity.

Let P denote an arbitrary linear wave property. All of the above formulas can be put into the standard form

$$P = \alpha K(\substack{\cos \\ \sin})(kx \cos\theta + ky \sin\theta - 2\pi ft + \phi) \tag{3}$$

where K is the appropriate function of f, z, d, and θ.

A representation of a confused sea with an intermixture of waves from a variety of directions with various frequencies can be stated as

$$P(x,y,z,t) =$$

$$\tag{4}$$

$$\sum_{m=1}^{M} \sum_{j=1}^{J} \alpha_{mj} K(f_m, z, d, \theta_j)(\substack{\cos \\ \sin})(k_m x \cos\theta_j + k_m y \sin\theta_j - 2\pi f_m t + \phi_{mj})$$

This representation can be made statistical if the phases, ϕ_{mj}, are taken as being uniformly distributed at random over the interval $(0, 2\pi)$ and independent of each other. The amplitude, α_{mj}, may be interrelated to the (two-sided) spectral density, S, of η by

$$\alpha_{mj} = 2\sqrt{S(f_m,\theta_j)\Delta f\Delta\theta} \tag{5}$$

where Δf and $\Delta\theta$ are increments on the frequency and angle axes for the intervals "occupied" by the (m,j) wave component.

STATISTICAL REPRESENTATIONS

For all the forms in Equation (1) the statistical expectation will be zero. The space-time covariance function for two wave properties, P_1 and P_2, is defined as

$$C_{12}(X,Y,\tau) = E[P_1(x,y,z,t)P_2(x+X, y+Y, z, t+\tau)] \tag{6}$$

where $E[\cdot]$ denotes the expectation operator.

For P_1 and P_2 either both cosine-type functions or sine-type functions

cos-cos or sin-sin.

$$C_{12}(X,Y,\tau) = 2\sum_{m=1}^{M}\sum_{j=1}^{J} S(f_m,\theta_j)K_1(f_m,z,d,\theta_j)K_2(f_m,z,d,\theta_j)$$
$$\cdot \cos(k_m X \cos\theta_j + k_m Y \sin\theta_j - 2\pi f_m\tau)\Delta\theta\Delta f \tag{7}$$

Similarly, if P_1 is a sine-type property and P_2 is a cosine-type

sin-cos.

$$C_{12}(X,Y,\tau) = 2\sum_{m=1}^{M}\sum_{j=1}^{J} S(f_m,\theta_j)K_1(f_m,z,d,\theta_j)K_2(f_m,z,d,\theta_j)$$
$$\cdot \sin(k_m X \cos\theta_j + k_m Y \sin\theta_j - 2\pi f_m\tau)\Delta\theta df \tag{8}$$

For P_1 a cosine-type property and P_2 a sine-type, $C_{12}(X,T,\tau)$ would be the same as Equation (8) except the sign would be reversed.

cos-sin.

$$C_{12}(X,Y,\tau) = -2\sum_{m=1}^{M}\sum_{j=1}^{J} S(f_m,\theta_j)K_1(f_m,z,d,\theta_j)K_2(f_m,z,d,\theta_j)$$
$$\cdot \sin(k_m X \cos\theta_j + k_m Y \sin\theta_j - 2\pi f\tau)\Delta\theta\Delta f \tag{9}$$

If P_1 and P_2 are actually the same property at the same location, then $X=0$, $Y=0$, and the pair are either sin-sin or cos-cos in type. Consequently, the usual covariance function may be written

$$C_{pp}(\tau) = 2 \sum_{m=1}^{M} \sum_{j=1}^{J} S(j_m, \theta_j) K^2(f_m, z, d, \theta_j) \cos(2\pi f \tau) \Delta\theta \Delta f \qquad (10)$$

A much less cluttered form results if Equations (7) through (10) are stated as integrals with the understanding that one can revert to the discrete formulations for computations with the fast Fourier transform. The integral formulations can also be regarded as the limit in quadratic mean as $\Delta f \to 0$ and $\Delta\theta \to 0$. Thus Equations (7) through (10) may be rewritten with some algebraic expansion as

cos-cos or sin-sin.

$$C_{12}(X,Y,\tau) =$$

$$2 \int_0^\infty \left[\int_0^{2\pi} S(f,\theta) K_1 K_2 \cos(kX\cos\theta + kY\sin\theta) d\theta \right] \cos 2\pi f \tau df$$

$$(11)$$

$$+ \int_0^\infty \left[\int_0^{2\pi} S(f,\theta) K_1 K_2 \sin(kX\cos\theta + kY\sin\theta) d\theta \right] \sin 2\pi f \tau df$$

sin-cos.

$$C_{12}(X,Y,\tau) =$$

$$2 \int_0^\infty \left[\int_0^{2\pi} S(f,\theta) K_1 K_2 \sin(kX\cos\theta + kY\sin\theta) d\theta \right] \cos 2\pi f \tau df$$

$$(12)$$

$$+ 2 \int_0^\infty \left[-\int_0^{2\pi} S(f,\theta) K_1 K_2 \cos(kX\cos\theta + kY\sin\theta) d\theta \right] \sin 2\pi f \tau df$$

cos-sin.

$$C_{12}(X,Y,\tau) =$$

$$2 \int_0^\infty \left[-\int_0^{2\pi} S(f,\theta) K_1 K_2 \sin(kX\cos\theta + kY\sin\theta) d\theta \right] \cos 2\pi f \tau df \qquad (13)$$

$$+ 2 \int_0^\infty [\int_0^{2\pi} S(f,\theta)K_1 K_2 \cos(kX\cos\theta + kY\sin\theta)d\theta]\sin2\pi f\tau \, df \qquad (13)$$

$$C_{pp}(\tau) = 2 \int_0^\infty [\int_0^{2\pi} S(f,\theta)K_1^2 \, d\theta]\cos2\pi f\tau \, df \qquad (14)$$

The theoretical relation between covariance functions and spectral densities is expressed by the Fourier integral.

$$\left. \begin{array}{l} C_{pp}(\tau) = \int_{-\infty}^\infty S_{pp}(f) \, e^{i2\pi f\tau} \, df \\[3mm] S_{pp}(f) = \int_{-\infty}^\infty C_{pp}(\tau) \, e^{-i2\pi f\tau} \, d\tau \end{array} \right\} \qquad (15)$$

$$\left. \begin{array}{l} C_{12}(\tau) = \int_{-\infty}^\infty S_{12}(f) \, e^{+i2\pi f\tau} \, df \\[3mm] S_{12}(f) = \int_{-\infty}^\infty C_{12}(\tau) \, e^{-i2\pi f\tau} \, d\tau \end{array} \right\} \qquad (16)$$

Here S with a single argument is understood to be the usual frequency spectra while S with two arguments denotes the directional spectra. The arguments, X and Y, in C_{12} are suppressed if they are regarded as fixed constants.

From basic definitions the following symmetries will hold if the sea surface is second-order, covariance-stationary.

$$\left. \begin{array}{l} C_{pp}(-\tau) = C_{pp}(\tau), \quad S_{pp}(-f) = S_{pp}(f) \\[3mm] C_{12}(-\tau) = C_{21}(\tau), \quad S_{12}(-f) = S_{12}(f) = S_{21}(f) \end{array} \right\} \qquad (17)$$

If the definitions of the exponentials in terms of sines and cosines are inserted and the symmetries are utilized, Equations (16) through (17) become

$$
\left.
\begin{aligned}
C_{pp}(\tau) &= 2 \int_0^\infty S_{pp}(f) \cos(2\pi f\tau)df \\
C_{12}(\tau) &= 2 \int_0^\infty c_{12}(f) \cos(2\pi f\tau)df + 2 \int_0^\infty q_{12}(f)\sin(2\pi f\tau)df
\end{aligned}
\right\} \quad (18)
$$

where

$$
S_{12}(f) = c_{12}(f) - iq_{12}(f) .
$$

The functions $c_{12}(f)$ and $q_{12}(f)$ are called the co- and quad-spectral densities, respectively, and have the symmetries

$$
\left.
\begin{aligned}
c_{12}(f) &= c_{21}(f) = c_{12}(-f) = c_{21}(-f) \\
q_{12}(f) &= -q_{21}(f) = -q_{12}(-f) = q_{21}(-f)
\end{aligned}
\right\} \quad (19)
$$

The co- and quad-spectral densities for wave properties P_1 and P_2 (in that order) can be immediately written by comparison of Equations (11) through (14) with Equation (18).

cos-cos or sin-sin.

$$
\left.
\begin{aligned}
c_{12}(f) &= \int_0^{2\pi} S(f,\theta)K_1K_2\cos(kX\cos\theta + kY\sin\theta)d\theta \\
q_{12}(f) &= \int_0^{2\pi} S(f,\theta)K_1K_2\sin(kX\cos\theta + kY\sin\theta)d\theta
\end{aligned}
\right\} \quad (20)
$$

sin-cos.

$$
\left.
\begin{aligned}
c_{12}(f) &= \int_0^{2\pi} S(f,\theta)K_1K_2\sin(kX\cos\theta + kY\sin\theta)d\theta \\
q_{12}(f) &= -\int_0^{2\pi} S(f,\theta)K_1K_2\cos(kX\cos\theta + kY\sin\theta)d\theta
\end{aligned}
\right\} \quad (21)
$$

cos-sin.

$$
c_{12}(f) = -\int_0^{2\pi} S(f,\theta)K_1K_2\sin(kX\cos\theta + kY\sin\theta)d\theta
$$

$$
q_{12}(f) = \int_0^{2\pi} S(f,\theta)K_1K_2\cos(kX\cos\theta + kY\sin\theta)d\theta
$$

(22)

$$
S_{pp}(f) = \int_0^2 S(f,\theta)K_1^2\, d\theta
$$

(23)

A UNIFIED APPROACH

Equations (20) through (23) form the basis of a simple straight-forward approach to estimating the directional wave spectra from an array of wave sensors. The spectral density for each sensor is estimated by usual procedures and substituted for the left side of Equation (23). Similarly, the co- and quad-spectra are estimated by each pair of wave gages and placed in the left-hand terms in Equation (22). If there are N sensors, there will be N equations of the form (23), and N(N-1)/2 pairs of equations of the form given in Equation (22). This represents a set of simultaneous integral equations which need to be solved for $S(f,\theta)$ by some means.

Thus, the estimation problem reduces to the inversion of a system of equations of the form

$$
\hat{c}_{12}(f) = \pm\int_0^{2\pi} S(f,\theta)K_1K_2\left(\genfrac{}{}{0pt}{}{\cos}{\sin}\right)(kX\cos\theta + kY\sin\theta)d\theta
$$

(24)

$$
\hat{q}_{12}(f) = \pm\int_0^{2\pi} S(f,\theta)K_1K_2\left(\genfrac{}{}{0pt}{}{\sin}{\cos}\right)(kX\cos\theta + kY\sin\theta)d\theta
$$

(25)

$$
\hat{S}_{pp}(f) = \int_0^{2\pi} S(f,\theta)K_p^2\, d\theta
$$

(26)

where the "^" denotes estimates of the functions from data.

PARAMETERIZATION OF S(f,θ)

Due to the cost and difficulty of maintaining instruments in the oceanic environment, arrays used for directional spectra estimation are quite sparse. Representations of $S(f,\theta)$ which involve only a few parameters are needed. Since completed general representations require many parameters, such representations are useful only in an initial step which gives preliminary indications of the proper modelling of $S(f,\theta)$.

Any collateral information indicating preferred choices for a model of $S(f,\theta)$ should be used to augment the data measurements. Thus, observations of aerial photographs may suggest that two unidirectional wave trains were involved. Then a model of the form

$$S(f,\theta) \quad = \quad S(f)[a_f\delta(\theta-\theta_1) + b_f\delta(\theta-\theta_2)] \tag{27}$$

might be used. Here a_f and b_f are coefficients which depend on f and $\delta(\theta)$ is the Dirac-delta function.

Waves from hurricanes seem to be dispersed over a relatively narrow sector of directions at a given time so that a model of the following form might be used.

$$S(f,\theta) \quad = \quad S(f)D_f(\theta) \tag{28}$$

with some choices for $D_f(\theta)$ being

von Mises.

$$D_f(\theta) \quad = \quad \frac{e^{a_f\cos(\theta-\mu_f)}}{2\pi I_0(a_f)} \tag{29}$$

wrapped normal.

$$D_f(\theta) \quad = \quad \begin{cases} \sum\limits_{k=1}^{1} \dfrac{e^{-(\theta-\mu-2\pi k)^2/(2\sigma^2)}}{\sqrt{2\pi}\ \sigma} \ , & \text{for} \quad \sigma < \dfrac{\pi}{3} \\[2em] \dfrac{1}{2\pi} + \sum\limits_{n=1}^{5} e^{-n^2\sigma^2/2} \cos(n\theta-n\mu), & \text{for} \quad \sigma > \dfrac{\pi}{3} \end{cases} \tag{30}$$

generalized cosine-squared.

$$D_f(\theta) \quad = \quad R\cos^{2\alpha}(\tfrac{\theta-\mu}{2}) \tag{31}$$

where

$$I_0(a) \ = \ \text{modified Bessel function of order zero}$$

$$R \ = \ \text{constant such that } \int_0^{2\pi} D_f(\theta)d\theta = 1.0 \ .$$

It is easy to come up with additional models combining several of these to yield unidirectional "lines" and multimodel angular distribution.

A convenient first parameterization of $S(f,\theta)$ is by the following steps.

(1) The frequency spectral density is estimated for each of the sensors and a combined best estimate of $\hat{S}(f)$ is obtained by

$$\hat{S}(f) \ = \ \left[\prod_{i=1}^{n} S_i(f) \right]^{1/n} \tag{32}$$

(i.e., the geometric mean).

(2) The spreading function, $D_f(\theta)$, is expanded in a Fourier series

$$D_f(\theta) \ = \ \frac{1}{2\pi} + \sum_{n=1}^{N} [a_n(f)\cos n\theta + b_n(f)\sin n\theta] \tag{33}$$

These expressions are substituted into Equations (24), (25), and (26). The coefficients $(\hat{a}_1, \hat{b}_1; \hat{a}_2, \hat{b}_2; \ldots; \hat{a}_N, \hat{b}_N)$ are obtained as least square solutions to the system of linear equations resulting from the substitution after integration. As large an integer N is used as possible consistent with the production of stable estimates.

The full-circle Fourier expansion has several disadvantages. The series must be truncated with relatively small N in order to develop a solvable system of linear equations. Secondly, the usual $D_f(\theta)$ function occurring in nature is zero for at least half of the circle of directions. A Fourier series which even approximately forces zeros on half of its period must have many coefficients. Thus, the Fourier series representation is not a model with just a few parameters.

SUB-CIRCLE EXPANSION OF $D_f(\theta)$

Often it may be known in advance that appreciable wave energy will be present only for $\theta_0 \leq \theta \leq \theta_0 + T$. This might happen, for example, if waves are refracted into a coastal measuring site so that the wave directions are all $\pm 20°$ from a perpendicular to the coast. In any case, it is usually known that the wave energy is all arriving on one-half of the directional circle. The back half would then have negligible energy.

The Fourier expansion of $D_f(\theta)$ can then be made over the interval $(\theta_0, \theta_0 + T)$.

$$D_f(\theta) = \frac{1}{T} + \sum_{n=1}^{\infty}\left[a_n(f)\cos\frac{2\pi n(\theta-\theta_0)}{T} + b_n(f)\sin\frac{2\pi n(\theta-\theta_0)}{T}\right] \quad (34)$$

for $\theta_0 \leq \theta \leq \theta_0 + T$. This equation reduces to Equation (33) if $\theta_0 = 0$ and $T = 2\pi$. The expansion in Equation (34) will require substantially fewer coefficients for a good fit than will that in Equation (33) because Equation (34) is not trying to force zeros on a major part of the circle. For example, suppose that $D_f(\theta) = \cos^2(3\theta)$ for $|\theta| < \pi/6$ and zero otherwise. This cosine-squared "bump" can be represented on the interval $(-\pi/6, \pi/6)$ with the two-coefficient series, $0.5 - 0.5\cos[2\pi(\theta+\pi/6)/T]$, where $T = \pi/3$. On the other hand, a Fourier series expansion on $(-\pi, \pi)$ would require many more coefficients in order to approximate the condition that $D_f(\theta) = 0$ for $\pi/6 < |\theta| < \pi$.

THE LINEAR SYSTEM OF EQUATIONS

The equations to be solved for the sub-circle Fourier coefficients are a set taken from the possible forms

$$\frac{\hat{c}_{12}(f)}{\hat{S}(f)} = \frac{1}{T}\int_{\theta_0}^{\theta_0+T} K_1 K_2 \binom{\cos}{\sin}(kX\cos\theta + kY\sin\theta)d\theta \quad (35)$$

$$+ \sum_{n=1}^{\infty} \hat{a}_n(f)\int_{\theta_0}^{\theta_0+T} K_1 K_2 \binom{\cos}{\sin}(kX\cos\theta + kY\sin\theta)\cos\frac{2\pi n(\theta-\theta_0)}{T}d\theta$$

$$+ \sum_{n=1}^{\infty} \hat{b}_n(f)\int_{\theta}^{\theta_0+T} K_1 K_2 \binom{\cos}{\sin}(kX\cos\theta + kY\sin\theta)\sin\frac{2\pi n(\theta-\theta_0)}{T}d\theta$$

$$\pm \; \frac{\hat{q}_{12}(f)}{\hat{S}(f)} \;\; = \;\; \frac{1}{T} \int_{\theta_0}^{\theta_0+T} K_1 K_2 \binom{\sin}{\cos} (kX\cos\theta + kY\sin\theta) d\theta \tag{36}$$

$$+ \; \sum_{n=1}^{\infty} \hat{a}_n(f) \int_{\theta_0}^{\theta_0+T} K_1 K_2 \binom{\sin}{\cos} (kX\cos\theta + kY\sin\theta) \cos \frac{2\pi n(\theta-\theta_0)}{T} d\theta$$

$$+ \; \sum_{n=1}^{\infty} \hat{b}_n(f) \int_{\theta_0}^{\theta_0+T} K_1 K_2 \binom{\sin}{\cos} (kX\cos\theta + kY\sin\theta) \sin \frac{2\pi n(\theta-\theta_0)}{T} d\theta$$

$$\frac{\hat{S}_{pp}(f)}{\hat{S}(f)} \;\; = \;\; \frac{1}{T} \int_{\theta_0}^{\theta_0+T} K_p^2 d\theta + \sum_{n=1}^{\infty} \hat{a}_n(f) \int_{\theta_0}^{\theta_0+T} K_p^2 \cos \frac{2\pi n(\theta-\theta_0)}{T} d\theta$$

$$+ \; \sum_{n=1}^{\infty} \hat{b}_n(f) \int_{\theta_0}^{\theta_0+T} K_p^2 \sin \frac{2\pi n(\theta-\theta_0)}{T} d\theta \tag{37}$$

For $\theta_0 = 0$ and $T = 2\pi$, the integrals all reduce to Bessel functions after some algebra (*Borgman*, 1969; *Panicker and Borgman*, 1970). Otherwise, the integrals can be evaluated by numerical integration, or, in many cases, explicitly.

Examples of these equations are given in the Appendix A for several typical cases. The equations achieve a particularly simple form when all the sensors are at the same (x,y) location, as for a buoy and for a (η, v_x, v_y) triple of sensors with the velocity gage nearly at the same location as the wave gage.

In any case, the sub-circle expansion reduces to a list of equations all of the form

$$\hat{U}(f) \;\; = \;\; A_0(f) + \sum_{n=1}^{\infty} [\hat{a}_n(f)A_n(f) + \hat{b}_n B_n(f)] \tag{38}$$

If the coefficients $a_n(f)$ and $b_n(f)$ are negligible for $n > N$, then Equation (38) can be rewritten as

$$\sum_{n=1}^{N} [\hat{a}_n(f)A_n(f) + \hat{b}_n(f)B_n(f)] \;\; = \;\; \hat{U}(f) \; - \; A_0(f) \tag{39}$$

The estimates of $a_n(f)$ and $b_n(f)$ can be obtained by least-square procedures if N is not too large.

A SEQUENTIAL ESTIMATION PROCEDURE

If the series is truncated too soon, Equation (39) should be written as

$$\sum_{n=1}^{N} [\hat{a}_n(f)A_n(f) + \hat{b}_n(f)B_n(f)] = \hat{U}(f) - A_0(f) - \sum_{n>N} [\hat{a}_n(f)A_n(f)$$

$$(40)$$

$$+ \hat{b}_n(f)B_n(f)]$$

Presumably, that part of the Fourier series which is on the right-hand side of the equation is relatively small compared with the terms on the left side of the equal sign. Consequently, $\{\hat{a}_n(f), \hat{b}_n(f), 1 \leq n \leq N\}$ could be estimated assuming $\{\hat{a}_n(f), \hat{b}_n(f)\} = \{0,0\}$ for $n > N$. Then, one of the parameterized models could be fitted using $\{a_n(f), b_n(f); 1 \leq n \leq N\}$ and $\{\hat{a}_n(f), \hat{b}_n(f); n > N\}$ could be computed for that fit. Equation (40) could be used to re-estimate $\{a_n(f), b_n(f); 1 \leq n \leq N\}$. The higher order coefficients could be recomputed, substituted into Equation (40), and still another estimate of the low order coefficients obtained. This cycle could be repeated until the process converged.

The advantages of this procedure are that the tail would be provided by a model fit, while the lower order coefficients would allow some departures from the overall model in terms of skewness, bimodality, and so forth.

Initially, the process could start with an estimate of the full circle Fourier coefficients. These would be used to estimate a preliminary set of values for the model parameters. Ordinarily, the dispersion parameter in the model would indicate that $D_f(\theta) \approx 0$ for a substantial sector of the directional circle. A sub-circle expansion would then be made on the remaining set of directions. The sequential procedure as indicated in Equation (40) and the subsequent discussion would then be followed to obtain the improved estimates of $D_f(\theta)$.

A reasonable model for the completion of the higher order Fourier coefficients is given in Appendix B.

SUMMARY

The statistical theory of linear waves leads to a system of
integral equations for a specified array of wave sensors. The
basic unknown is the directional spectral density at each frequency, which is contained in the kernal of the integral in each
equation. A unified standard procedure is given for developing
the integral equations assocated with any particular pair of wave
sensors.

In principle, there are two real-valued equations for each
pair. These are the co-spectral equations, and the quad-spectral
equation. In addition, there is a frequency spectral density
equation associated with each sensor. Thus, there would be n^2
real equations for an n-gage array. In practice, some of these
equations may be theoretically zero. As a result there are usually
fewer than n^2 equations in the actual system.

The system of integral equations can be solved for the
directional spectrum by several procedures. The method outlined
in the text is based on representing the directional spectrum as a
frequency spectrum times a unit area frequency-dependent spreading
function. The frequency spectrum may be estimated easily from
standard procedures. If the spreading function in the integral
equation is expanded in a Fourier series over all directions or
some selected sector of directions, the integral equations reduce
to a linear system solvable for the Fourier coefficients by least-
squares procedures.

If a sufficient number of gage pairs are available, the set of
estimated Fourier coefficients may provide all non-negligible coefficients for the spreading function, and thus give a complete
solution. Usually though, the sparseness of the sensor array
requires the deletion of non-negligible coefficients from the system
of equations. Hence, the truncated Fourier series produced from the
solution is typically a distorted version of the spreading function.

The spreading function estimate may be improved or enhanced by
further processing. On the supposition that the first few harmonics
are the least distorted in the truncation, they can be used to fit
a subjectively selected, parameterized shape formula to the spreading function. A variety of such formulas or models are available.
A few of the standard unimodal types are given in the text.

Any particular model fitted to the low-order Fourier coefficients of the spreading function implies values for the high-
order coefficients truncated out in the original reduction to a
linear equation system. Consequently, the high-order Fourier
coefficients consistent with the fitted model can be inserted into

the linear system and the low-order coefficients recomputed. The
model is refitted and the cycle continues until convergence occurs.

     The equations for various examples of sensor arrays are given
in Appendix A. A sub-circle wrapped-normal model is outlined in
Appendix B. A set of example calculations based on realistic, but
artificial, data is given in Appendix C.

APPENDIX A

SOME EXAMPLES

(See Equations (20), (21), and (22) for background formulas.)

1. <u>Space array of water level elevations.</u>

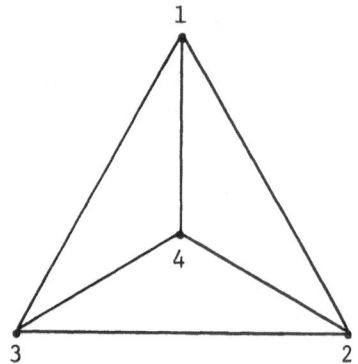

<u>Basic measurements</u>: (x_i, y_i, t) for $i=1,2,3,4$

<u>Fundamental equations</u>:

$$\hat{S}(f) = \sqrt[4]{\hat{S}_1(f)\ \hat{S}_2(f)\ \hat{S}_3(f)\ \hat{S}_4(f)} \tag{A-1}$$

$$\frac{\hat{c}_{ij}}{\hat{S}(f)} = \int_0^{2\pi} D_f(\theta)\ \cos(kX_{ij}\cos\theta + kY_{ij}\sin\theta)\,d\theta \tag{A-2}$$

$$\frac{\hat{q}_{ij}}{\hat{S}(f)} = \int_0^{2\pi} D_f(\theta) \sin(kX_{ij}\cos\theta + kY_{ij}\sin\theta)d\theta \qquad \text{(A-3)}$$

$$\text{for } 1 \leq i \leq j \leq 4$$

Sub-circle expansion:

$$\frac{\hat{c}_{ij}}{\hat{S}(f)} = \frac{1}{T}\int_{\theta_0}^{\theta_0+T} \cos(kX_{ij}\cos\theta + kY_{ij}\sin\theta)d\theta$$

$$+ \sum_{n=1}^{N} \hat{a}_n(f) \int_{\theta_0}^{\theta_0+T} \cos(kX_{ij}\cos\theta + kY_{ij}\sin\theta)\cos\frac{2\pi n(\theta-\theta_0)}{T} d\theta$$

$$+ \sum_{n=1}^{N} \hat{b}_n(f) \int_{\theta_0}^{\theta_0+T} \cos(kX_{ij}\cos\theta + kY_{ij}\sin\theta)\sin\frac{2\pi n(\theta-\theta_0)}{T} d\theta$$

$$\text{(A-4)}$$

$$\hat{q}_{ij}(f)/\hat{S}(f) =$$

$$\frac{1}{T}\int_{\theta_0}^{\theta_0+T} \sin(kX_{ij}\cos\theta + kY_{ij}\sin\theta)d\theta$$

$$+ \sum_{n=1}^{N} \hat{a}_n(f) \int_{\theta_0}^{\theta_0+T} \sin(kX_{ij}\cos\theta + kY_{ij}\sin\theta)\cos\frac{2\pi n(\theta-\theta_0)}{T} d\theta \quad \text{(A-5)}$$

$$+ \sum_{n=1}^{N} \hat{b}_n(f) \int_{\theta_0}^{\theta_0+T} \sin(kX_{ij}\cos\theta + kY_{ij}\sin\theta)\sin\frac{2\pi n(\theta-\theta_0)}{T} d\theta$$

(Note: If $\theta = \theta_0$ and $T = 2\pi$, the integrals reduce to Bessel functions (*Panicker and Borgman*, 1974.))

2. <u>Line array of water level elevations.</u>

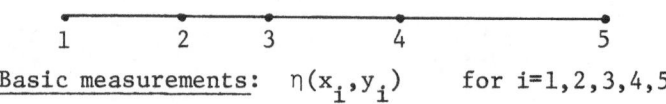

Basic measurements:$\quad \eta(x_i, y_i) \quad$ for $i = 1,2,3,4,5$

<u>Fundamental equations:</u> same as for Example (1).

3. <u>Wave staff and velocity meters.</u>

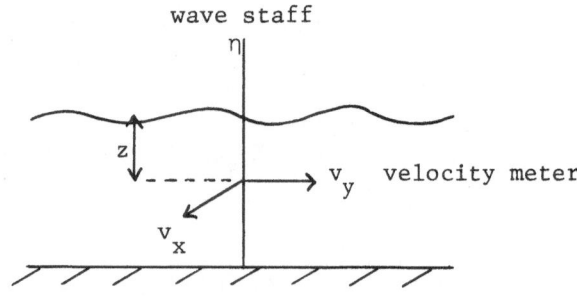

<u>Basic measurements:</u> $\eta(t)$, $v_x(z,t)$, $v_y(z,t)$ all at the same space position (i.e., $X_{ij}=0$, $Y_{ij}=0$ for all sensor pairs)

<u>Fundamental equations:</u>

$\hat{S}(f)$ estimated from $\eta(t)$ data

$$\frac{\hat{S}_{v_x v_x}(f)}{\hat{S}(f)} = (2\pi f)^2 \frac{\cosh^2 k(d-z)}{\sinh^2 kd} \int_0^{2\pi} D_f(\theta) \cos^2\theta \, d\theta \qquad (A\text{-}6)$$

$$\frac{\hat{S}_{v_y v_y}(f)}{\hat{S}(f)} = (2\pi f)^2 \frac{\cosh^2 k(d-z)}{\sinh^2 kd} \int_0^{2\pi} D_f(\theta) \sin^2\theta \, d\theta \qquad (A\text{-}7)$$

The sum of these last two equations gives a natural estimate of the attenuation term.

$$\frac{\hat{S}_{v_x v_x}(f) + \hat{S}_{v_y v_y}}{\hat{S}(f)} = \frac{(2\pi f)^2 \cosh^2 k(d-z)}{\sinh^2 kd} = [\hat{R}(f)]^2 \qquad (A\text{-}8)$$

(where $\hat{R}(f)$ is defined by the above expression).

The cross spectral terms are

$$\frac{\hat{c}_{\eta v_x}(f)}{\hat{S}(f)\hat{R}(f)} = \int_0^{2\pi} D_f(\theta) \cos\theta \ d\theta \tag{A-9}$$

$$\hat{q}_{\eta v_x}(f) = 0$$

$$\frac{\hat{c}_{\eta v_y}(f)}{\hat{S}(f)\hat{R}(f)} = \int_0^{2\pi} D_f(\theta) \sin\theta \ d\theta \tag{A-10}$$

$$\hat{q}_{\eta v_y}(f) = 0$$

$$\frac{\hat{c}_{v_x v_y}(f)}{\hat{S}(f)\hat{R}^2(f)} = \int_0^{2\pi} D_f(\theta) \sin(\theta) \cos(\theta) \ d\theta \tag{A-11}$$

$$\hat{q}_{v_x v_y}(f) = 0$$

The quad-spectra are all zero because X=Y=0 in Equation (20).

Sub-circle expansion:

$$\frac{\hat{c}_{\eta v_x}(f)}{\hat{S}(f)\hat{R}(f)} = \frac{1}{T} \int_{\theta_0}^{\theta_0+T} \cos\theta \ d\theta$$

$$+ \sum_{n=1}^{N} \hat{a}_n(f) \int_{\theta_0}^{\theta_0+T} \cos\theta \ \cos \frac{2\pi n(\theta-\theta_0)}{T} \ d\theta \tag{A-12}$$

$$+ \sum_{n=1}^{N} \hat{b}_n(f) \int_{\theta_0}^{\theta_0+T} \cos\theta \ \sin \frac{2\pi n(\theta-\theta_0)}{T} \ d\theta$$

$$\frac{\hat{c}_{\eta v_y}(f)}{\hat{S}(f)\hat{R}(f)} = \frac{1}{T}\int_{\theta_0}^{\theta_0+T}\sin\theta\ d\theta$$

$$+ \sum_{n=1}^{N}\hat{a}_n(f)\int_{\theta_0}^{\theta_0+T}\sin\theta\ \cos\frac{2\pi n(\theta-\theta_0)}{T}\ d\theta \qquad (A-13)$$

$$+ \sum_{n=1}^{N}\hat{b}_n(f)\int_{\theta_0}^{\theta_0+T}\sin\theta\ \sin\frac{2\pi n(\theta-\theta_0)}{T}\ d\theta$$

$$\frac{\hat{c}_{v_x v_y}(f)}{\hat{S}(f)\hat{R}(f)} = \frac{1}{T}\int_{\theta_0}^{\theta_0+T}\sin\theta\ \cos\theta\ d\theta$$

$$+ \sum_{n=1}^{N}\hat{a}_n(f)\int_{\theta_0}^{\theta_0+T}\sin\theta\ \cos\theta\ \cos\frac{2\pi n(\theta-\theta_0)}{T}\ d\theta \qquad (A-14)$$

$$+ \sum_{n=1}^{N}\hat{b}_n(f)\int_{\theta_0}^{\theta_0+T}\sin\theta\ \cos\theta\ \sin\frac{2\pi n(\theta-\theta_0)}{T}\ d\theta$$

$$\frac{\hat{S}_{v_x v_x}(f)}{\hat{S}(f)\hat{R}^2(f)} = \frac{1}{T}\int_{\theta_0}^{\theta_0+T}\cos^2\theta\ d\theta$$

$$+ \sum_{n=1}^{N}\hat{a}_n(f)\int_{\theta_0}^{\theta_0+T}\cos^2\theta\ \cos\frac{2\pi n(\theta-\theta_0)}{T}\ d\theta \qquad (A-15)$$

$$+ \sum_{n=1}^{N}\hat{b}_n(f)\int_{\theta_0}^{\theta_0+T}\cos^2\theta\ \sin\frac{2\pi n(\theta-\theta_0)}{T}\ d\theta$$

$$\frac{\hat{S}_{v_y v_y}(f)}{\hat{S}(f)\hat{R}^2(f)} = \frac{1}{T}\int_{\theta_0}^{\theta_0+T}\sin^2\theta\ d\theta$$

$$+ \sum_{n=1}^{N}\hat{a}_n(f)\int_{\theta_0}^{\theta_0+T}\sin^2\theta\ \cos\frac{2\pi n(\theta-\theta_0)}{T}\ d\theta \qquad (A-16)$$

$$+ \sum_{n=1}^{N} \hat{b}_n(f) \int_{\theta_0}^{\theta_0+T} \sin^2\theta \, \sin \frac{2\pi n(\theta-\theta_0)}{T} \, d\theta \qquad \text{(A-16)}$$

For full-circle expansion ($\theta_0=0$, $T=2\pi$) these formulas reduce to

$$\frac{\hat{c}_{\eta v_x}(f)}{\hat{S}(f)\hat{R}(f)} = \pi\hat{a}_1(f) \qquad \text{(A-17)}$$

$$\frac{\hat{c}_{\eta v_y}(f)}{\hat{S}(f)\hat{R}(f)} = \pi\hat{b}_1(f) \qquad \text{(A-18)}$$

$$\frac{\hat{c}_{v_x v_y}(f)}{\hat{S}(f)\hat{R}^2(f)} = \frac{\pi}{2}\,\hat{b}_2(f) \qquad \text{(A-18)}$$

$$\frac{\hat{s}_{v_x v_x}(f)}{\hat{S}(f)\hat{R}^2(f)} = \frac{1}{2} + \frac{\pi}{2}\,\hat{a}_2(f) \qquad \text{(A-20)}$$

$$\frac{\hat{s}_{v_y v_y}(f)}{\hat{S}(f)\hat{R}^2(f)} = \frac{1}{2} - \frac{\pi}{2}\,\hat{a}_2(f) \qquad \text{(A-21)}$$

4. <u>Floating buoy</u>

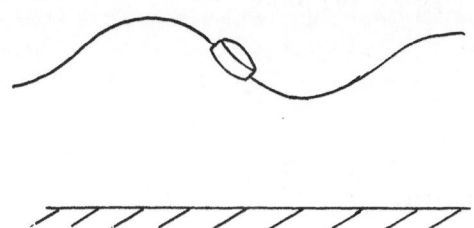

For general interest, we could note that wave height and slope measurements are equivalent to measurements of a buoy's heave, pitch and roll.

Basic measurements: $\eta(t)$, $\eta_x(t)$, $\eta_y(t)$ at same (x,y) location (i.e.), $X_{ij} = Y_{ij} = 0$ for all pairs of sensors)

Fundamental equations:

$\hat{S}(f)$ estimated from $\eta(t)$ data

$$\frac{\hat{S}_{\eta_x \eta_x}(f)}{\hat{S}(f)k^2} = \int_0^{2\pi} D_f(\theta) \cos^2\theta \, d\theta \tag{A-22}$$

$$\frac{\hat{S}_{\eta_y \eta_y}(f)}{\hat{S}(f)k^2} = \int_0^{2\pi} D_f(\theta) \sin^2\theta \, d\theta \tag{A-23}$$

$$\frac{q_{\eta \eta_x}(f)}{\hat{S}(f)k} = \int_0^{2\pi} D_f(\theta) \cos\theta \, d\theta \tag{A-24}$$

$$q_{\eta \eta_x}(f) = 0$$

$$\frac{q_{\eta \eta_y}(f)}{\hat{S}(f)k} = \int_0^{2\pi} D_f(\theta) \sin\theta \, d\theta \tag{A-25}$$

$$c_{\eta \eta_y}(f) = 0$$

$$\frac{c_{\eta_x \eta_y}(f)}{\hat{S}(f)k^2} = \int_0^{2\pi} D_f(\theta) \sin\theta \cos\theta \, d\theta \tag{A-26}$$

$$q_{\eta_x \eta_y}(f) = 0$$

Sub-circle expansion:

$$\frac{\hat{S}_{\eta_x \eta_x}(f)}{\hat{S}(f)k^2} = \frac{1}{T} \int_{\theta_0}^{\theta_0 + T} \cos^2\theta \, d\theta \tag{A-27}$$

$$+ \sum_{n=1}^{N} \hat{a}_n(f) \int_{\theta_0}^{\theta_0+T} \cos^2\theta \, \cos\frac{2\pi n(\theta-\theta_0)}{T} \, d\theta \qquad (A-27)$$

$$+ \sum_{n=1}^{N} \hat{b}_n(f) \int_{\theta_0}^{\theta_0+T} \cos^2\theta \, \sin\frac{2\pi n(\theta-\theta_0)}{T} \, d\theta$$

$$\frac{\hat{S}_{\eta_y\eta_y}(f)}{\hat{S}(f)k^2} = \frac{1}{T} \int_{\theta_0}^{\theta_0+T} \sin^2\theta \, d\theta$$

$$+ \sum_{n=1}^{N} \hat{a}_n(f) \int_{\theta_0}^{\theta_0+T} \sin^2\theta \, \cos\frac{2\pi n(\theta-\theta_0)}{T} \, d\theta \qquad (A-28)$$

$$+ \sum_{n=1}^{N} \hat{b}_n(f) \int_{\theta_0}^{\theta_0+T} \sin^2\theta \, \sin\frac{2\pi n(\theta-\theta_0)}{T} \, d\theta$$

$$\frac{\hat{q}_{\eta\eta_x}(f)}{\hat{S}(f)k} = \frac{1}{T} \int_{\theta_0}^{\theta_0+T} \cos^2\theta \, d\theta$$

$$+ \sum_{n=1}^{N} \hat{a}_n(f) \int_{\theta_0}^{\theta_0+T} \cos\theta \, \cos\frac{2\pi n(\theta-\theta_0)}{T} \, d\theta \qquad (A-29)$$

$$+ \sum_{n=1}^{N} \hat{b}_n(f) \int_{\theta_0}^{\theta_0+T} \cos\theta \, \sin\frac{2\pi n(\theta-\theta_0)}{T} \, d\theta$$

$$\frac{q_{\eta\eta_y}(f)}{\hat{S}(f)k} = \frac{1}{T} \int_{\theta_0}^{\theta_0+T} \sin\theta \, d\theta$$

$$+ \sum_{n=1}^{N} \hat{a}_n(f) \int_{\theta_0}^{\theta_0+T} \sin\theta \, \cos\frac{2\pi n(\theta-\theta_0)}{T} \, d\theta \qquad (A-30)$$

$$+ \sum_{n=1}^{N} \hat{b}_n(f) \int_{\theta_0}^{\theta_0+T} \sin\theta \, \sin\frac{2\pi n(\theta-\theta_0)}{T} \, d\theta$$

$$\frac{\hat{c}_{\eta_x\eta_y}(f)}{\hat{S}(f)k} = \frac{1}{T} \int_{\theta_0}^{\theta_0+T} \sin\theta \cos\theta \, d\theta$$

$$+ \sum_{n=1}^{N} \hat{a}_n(f) \int_{\theta_0}^{\theta_0+T} \sin\theta \cos\theta \cos \frac{2\pi n(\theta-\theta_0)}{T} \, d\theta \qquad (A-31)$$

$$+ \sum_{n=1}^{N} \hat{b}_n(f) \int_{\theta_0}^{\theta_0+T} \sin\theta \cos\theta \sin \frac{2\pi n(\theta-\theta_0)}{T} \, d\theta$$

For a full circle expansion ($\theta_0=0$, $T=2\pi$)

$$\frac{\hat{S}_{\eta_x\eta_x}(f)}{\hat{S}(f)k^2} = \frac{1}{2} + \frac{\pi\hat{a}_2(f)}{2}$$

$$\frac{\hat{S}_{\eta_y\eta_y}(f)}{\hat{S}(f)k^2} = \frac{1}{2} - \frac{\pi\hat{a}_2(f)}{2}$$

$$\frac{\hat{q}_{\eta\eta_x}(f)}{\hat{S}(f)k} = \pi\hat{a}_1(f)$$

$$\frac{\hat{q}_{\eta\eta_y}(f)}{\hat{S}(f)k} = \pi\hat{b}_1(f)$$

$$\frac{\hat{c}_{\eta_x\eta_y}(f)}{\hat{S}(f)k^2} = \frac{\pi}{2}\hat{b}_2(f)$$

APPENDIX B

A SUB-CIRCLE NORMAL APPROXIMATION

Suppose $D(\theta)$ is nonzero only for $\theta_0 \leq \theta \leq \theta_0+T$. Let $f_W(w)$ be any probability density defined for $-\infty < w < \infty$ which is also nonzero only for $\theta_0 \leq w \leq \theta_0+T$. Clearly, $f_W(\theta)$ may be taken as a possible model for $D(\theta)$.

The Fourier series for $D(\theta)$ is

$$D(\theta) = \frac{a_0}{2} + \sum_{n=1}^{\infty} [a_n \cos \frac{2\pi n(\theta-\theta_0)}{T} + b_n \sin \frac{2\pi n(\theta-\theta_0)}{T}] \qquad (B-1)$$

where

$$a_n = \frac{2}{T} \int_{\theta_0}^{\theta_0+T} D(\theta) \cos \frac{2\pi n(\theta-\theta_0)}{T} d\theta \qquad (B-2)$$

$$b_n = \frac{2}{T} \int_{\theta_0}^{\theta_0+T} D(\theta) \sin \frac{2\pi n(\theta-\theta_0)}{T} d\theta \qquad (B-3)$$

Consequently, if $D(\theta) \approx f_W(\theta)$

$$(a_n + ib_n) \frac{T}{2} = \int_{\theta_0}^{\theta_0+T} f_W(\theta) e^{i2\pi n(\theta-\theta_0/T} d\theta$$

$$= e^{-i2\pi n\theta_0/T} \int_{-\infty}^{\infty} f_W(w) e^{i2\pi nw/T} dw \qquad (B-4)$$

$$= e^{i2\pi n\theta_0/T} \phi_W(\frac{2\pi n}{T})$$

where

$$\phi_W(u) = E[e^{iuW}] \qquad (B-5)$$

is the characteristic function for $f_W(w)$.

For a normal probability density

$$f_W(w) = \frac{e^{-(w-\mu)^2/2\sigma^2}}{\sqrt{2\pi} \sigma} , \quad -\infty < w < \infty \qquad (B-6)$$

the characteristic function is

$$\phi_W(u) = e^{i\mu u - \sigma^2 u^2/2} \qquad (B-7)$$

Therefore, if a normal model is used for $D(\theta)$,

$$(a_n + ib_n) \frac{T}{2} = e^{-i2\pi n\theta_0/T} e^{i2\pi n\mu/T} e^{-\sigma^2(\frac{2\pi n}{T})^2/2} \qquad (B-8)$$

or $\quad a_n = \frac{2}{T} e^{-2\pi^2 n^2 \sigma^2 / T^2} \cos[2\pi n(\mu - \theta_0)/T]$

$\quad\quad b_n = \frac{2}{T} e^{-2\pi^2 n^2 \sigma^2 / T^2} \sin[2\pi n(\mu - \theta_0)/T]$ $\quad\quad\quad\quad\quad$ (B-9)

A fit to the model may be obtained from

$$\sqrt{a_1^2 + b_1^2} = \frac{2}{T} e^{-2\pi^2 \hat{\sigma}^2 / T^2} \quad\quad\quad\quad\quad (B\text{-}10)$$

or

$$\hat{\sigma}^2 = -\frac{T^2}{2\pi^2} \ell n \left[\frac{T}{2} \sqrt{a_1^2 + b_1^2}\right] \quad\quad\quad\quad\quad (B\text{-}11)$$

and

$$\frac{2\pi(\hat{\mu} - \theta_0)}{T} = \text{arc } \tan(b_1/a_1) \quad\quad\quad\quad\quad (B\text{-}12)$$

or

$$\hat{\mu} = \theta_0 + \frac{T}{2\pi} \text{arc } \tan(b_1/a_1) \quad\quad\quad\quad\quad (B\text{-}13)$$

Higher order coefficients may similarly be used to get estimates or some type of weighted average or least-square technique may be introduced.

APPENDIX C

A NUMERICAL EXAMPLE

Values of $S(f)$, a_1, b_1, b_2, μ, and σ for an artificial data are shown in Table 1. Although the values are not taken from actual data because the immediately available examples are still proprietary, the general range and scatter are consistent with cross-spectral estimates with 20 degrees of freedom and a resolution of .01 Hz.

The pre-supposed instrument array for the example consists of a wave staff and three velocity gages at various depths directly below the wave staff measuring horizontal water particle velocity components.

For the velocity meter at a given position from Equation (A-8)

$$[\hat{S}_{v_x v_x}(f) + \hat{S}_{v_y v_y}(f)]/\hat{S}(f) = (\hat{R})^2$$

From the definition of the full circle Fourier coefficients

$$\binom{a_m}{b_m} = \frac{1}{\pi}\int_0^{2\pi} D_f(\theta) \frac{\cos}{\sin} m\theta \, d\theta$$

It follows from Equation (A-9), together with a similar relation for v_x with v_x and v_y with v_y, that for all gage pairs of the type indicated

$$a_1 = \hat{c}_{\eta v_x}(f)/[\pi\hat{S}(f)\hat{R}]$$

$$b_1 = \hat{c}_{\eta v_y}(f)/[\pi\hat{S}(f)\hat{R}]$$

$$a_2 = (2/\pi)\{(\hat{c}_{v_x v_x}(f)/[\hat{S}(f)\hat{R}_1\hat{R}_2]) - 05\}$$

$$= (2/\pi)\{0.5 - ([\hat{c}_{v_y v_y}(f)]/[\hat{S}(f)\hat{R}_1\hat{R}_2])\}$$

$$b_2 = (2/\pi)\{[\hat{c}_{v_x v_y}(f)]/[\hat{S}(f)\hat{R}_1\hat{R}_2]\}$$

As an example of the use of these formulas, a typical set of possible values for cross-spectral density estimates at f = 0.1 Hz is shown in Table 2. Similar tables would be prepared for each analyzed frequency. The values in Table 2 may be substituted into the formula to obtain the coefficient values in Table 3. The scatter shown is typical of real data with spectral estimates having the resolution and degrees of freedom previously mentioned.

Various methods could be used to distill a single set of coefficients from all those displayed in Table 3. Least-square methods or some type of weighted average could be used. However, for the artificial example given, a simple average is listed for the coefficients at f = 0.1 Hz in Table 1.

The formulas for the fit of the wrapped-normal follow from Equation (B-9)

$$a_1 = (1/\pi)e^{-\sigma^2/2}\cos\mu$$

TABLE 1. An illustrative example (artificial data) of
spectral density, Fourier coefficients, and
wrapped-normal parameter estimates.

f	$\hat{S}(f)$	$\hat{a}_1$	$\hat{b}_1$	$\hat{a}_2$	$\hat{b}_2$	μ	σ
.07	5.1	-.294	-0.64	.239	.082	192°	19°
.08	40.3	-.301	-.075	.248	.166	194°	13°
.09	128.6	-.302	-.081	.232	.175	195°	11°
.10	162.8	-.290	-.110	.225	.189	201°	13°
.11	100.0	-.279	-.129	.217	.196	205°	15°
.12	50.2	-.250	-.158	.128	.173	212°	22°
.13	44.7	-.211	-.195	.048	.193	223°	26°
.14	35.2	-.202	-.216	.026	.192	227°	22°
.15	31.9	-.223	-.207	.044	.211	223°	17°
.16	23.5	-.221	-.192	.050	.175	221°	23°
.17	20.8	-.218	-.154	.076	.130	215°	34°
.18	19.8	-.220	-.174	.063	.143	218°	29°
.19	15.6	-.188	-.189	.041	.151	225°	34°
.20	12.2	-.164	-.178	.051	.107	227°	42°

$$b_1 = (1/\pi)e^{-\sigma^2/2}\sin\mu$$

Thus

$$\mu = \text{arc tan}(b_1/a_1)$$

$$\sigma^2 = -2 \ln[\pi\sqrt{a_1^2 + b_1^2}]$$

Values of μ and σ stated in degrees are listed in the last column
of Table 1. A similar set of equations for μ and σ^2 can be obtained
from a_2 and b_2, or, alternately, various more sophisticated methods
using all four coefficients can be introduced. The simplest approach
using only a_1 and b_1 was followed in the listed example.

TABLE 2. An illustrative example (artifical data) of
cross-spectral densities at f = 0.1 Hz.
(Spectrum on main diagonal, co-spectrum above
diagonal, quad-spectrum below diagonal).

	η	v_{x1}	v_{y1}	v_{x2}	v_{y2}	v_{x3}	v_{y3}
η	162.8	−72.2	−26.7	−64.6	−22.9	−62.3	−25.6
v_{x1}	− .40	32.8	10.7	28.9	9.9	28.0	10.7
v_{y1}	.60	− .52	5.4	10.1	4.5	9.6	3.1
v_{x2}	− .06	− .11	.36	26.2	8.8	24.9	8.7
v_{y2}	− .90	.31	.40	.60	4.2	8.4	4.5
v_{x3}	− .80	.95	.81	.93	− .22	24.4	9.3
v_{y3}	− 1.12	.42	.23	.71	.23	.67	5.1

Studies in which the above general procedures were applied to
ocean measurements are described by *Forristall, et al.* (1978a,b).

The Fourier series, least-square approach with subsequent
model-fitting as outlined in the preceeding, depend critically on
the stability and accuracy of the cross-spectral estimates. As
many degrees of freedom as possible should be obtained in the
spectral estimation. This, in turn, depends on the length of the
data and the resolution imposed. Let T be the length of record
which is digitized at increment Δt. Then $N = T/\Delta t$ = number of
measurements in the data sequence and the basic frequency increment
will be $\Delta f = 1/T$. The number of FFT spectral lines in a resolution
interval (or interval of spectral averaging) R will be $R/\Delta f = RT$.
There are two degrees of freedom per line, so the degrees of freedom per spectral estimate will be 2RT. Thus, statistical stability
increases with record length and coarseness of resolution.

TABLE 3. An illustrative example (artifical data) of
coefficient estimates consistent with
Table 2 numbers.

(a) Separate estimates of a_1 at f = 0.1 Hz.

	v_{x1}	v_{x2}	v_{x3}	
η	−.292	−.292	−.286	$\longrightarrow$

TABLE 3.   An illustrative example (artificial data) of
coefficient estimates consistent with
Table 2 numbers. (Continued.)

(b)   Separate estimates of b_1 at f = 0.1 Hz.

	v_{y1}	v_{y2}	v_{y3}
η	-.108	-.104	-.117

(c)   Separate estimates of a_2 at f = 0.1 Hz.

	v_{x1}	v_{y1}	v_{x2}	v_{y2}	v_{x3}	v_{y3}
v_{x1}	.299					
v_{y1}		.228				
v_{x2}	.222		.231			
v_{y2}		.234		.230		
v_{x3}	.213		.211		.207	
v_{y3}		.260		.223		.208

(d)   Separate estimates of b_2 at f = 0.1 Hz.

	v_{x1}	v_{y1}	v_{x2}	v_{y2}	v_{x3}
v_{x1}					
v_{y1}	.200				
v_{x2}		.189			
v_{y2}	.185		.184		
v_{x3}		.182		.174	
v_{y3}	.202		.185		.200

REFERENCES

Borgman, L. E. 1969. Directional spectra models for design use. *Offshore Technology Conference*, Houston, Texas.

Bowden, K. F. and R. A. White. 1966. Measurements of the orbital velocities of sea waves and their use in determining the directional spectrum, *Geophys. J. Roy. Astr. Soc., 12*, 33-54.

Burg, J. P. 1972. The relationship between maximum entropy spectra and maximum likelihood spectra, *Geophys., 37*, 33-54.

Capon, J. 1969. High-resolution frequency-wave number spectrum analyses, *Proc. IEEE, 57*, 1408-1418.

Cartwright, D. H. and N. D. Smith. 1964. Buoy techniques for obtaining directional wave spectra, *Buoy Technology*, Marine Technology Society, 112-121.

Chakrabarti, S. K. 1971. Wave train direction analysis. *J. Waterways, Harbors and Coastal Engineering Div., Am. Soc. Civ. Engineers, 97*, 787-792.

Chakrabarti, S. K. and R. H. Snider. 1973. Two-dimensional wave energy spectra, *Underwater Tech., 2*, 80-85.

_____. 1974. Multidirectional analysis of monochromatic wave trains, Tech. Note, *J. Waterways, Harbors and Coastal Engineering Div., Am. Soc. Cil. Engineers, 100*, 61-67.

Cummins, W. E. 1959. The determination of directional wave spectra in the TMB maneuvering-seakeeping basin, U.S. Navy David Taylor Model Basin, Report 1362, 12th American Towing Tank Conf., University of California, Berkeley.

Ewing, J. A. 1969. Some measurements of the directional wave spectrum, *J. Marine Sci., 27*, 163-171.

Forristall, G. Z., E. G. Ward, L. E. Borgman, and V. J. Cardone. 1978a. Storm wave kinematics, Offshore Technology Conference, Houston, Texas, paper 3227.

_____. 1978b. The directional spectra and kinematics of surface waves in tropical storm Delia, submitted to *J. Phys. Oceanogr.*

Hasselmann, K., T. P. Barnett, E. Bouws, H. Carlson, D. E. Cartwright, K. Enke, J. A. Ewing, H. Gienapp, D. E. Hasselmann, P. Kruseman, A. Meerburg, P. Müller, D. J. Olbers, K. Richter, W. Sell, and H. Walden. 1973. Measurement of wind-wave growth and swell decay during the Joint North Sea Wave Project (JONSWAP), *Erganzungshaft zur Deutschen Hydrographischen Zeitschrift, Reihe A (8°)*, No. 12.

Lacoss, R. T. 1971. Data adaptive spectral analysis methods, *Geophys. 36*, 661-675.

Longuet-Higgins, M. S., D. E. Cartwright, and N. D. Smith. 1963. Observations of the directional spectrum of sea waves using the motions of a floating buoy, *Ocean Wave Spectra*, Prentice-Hall, Englewood Cliffs, N.J., 111-136.

Lowe, R., D. L. Inman, and B. M. Brush. 1972. Simultaneous data system for instrumenting the shelf, *Proc. Thirteenth International Coastal Engineering Conference*, Vancouver, B.C., 95-112.

Mitsuyasu, H., F. Tasai, T. Suhara, S. Mizuno, M. Ohkusu, T. Honda, and K. Rikiishi. 1974. Observations of the directional spectrum of ocean waves using a Cloverleaf Buoy, *Report of Research Institute for Applied Mechanics*, Kyushu University, Fukuoka, Japan.

Mobarek, I. E. 1965. Directional spectra of laboratory wind waves, *J. Waterways, Harbors and Coastal Engineering Div., Am. Soc. Civ. Engineers, 91*, 91-116.

Nagata, Y. 1964. The statistical properties of orbital wave motions and their applications for the measurement of directional wave spectra, *J. Oceanogr. Soc. of Japan, 19*, 169-181.

Panicker, N. N. 1974. Review of techniques for directional wave spectra, *Proc. International Symposium on Ocean Wave Measurement and Analysis*, Am. Soc. Civ. Engineers, New Orleans, 669-688.

Panicker, N. N. and L. E. Borgman. 1970. Directional spectra from wave-gage arrays, *Proc. Twelfth International Coastal Engineering Conference*, Washington, D.C., 117-136.

Panicker, N. N. and L. E. Borgman. 1974. Enhancement of directional wave spectrum estimates, *Proc. Fourteenth International Coastal Engineering Conference*, Copenhagen, Denmark.

Ploeg, J. 1972. Some results of a directional wave recording station, *Proc. Thirteenth International Coastal Engineering Conference*, Vancouver, B.C., 131-141.

Rikiishi, K. 1977. A study on the measurement of the directional spectrum and phase velocity of laboratory waves, *Report of Research Institute for Applied Mechanics*, Kyushu University, Fukuoka, Japan.

Simpson, J. H. 1969. Observations of the Directional characteristics of sea waves, *Geophys. J. Roy. Astr. Soc., 17*, 93-120.

Stevens, R. G. 1965. On the measurement of the directional spectra of wind generated waves using a linear array of surface elevation detectors, unpublished manuscript, Technical Report Ref. No. 65-20, Woods Hole Oceanographic Institution, Woods Hole, Massachusetts.

Suzuki, Y. 1969. Determination of approximate directional spectra for coastal waves, *Report of the Port and Harbor Research Institute, 8*, 43-101, Japan.

Ulrych, T. J. 1972. Maximum entropy power spectrum of truncated sinusoids, *J. Geophys. Res., 77*, 1396-1400.

DATA BUOY WAVE MEASUREMENTS

Kenneth Steele and Andrew Johnson, Jr.

NOAA Data Buoy Office

ABSTRACT

The Data Buoy Office of the National Oceanic and Atmospheric Administration (NOAA) develops and deploys data buoy systems for the acquisition and reporting of marine data. Various types of these systems acquire wave data. The quality of wave spectra from these systems depends upon hull and hardware configurations. This paper provides an overview of existing and planned systems for the measurement of wave data.

DESCRIPTION OF WAVE MEASUREMENT SYSTEMS

The NOAA Data Buoy Office (NDBO) is acquiring wave measurements from a number of ocean stations, the locations of which are shown in Figure 1. The stations fall into three categories: deep ocean stations, continental shelf stations, and an experimental station. Buoys moored at these stations have on-board data acquisition and reporting systems, commonly referred to as "payloads." A payload comprises meteorological and oceanographic sensors; electronics for the acquisition, processing and formatting of data from these sensors; and a communication system to relay the formatted data to shore. The deep ocean and the continental shelf buoys routinely report wave data to users. Data from two payloads aboard a buoy known by the acronym "XERB," which is located at the experimental station shown in Figure 1, are used for engineering development purposes. The data from this buoy are not normally reported to users.

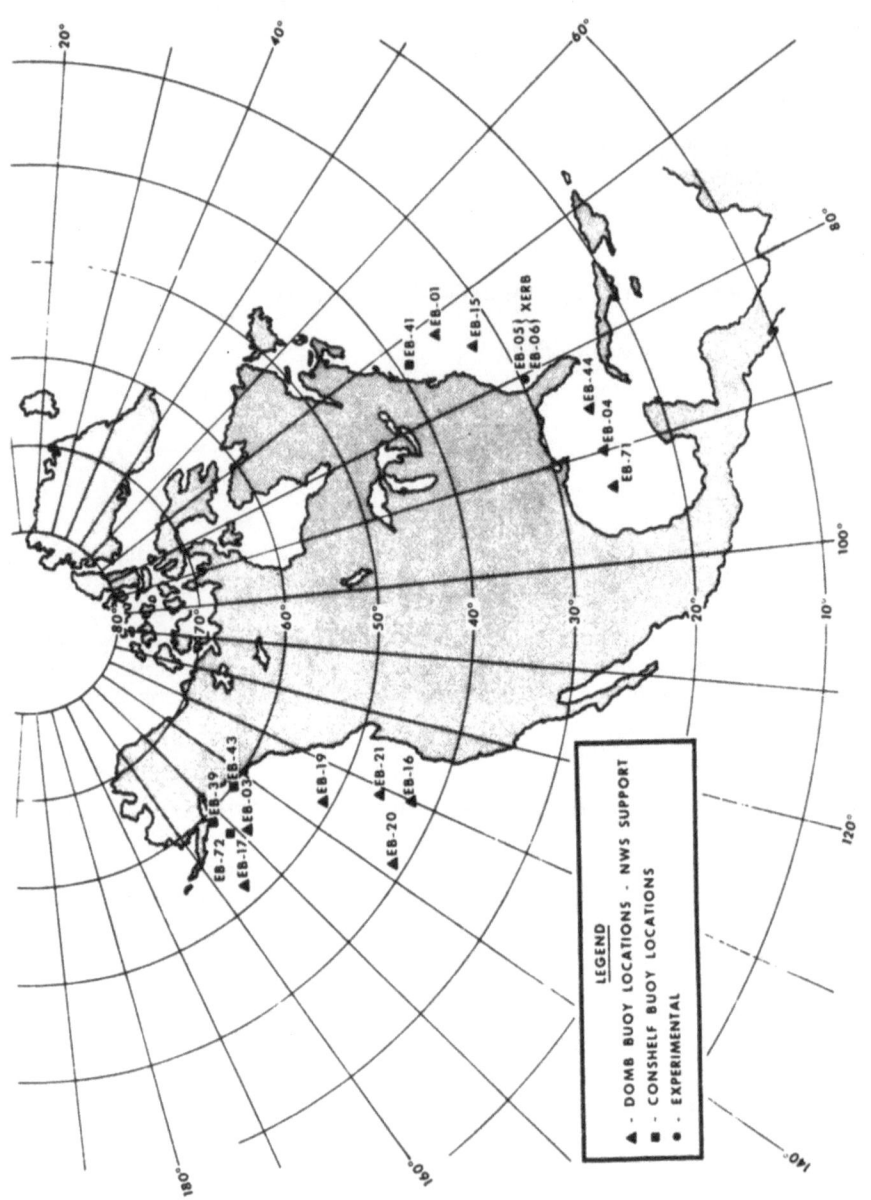

Fig. 1. Wave measurement stations (as of July 1, 1977)

Figure 2 contains a table showing the types of systems used at these stations. Several combinations of platforms, sensors, sensor mountings, signal conditioning, and data processing techniques make up the systems that routinely report wave data. This variety has arisen due to the necessity of making use of existing hulls and payloads; however, all systems producing wave data have the capability to produce one-dimensional displacement spectra, significant wave height ($H_{1/3}$), and average zero-up-crossing wave period by Rice's theory (*Goda*, 1974). Of the wave systems shown in Figure 2, the earliest is the one installed in the Engineering Experimental Phase (EEP) type buoy, and described in a paper by *Steele et al.* (1975). This system performed well despite the relatively high noise levels associated with it. However, the hardware was not operationally reliable and NDBO now has only one such system at sea.

The type of wave measurement system most used by NDBO at this time (1977) is the Wave Spectrum Analyzer (WSA) (*Remond*, 1976). This system, in two slightly different hardware configurations, can be used with either the Limited Capability Buoy payload (*NOAA Data Buoy Office*, 1973), or the Prototype Environmental Buoy payload (*General Dynamics*, 1975). A schematic diagram of the system is shown in Figure 3.

The WSA consists of 12 separate parallel analog filters, each of which has an adjustable center frequency, bandwidth, and gain. An accelerometer output is fed into each filter. The near-DC output voltages from the channels are sampled and the mean values of these samples are sent to shore, where the heave displacement of the sea surface is reconstructed.

Figure 4 shows the power gain curves for the 12 filters, each normalized to unity at the center frequency. The near-DC voltage output of each filter is the result of the product of the frequency dependent filtering function and the voltage spectrum. The voltage output of a filter will be very nearly proportional to the center frequency spectral density only if the bandwidth of the filter is small compared to the variation of the spectrum near the center frequency. The entire spectrum cannot be covered effectively with only 12 very narrow filters. Even if the use of more than 12 filters was practical, very narrow-band filters have a tendency to be unstable. For these reasons, the bandwidths of the filters are made relatively large. With broader bands, the "skirts" of each filter, in which the spectral density may be very different from the spectral density at the center frequency, will contribute significantly to the output of the filter. To compensate for the spectral density errors that normally occur due to the filter skirts, the frequency axis is divided into contiguous intervals equal to the filter bandwidths and centered on the filters. The voltage output

Wave System Type	On-Board Hardware	Analog Filter Cutoff (Hz)	Spectral Estimates Produced (Number/fr, .---Hz)	Payload Type	RF Type	EB-SITE NUMBER AND/OR (BUOY NAME)
Wave Spectrum Analyzer (WSA)	12 Parallel Analog Filter	0.3	(12/f = .05, .06, .07, .08, .09, .10, .1225, .16, .20, .24, .28, .33 Hz)	LCB	HF	41 ... 39
		None		PEB	UHF	43
Experimental EEP	Digital Minicomputer		(48/f = .03, .04, - - .49, .50 Hz)	Modified *EEP	HF	06 (XERB)
Wave Data Analyzer (WDA)	Digital Micro Computer Producing Covariances	0.5	(95/f = .030, .035, - - .495, .500 Hz)	*GDTT	UHF	(WRANSAC)
			(48/f = .03, .04, - - .49, .50 Hz)	Magnavox Phase II	HF	71, 72 ... 01, 44
		0.25	(45/f = .030, .035, - - .245, .250 Hz)		UHF	05 (XFRB)
			(23/f = .03, .04, .05 - - .24, .25 Hz)	*CSBP	HF	
					UHF	

Wave Sensors					
Accelerometer Output's Signal Conditioning	None		Double Integration Produces Displacement		None
Accelerometer's Mounting	Fixed Parallel to Vertical Mast		Vertically Stabilized Earth Frame		Fixed ll to Mast

Buoy Hulls					
Hull Dimensions	5 Meter Diameter	10 Meter Diameter	12 Meter Diameter	0.9 Meter Diameter	6.22 Meter Length 2.95 Meter Beam
Hull Type	Discus			Waverider	NOMAD (Boat Shape)

* EEP – Engineering Experimental Phase; GDTT – "GOES" Data Transmit Terminal; WRANSAC – Waverider Analyzer Satellite Communicator; CSBP – Continental Shelf Buoy Payload

Fig. 2. Types of NDBO one dimensional wave measurement systems (as of July 1, 1977). Measurement, sensor and buoy hull characteristics are found by reading across or downward from a given buoy number.

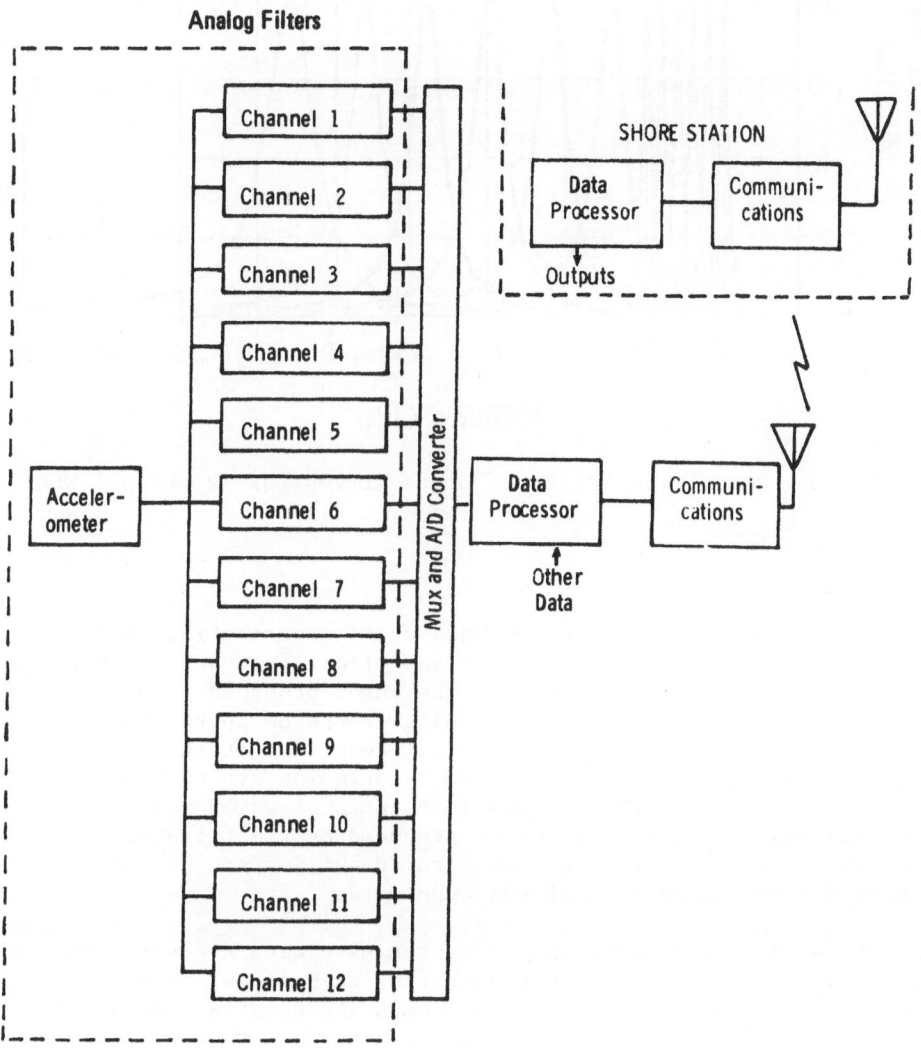

Fig. 3. Wave spectrum analyzer system schematic. Accelerometer data is filtered by twelve analog filters before sampling
and transmission to shore where sea surface displacements are
calculated.

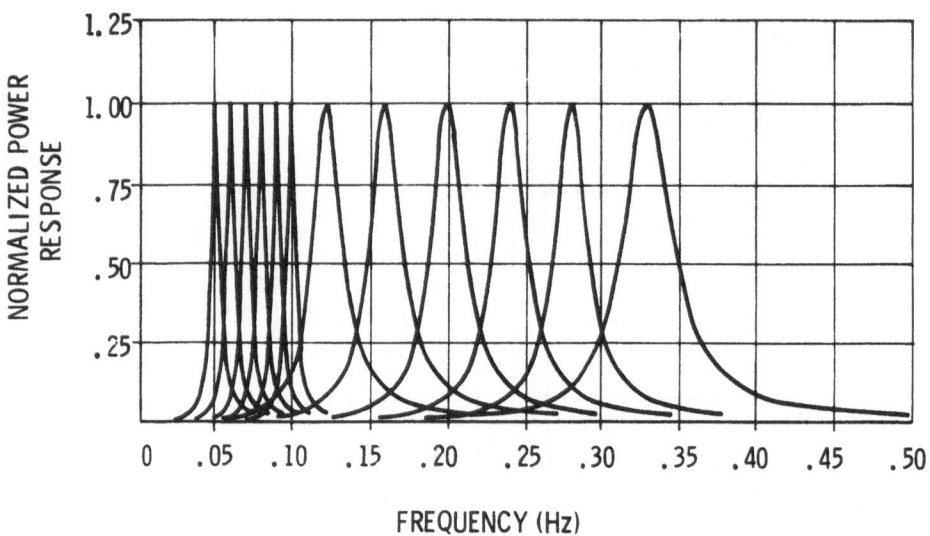

Fig. 4. Normalized filter response curves for 12 channel WSA.

of each filter is assumed to be made up of contributions made
separately by each interval. This analytical approach leads to
the voltage from each filter, and the count produced from it by
the analog-to-digital converter, being expressed approximately as
a linear sum of the (unknown) spectral densities at the centers of
the filters. These 12 equations in 12 unknowns are solved simultaneously to produce 12 new equations, each of which relates one
spectral density to all 12 counts produced by the filters. It is
from these equations that the spectral densities are produced.
Figure 5 shows an example of WSA type data.

Wave data from this type of system have been evaluated (*Hubert*,
1976a, 1976b, 1977a) through hindcasting techniques. Favorable
results have also been obtained by comparing spectra from a WSA-
equipped buoy with spectra obtained from wavefider buoy measurements. Examples of comparisons are shown in Figures 6, 7, and 8.

Although the WSA systems are generally reliable and produce
useful data, their spectral range is very limited and resolution is
too coarse. It is NDBO's intent to replace those in service with
superior systems as soon as possible. Since the WSAs were produced
in response to the need for systems that would be compatible with
payloads that had only analog sensor interfaces, the retirement of
the WSAs will depend on the retirement of these payloads.

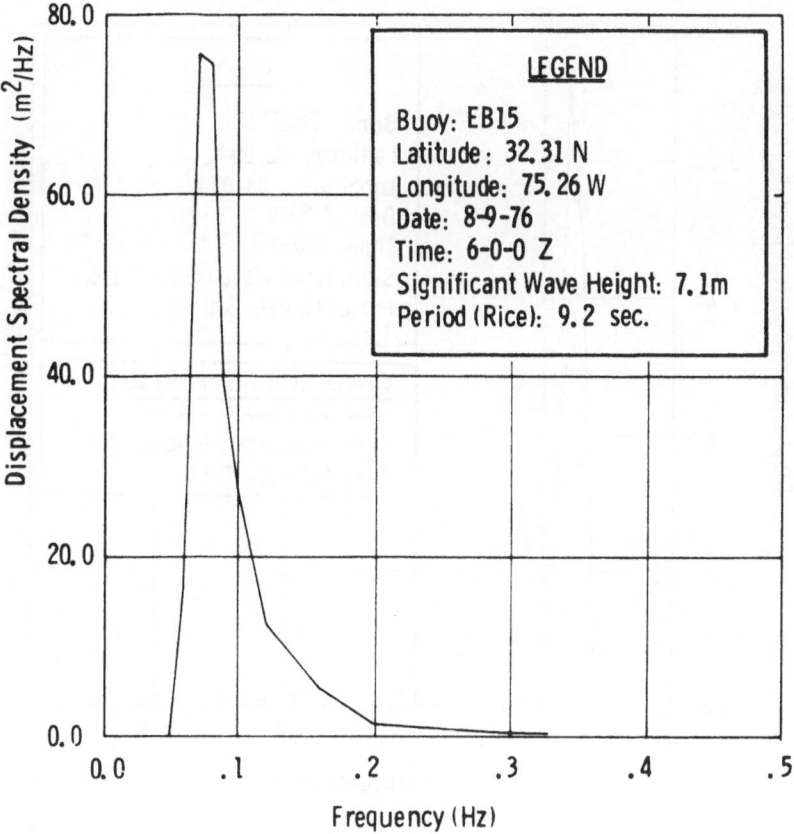

Fig. 5. Example of Wave Spectrum Analyzer (WSA). This spectrum was measured during a heavy sea state off the U.S. east coast.

The most advanced type of system used by NDBO for wave measurements is the Wave Data Analyzer (WDA). This system is part of the Phase II payload system (*Magnavox*, 1976, and *Steele et al.*, 1976). The WDA consists of a vertical linear accelerometer whose axis is fixed in the buoy and whose output voltage is filtered, sampled, and transformed into the equivalent of covariances. Figure 9 shows the key spectral parameters for two different versions of the WDA. The data produced by the WDA are telemetered by the payload to shore, where they are transformed into the spectrum. An example is shown in Figure 10. Evaluations of WDA type data have been performed (*Hubert*, 1977b) with positive results. Based on *Hubert's* analysis and the WDA design features, the wave data

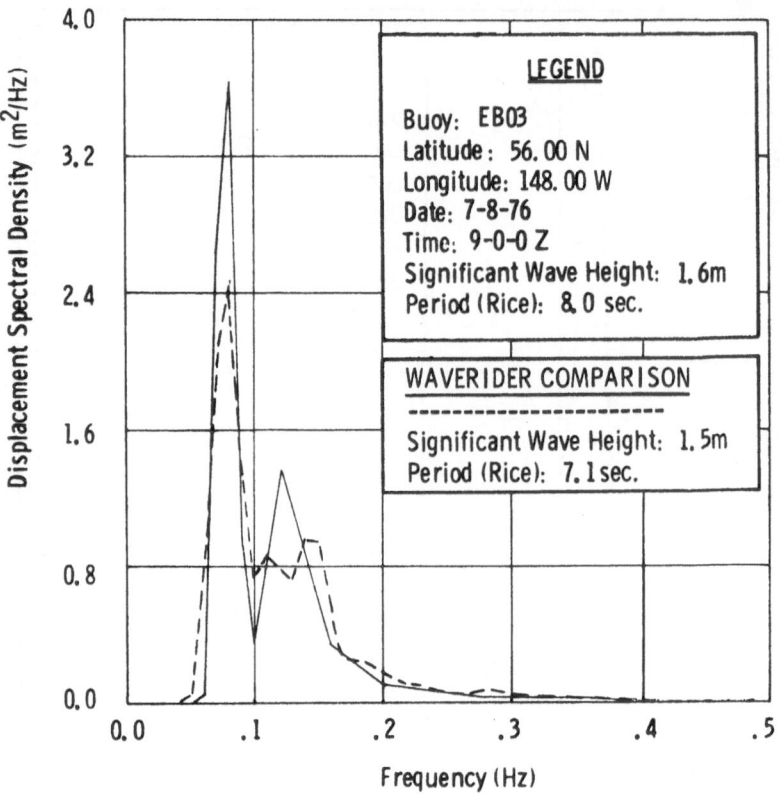

Fig. 6. Wave Spectrum Analyzer (WSA) data compared to Waverider data for a double peaked spectrum.

produced by the WDAs are considered the best of any NDBO measurement system. The accuracy of significant wave heights produced by this system is estimated to be ±0.3 meters. The quality of WDA data will be further refined after sufficient comparison data are collected from Waverider systems. Until recently, no great demand by the user community for high accuracy had been communicated to NDBO. In recent months, interest in wave data has risen and NDBO has attached a higher priority to upgrading the calibrations of both the WDAs and the WSAs.

COMMUNICATION AND DISSEMINATION OF THE DATA

Spectral wave data are acquired by each operational buoy every 3 hours at standard synoptic times (hourly data are available on command from certain buoys). No more than 30 minutes after the synoptic hour, the data are transmitted via HF to NDBO's Shore

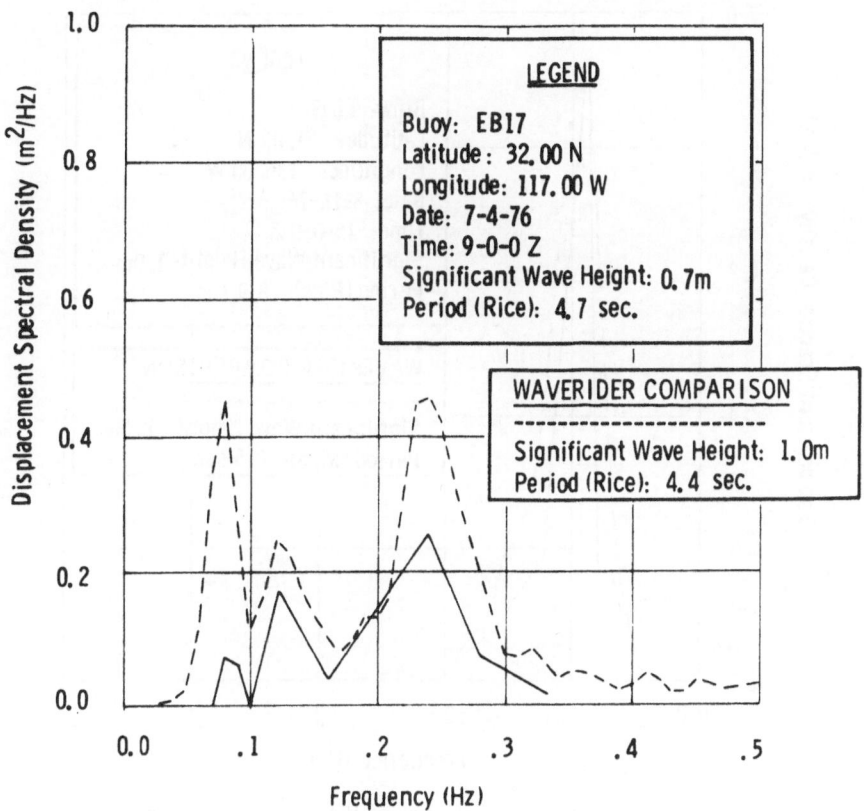

Fig. 7. Wave Spectrum Analyzer (WSA) data compared to
Waverider data for low wave heights.

Collection Station in Miami as shown in Figure 11. In Miami, these
data are coded and used to compute spectral densities of vertical
displacement of the sea surface. The spectral densities are used
to calculate significant wave height and average wave period.
These are sent to the National Meteorological Center of the National
Weather Service via teletype, along with meteorological data obtained from the buoy in standard ship's weather message format.
The displacement spectral densities are then coded into a special
format (Figure 12) for transmission from Miami to the National
Meteorological Center for further dissemination to users involved
with marine forecasting, operational forecast verification, and the
development of forecast models.

Once each day all raw and processed buoy data are sent from
Miami to NDBO for evaluation and further processing. All data from
the operational buoys are examined for time continuity and internal
consistency. They are also checked to determine if they are

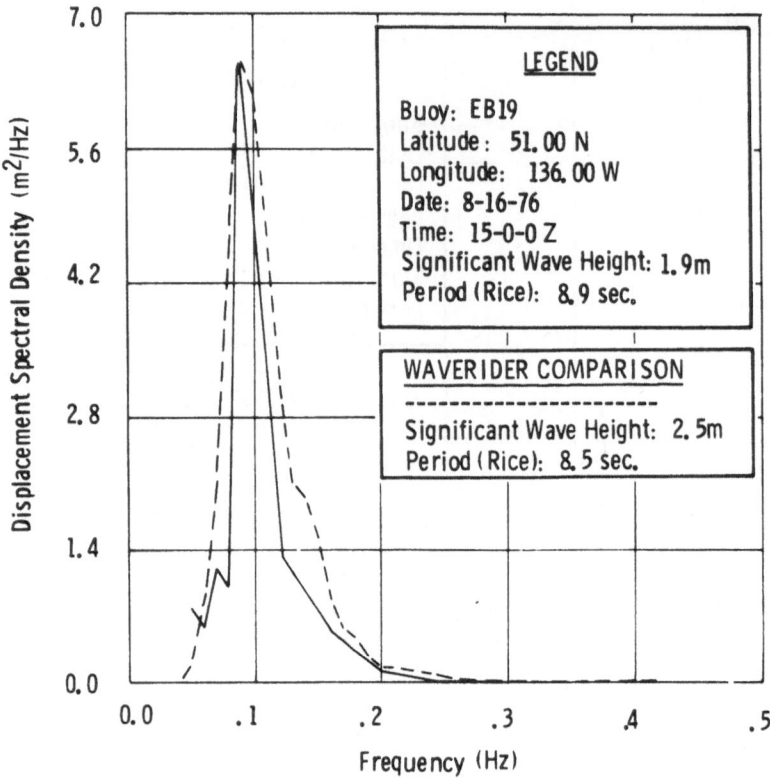

Fig. 8. Wave Spectrum Analyzer (WSA) data compared to Waverider data for a typical single peaked wave spectrum.

representative of environmental situations documented by other data sources. Reports of the status of all operational data are mailed weekly to interested users. When errors are detected in the data, they are flagged on NDBO's data base and deleted from archival tapes which are prepared monthly and sent to the National Climatic Center and the National Oceanographic Data Center.

PRESENT AND FUTURE DEVELOPMENT ACTIVITIES

NDBO's present and future wave measurement development activities can be classified into the following categories: (1) reduction of procurement and operating costs, (2) development of a small buoy for measuring waves only, (3) development of systems to measure directional wave properties, and (4) improvement of routine operations. The latter development activity encompasses

System Design Parameters	Quantitative Values, Two Different System Variations	
Half-Power Frequency, 6-Pole Butterworth Filter (Hz)	0.500	0.250
Nyquist Frequency, f_N (Hz)	0.750	0.300
Time Between Samples, $\Delta t = (1/2f_N)$ (sec)	2/3	5/3
Frequency Separation, Spectral Estimates, Δf (Hz)	0.005	0.005
Number of Lags Set Into WDA, $M = (f_N/\Delta f)$ (#)	150	60
Total Lag Time (sec)	100	100
Duration of Sampling, T (min)	20	15
Number of Samples, $N = (T/\Delta t)$ (#)	1800	540
Number of Degrees of Freedom, $DOF = (2N/M)$ (#)	24	18
16-Bit Words Sent to Shore, $(M+5)$ in WDA's (#)	155	65

Fig. 9. Key parameters for Wave Data Analyzer (WDA) type systems. Accelerometer data is transformed into the equivalent of covariances on board the buoy before transmission to shore.

several elements including the improvement of the calibration of each system type and the replacement of WSAs with WDAs as these become available.

An important part of these plans is the reduction of the cost of wave measurement systems. One form of the cost reduction effort is an attempt to develop a small, easy-to-handle buoy that will measure only waves. A Waverider buoy is now being modified to report spectral wave data via the Geostationary Operational Environmental Satellite (GOES). This buoy, named the Waverider Analyzer Satellite Communicator (WRANSAC), is described in Figure 2. The WRANSAC has great flexibility and can serve in a variety of applications. One such application is to provide ground truth data for the

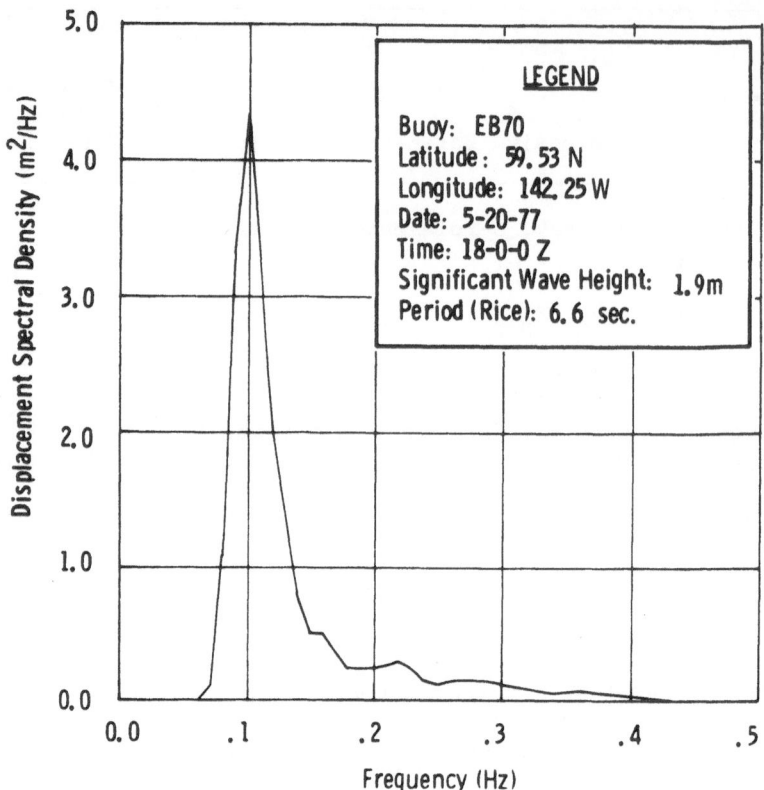

Fig. 10.   Example of Wave Data Analyzer (WDA) data.   This spectrum was measured during a medium sea state in the Gulf of Alaska.

calibration of larger buoys. With the help of the WRANSAC and conventional Waveriders, the quality of wave data will be systematically improved. Satellite telemetering capability for Waveriders will be very useful, because with it they can be used far from shore and near NDBO buoy locations. The present Wave-rider radio link requires that the Waveriders be used near ship or coastal receiving stations.

Another of NDBO's activities is the development of systems to measure directional spectra, or to measure some other direc-tional properties of the sea. Directional measurements can be implemented operationally at a cost only slightly greater than that for one-dimensional spectra. NDBO expects to deploy discus hull systems capable of measuring directional properties of the sea in the next 2 years. The first five coefficients of the

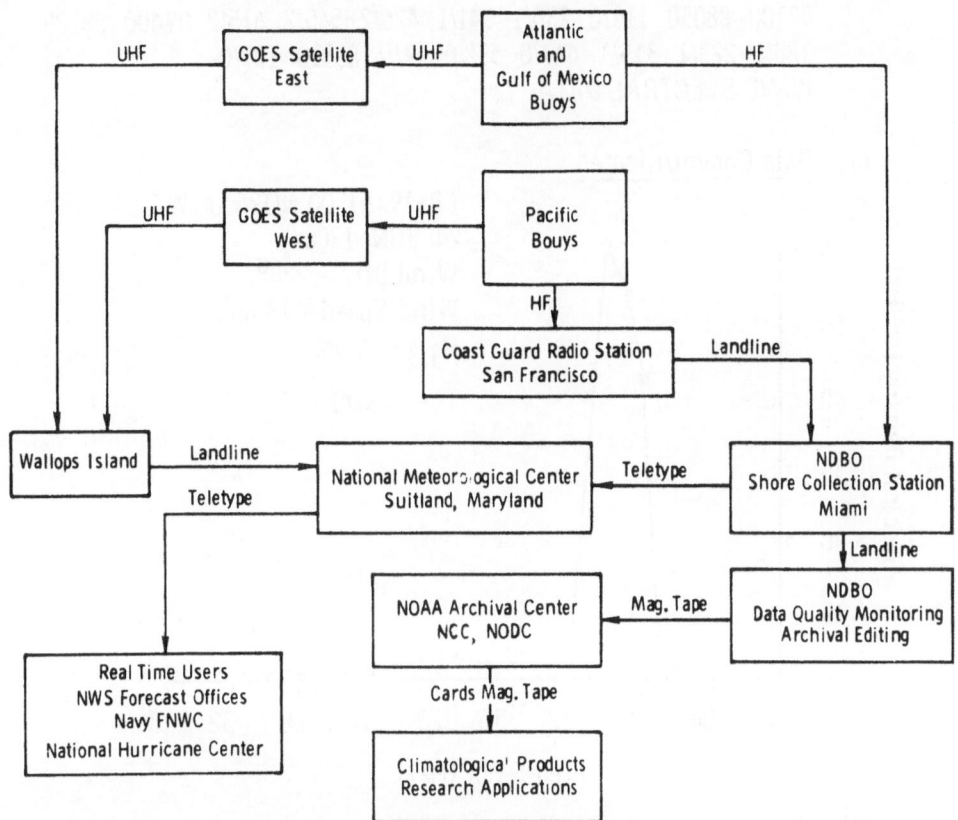

Fig. 11. Communications and dissemination of buoy data. Data can be obtained from the National Climatic Center (NCC) and the National Oceanographic Data Center (NODC). Data is used for wave forecast applications by National Weather Service (NWS) forecast offices, Fleet Numerical Weather Central (FNWC) and the National Hurricane Center.

I. <u>Sample Coded Message</u>

NNNNAB
ZCZC
SXVD10 PANC 241200
EB19 //510 71360 24121 /2213 30809
99100 88050 119/0 230/1 341/1 426/2 540/2 615/2 99400 88120
188/1 223/1 315/1 447/0 538/0 99/// 88330 125/0
WAVE SPECTRAL DATA;

II. <u>Data Communicated</u>

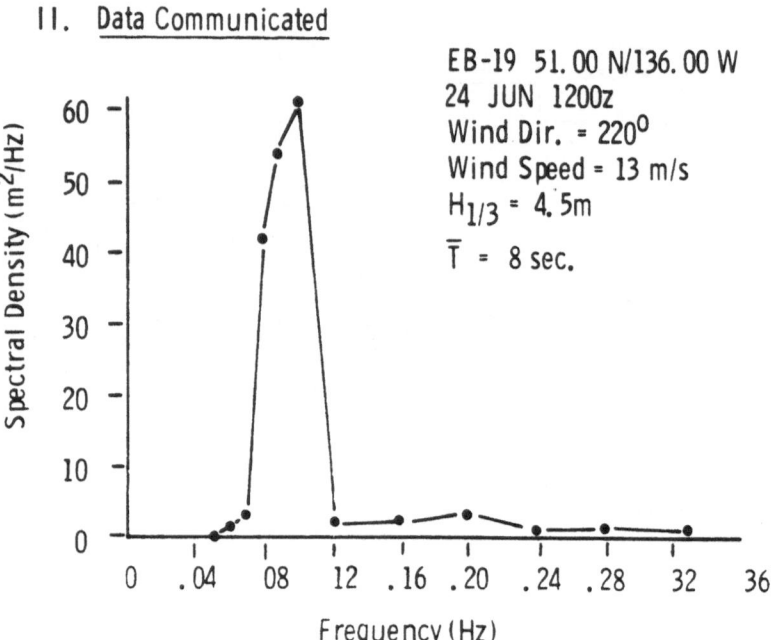

EB-19 51.00 N/136.00 W
24 JUN 1200z
Wind Dir. = 220°
Wind Speed = 13 m/s
$H_{1/3}$ = 4.5m

$\bar{T}$ = 8 sec.

Fig. 12. Example of NDBO coded spectral wave data report. The coded data is plotted in the lower half of the figure. This type of data is available to real-time users.

Fourier series expansion of directional spectra in azimuthal angle at a particular frequency can be determined through the measurement of the time series records of buoy heave, heading, pitch, and roll angles. The measurement method has been applied successfully in the past (*Longuet-Higgins et al.*, 1963), but the adaptation of it to operational use presents significant practical problems. This program is being actively pursued.

NDBO has achieved some success in the verification of wave
measurements from one buoy by relating them to measurements from
one or more other buoys. Substantial swells appearing in a particular narrow frequency band at one buoy will often arrive at other
buoys in the geographical area at other times depending on the
location of the source. More wave information can almost certainly
be extracted from a network of buoys than from the same buoys
treated as individual measurement stations. NDBO is now evaluating
the feasibility and desirability of systematically relating data
from several buoys to monitor the proper functioning of hardware
on individual buoys. The spectral wave data collected from a network of buoys in an area (such as the Gulf of Alaska) could be used
to produce a report of conditions in the area.

SUMMARY

NDBO's operational buoys produce useful wave data for the
National Weather Service and for other users. The quality of these
data is adequate for operational use. The accuracy of the wave
data analyzer systems is ±0.3 meters. Systems with the capability
for measuring directional spectra are under development. Additional
effort is being expended to refine the buoy transfer functions.

REFERENCES

General Dynamics. 1975. Technical manual for the Model 313 data
    acquisition and control system for the prototype environmental
    buoy. TM-75-PEB-3.
Goda, Y. 1974. Estimation of wave statistics from spectral infor-
    mation, International Symposium on Ocean Wave Measurements
    and Analysis, New Orleans, Am. Soc. Civ. Engs., *1*. 320-337
Hubert, W. E. 1976a. Evaluation of spectral wave data from EB-41,
    Tech. Report 6, Ocean Data Systems, Inc.
Hubert, W. E. 1976b. An operational evaluation of PEB buoy systems,
    Ocean Data Systems, Inc.
Hubert, W. E. 1977a. Evaluation of environmental data acquired by
    Magnavox Phase II Buoy Systems, Final Technical Report, Ocean
    Data Systems, Inc.
Hubert, W. E. 1977b. Evaluation of spectral wave data acquired
    by NOMAD buoys, Final Technical Report, Ocean Data Systems, Inc.
Longuet-Higgins, M.S., D. E. Cartwright, and N. D. Smith. 1963.
    Observations of the directional spectrum of sea waves using the
    motions of a floating buoy, *Ocean Wave Spectra,* Prentice Hall,
    N.Y. 111-136.
Magnavox. 1976. Equipment operations manual for Phase II payload
    system.
NOAA Data Buoy Office. 1973. Practical experience with buoys,
    internal NDBO report.

Remond, F. X. 1976. Ocean spectrum measurement with analog filters,
   In: *Proceedings of 22nd International Instrumentation Symposium,*
   San Diego, Calif.

Steele, K. E., J. M. Hall, and F. X. Remond. 1975. Routime measurements of heave displacement spectra from large discus buoys in the
   deep ocean. In: *Proceedings of the First Combined IEEE Conference on Engineering in the Ocean Environment and the Annual
   Meeting of the Marine Technology Society* (OCEANS '75), San Diego.

Steele, K. E., P. A. Wolfgram, A. Trampus, and B. S. Graham. 1976.
   An operational high resolution wave data analyzer system for
   buoys. In: *Proceedings of the Second Annual Combined Conference Sponsored by the Marine Technology Society and the IEEE*
   (OCEANS '76), Washington, D.C.

Measuring the Nearshore Wave Climate:
California Experience

Richard J. Seymour

California Department of Navigation and Ocean Development

ABSTRACT

A nearshore wave climate measurement network for the California coastline has been expanded to 16 stations since its inauguration in 1975. The system utilizes a compact slope array to measure wave directionality in shallow water as well as offshore buoys and nearshore non-directional gages. Data are collected and analyzed automatically by a central computer. Plans are being developed to expand the system to provide high density coverage of the whole state coastline.

INTRODUCTION

The need for characterizing the nearshore or coastal wave climate follows the experience of more conventional meteorological climate measurement programs: the higher the population density, the more intense the usage of the resource, the greater the penalty for ignorance -- then the greater the need for local detail and accuracy in the climate statistics. California has a coastline roughly as long as the stretch from Massachusetts to Florida. Much of it is heavily developed. The continental shelf has an extremely convoluted bathymetry and much of the state is partially sheltered by a long string of offshore islands. The coastline is predominantly rocky, punctuated by long reaches of sandy beaches in delicate dynamic equilibrium with the waves and currents, and fed by occasionally swiftly flowing rivers which dump sediment directly into the ocean. These factors combine to produce a situation in which there is a pronounced spatial inhomogeneity in the nearshore wave climate and strong political and economic pressures to define

317

it accurately. This has led the State, through its Department
of Navigation and Ocean Development (DNOD), to test the feasibility
of constructing a coastal wave data network to automatically acquire
coastal wave climate statistics by direct measurement. The principal purpose for this system is to provide information on which to
base decisions on coastal erosion protection programs. A secondary
purpose is to gather data on wave climates that impact boating and
other navigational activities.

THE CALIFORNIA WAVE NETWORK

     The system was inaugurated in 1975 by the Scripps Institution
of Oceanography (SIO) with overall direction and financial support
by DNOD, and with additional funding support by the Sea Grant program during the second year. The expansion of the system beyond
the initial seven stations funded by DNOD has been supported by the
South Pacific Division of the Army Corps of Engineers. A total of
16 stations were installed by the summer of 1978. These include
four directional measurement stations. Three of the stations have
"Waverider" wave measuring buoys offshore at a depth of approximately 80 meters in addition to the nearshore wave energy measurement capability. A system map is shown in Figure 1.

     The first station, at Imperial Beach near the Mexican Border,
has been reporting data since December 1975. With substantial
support from the Army Corps of Engineers, the network is expected
to be expanded to include well over 100 new stations along the
California coast in the period 1979-1982. The basic station consists of one or more bottom-resting pressure transducers which are
placed at a depth of about 10 meters. This results in their remaining outside the surf zone under all conditions in the present
locations. The transducers are hard wired to a terminal on the
beach or on a pier head which contains a telephone connection.
When offshore buoys are deployed, a radio telemetry link connects
the wave measuring buoy and the terminal on shore. Each channel
of information is sampled, digitized and stored in memory in the
terminal. The standard sampling rate for ocean waves is 1 Hz and
the sampling interval is 1024 secs (approximately 17 minutes).
The most recent data is maintained in memory. Each station is
automatically polled every ten hours by a central minicomputer at
SIO using ordinary dialup telephone lines. A schematic of the
system is shown in Figure 2. Details of the data transmission
system are described in *Seymour and Sessions* (1976). On certain
stations, either the wave buoy or a nearshore pressure gage is
monitored continuously. The number of zero upcrossings and the
mean squared value of surface elevation or bottom pressure are
calculated over 30 seconds and stored in memory. This provides a
nearly continuous record of the variance from which data on wave

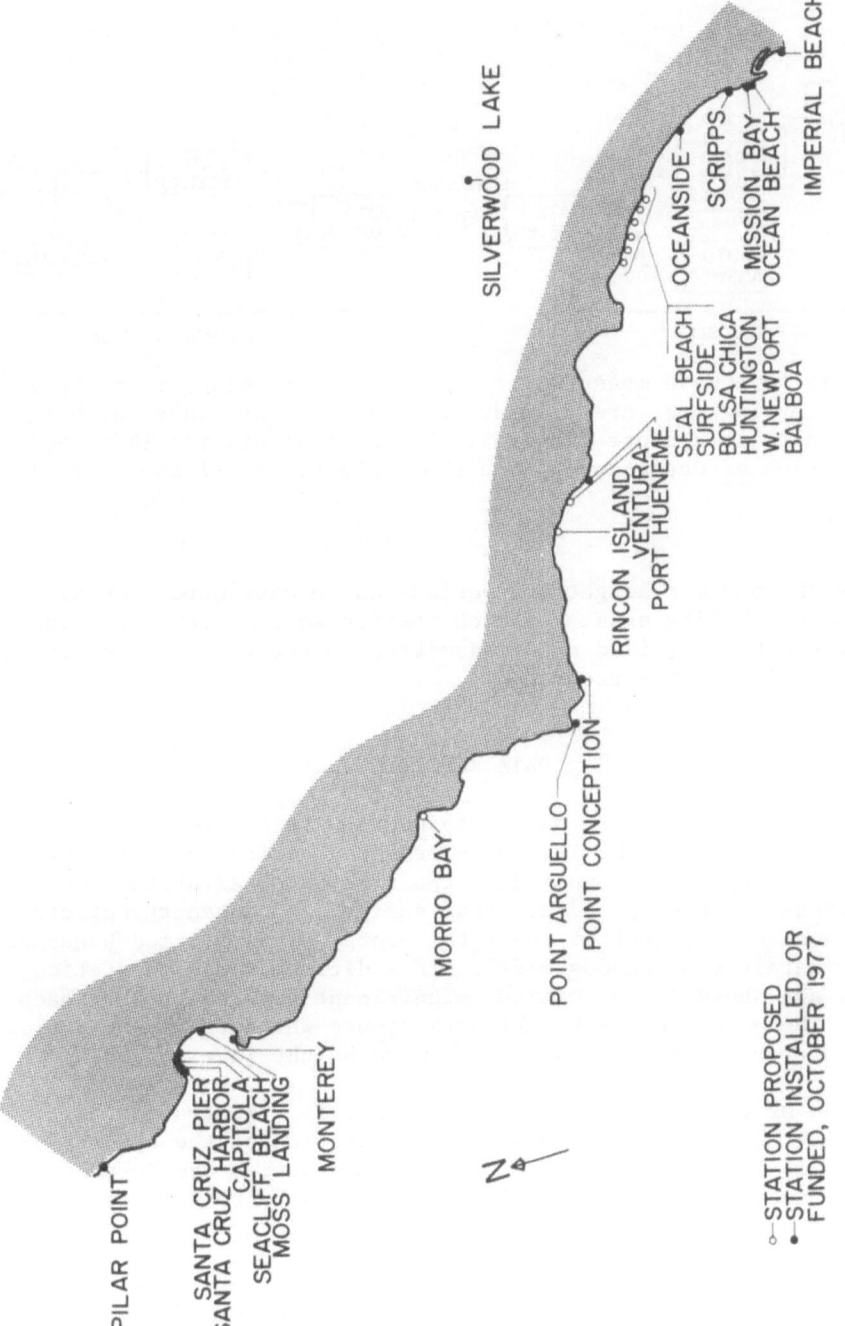

Fig. 1. California Coastal Engineering Data Network wave measurement stations. The indicated stations were installed or proposed as of October 1977.

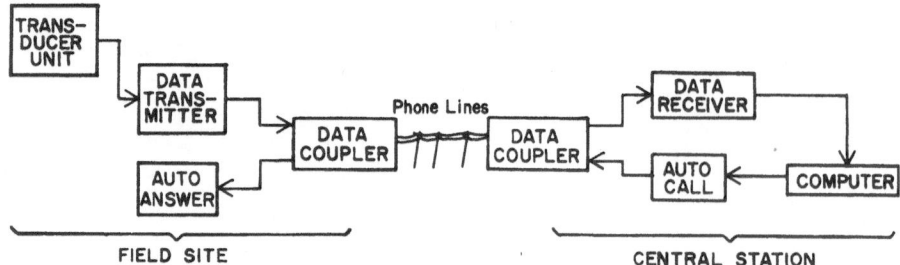

Fig. 2. Wave measurement system. The transducer can be a
bottom-resting pressure transducer or an offshore buoy which tele-
meters data to a shore terminal. The central station is at Scripps
Institution of Oceanography and the field sites are shown in Fig. 1.

groups and maximum height and period can be extracted. Multiple
channel capability exists at each station so that any other analog
signal can be digitized and transmitted to the central station in
addition to the wave records.

DATA REPORTING

Monthly reports are mailed to users within a few days after
the end of every month. For each ten-hour period and for each
station, these reports contain: the time of observation, the
significant wave height, the total energy, and an energy spectrum
expressed as a percent of the total energy within period bands of
approximately two seconds width. In addition, for each station,
tables are shown of the maximum significant wave height for each
day in the month and of height persistence which is the number of
consecutive days that a particular wave height is exceeded.

One of the most useful data displays is illustrated in
Figure 3. It shows a pictorial representation of the spectra for
a month at each station so that relative energies can be evaluated
and the decay of swell following a storm can be readily observed.
Although the energy relationships shown are largely qualitative,
they provide a concise representation of many pages of tabular data
and can rapidly direct the attention of a user to significant
events. For direction measuring stations, representative directions are reported for each period band using the methodology
described below.

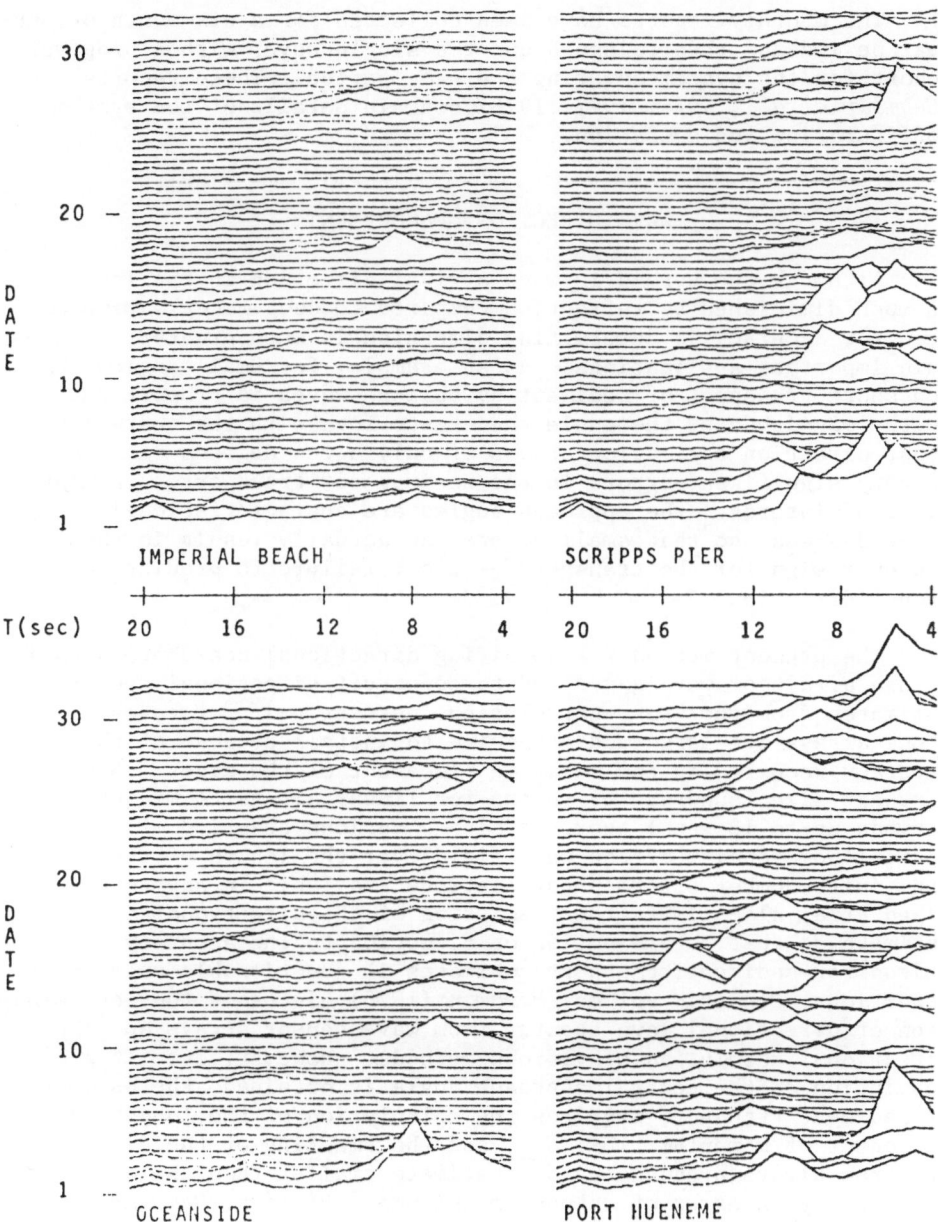

Fig. 3. Qualitative energy spectra for four stations during
the month of March 1977. Each line represents a measurement taken
every ten hours. Time moves upwards in the figure. The ordinates
of each plotted line are proportional to the wave energy at the
indicated period. Note that plots are linear in period, rather
than frequency.

The monthly reports have been collected for a six-month or one-year period and then reissued under a single cover with a progress report for the system covering the same period. These reports (*Seymour et al.*, 1976, 1977, 1978) contain details of the system configuration and operation.

WAVE DIRECTION MEASUREMENT

Although the spread in wave arrival direction in shallow water is much diminished by refraction compared to deep water waves, the need for accuracy in determining direction is much greater in certain important applications. As an example, for small angles the longshore transport of sediment by oblique waves is proportional to the magnitude of the angle the crests form with the shoreline. Thus, direction estimation errors of only a few degrees can result in very significant errors in the estimate of transport. In many cases of interest, the approach angles are characteristically only a few degrees, so that small errors can actually result in the improper sign for the transport -- i.e., failure to predict the proper direction.

The present method for acquiring directional nearshore data is to use arrays of wave gages and to construct directional spectrum estimates from the cross-correlations between pairs of gages. These arrays are large and expensive to install because of the accuracy required in the relative placement of the gages. Because of their size (hundreds of meters for linear arrays to measure periods up to 18 secs) there is also a serious question of the spatial homogeneity of the wave field over the length of the array. The finite number of cross-correlations available in any reasonable sized array also limits the resolution of the directional spectrum estimate so that determination of direction to the accuracy required for sediment transport estimates is questionable under many conditions. In this volume, *Borgman* (1978) discusses the determination of directional wave spectra from wave sensor arrays and the directional resolution of various methods. However, even if good resolution of the central approach angle is obtained, the estimate of sediment transport requires the calculation of the longshore component of momentum flux, S_{xy}, and the smearing of the directional spectrum always produces an S_{xy} estimate smaller than the actual quantity by an error of unknown magnitude. *Higgins, Seymour, and Pawka* (1978) show the analytical proof for this and describe a series of verifying field experiments.

Since the principal motivation of wave direction measurement in this program is to predict sediment transport, it was decided to measure S_{xy} as a function of frequency directly. Following *Longuet-Higgins* (1970)

$$S_{xy}(f) = \overset{\theta}{\Sigma} \; E(f, \; \theta) \; n \; sin \; \theta \; cos \; \theta$$

$$(-90° \leq \theta \leq + 90°)$$

(1)

where S_{xy} = longshore component of shoreward directed momentum flux

$E(f, \theta)$ = energy as a function of frequency f and direction θ

n = ratio of group and phase velocities

θ = angle between wave arrival direction and the normal to the shoreline

Equation (1) provides a conceptual picture of the longshore component of momentum flux showing its dependence upon approach angles. However, in longshore sediment transport calculations using current models, the transport rate is proportional to the magnitude of S_{xy} and its direction is determined by the sign of this quantity. Therefore, the component approach angles need not be measured if a means exists for direct measurement of S_{xy}. An estimate of S_{xy} is obtained from this system using the approach suggested in *Longuet-Higgins et al.* (1963) of utilizing the time averaged value of the product of the orthogonal sea surface slope components.

$$S_{xy}(f) = \frac{n}{k^2} \; <n_x n_y>$$

(2)

where k = wave number

and $<n_x n_y>$ = time average of slope components in the offshore and alongshore directions

The averaging of the slope components is done in frequency space by computing the cospectrum between the slope component time histories. The slopes are measured by using the difference between the two pairs of bottom-mounted pressure transducers. The Fourier coefficients are corrected by standard linear wave theory for the pressure attenuation effects.

A more complete treatment of the theory and a discussion of the magnitude of the errors inherent in this estimation scheme are found in *Seymour and Higgins* (1978A) and *Higgins, Seymour, and Pawka* (1978).

Measurement of S_{xy} for ocean waves to acceptable engineering
accuracy can be made with relatively small arrays -- on the order
of six meters on a side. This allows the array to be built into a
single rigid frame. The alignment of the frame is established
within approximately one-half degree by divers using a waterproof
magnetic compass. Coordinate rotation is employed to correct the
slopes to be referenced to the local trend in bottom contours. The
estimation of longshore sediment transport using data obtained
through this system is described in *Seymour and Higgins* (1978B).
It can be seen that employing the values of $S_{xy}(f)$ from Equation (2)
in Equation (1) allows the estimation of a representative approach
angle for each frequency interval.

INTERACTIONS WITH DEEP WATER MEASUREMENT PROGRAMS

Since the principal concern with wave statistics is their
application to the nearshore environment, any scheme for producing
deepwater statistics -- either using direct measurement or hindcasting techniques -- requires a test of its ability to allow
projection into shallow water. Synoptic data from large numbers
of nearshore measurement stations provides the ultimate test for
any deep water system. The usefulness of deep water statistics to
the coastal engineer depends upon using refraction analyses
and upon being able to predict the nearshore wave environment to
reasonable accuracy.

If comparisons of nearshore measured climates and climates
predicted from deep water data show that the deep water directions
cannot be measured or inferred with sufficient accuracy to allow
such projections into shallow water, nearshore measurements can
possibly provide a reasonable means for estimating deep water
direction. This approach, suggested by M. P. O'Brien of
Berkeley, is based upon using non-directional nearshore wave gages
spaced many kilometers apart along a coastline as an array to
determine deep water wave direction. This large-scale array could
be used in two modes. For a given frequency band, each deep water
wave approach direction can produce a unique fingerprint of spectral
intensities at the various elements in the array, provided that the
intervening bathymetry is sufficiently irregular to cause non-
homogenous refraction, so that a deep water direction can be assigned.
The second mode of operation would be to use the shore stations as
a directional array which could be considered coherent for events
of sufficiently long time and space scales. *Thompson and Smith*
(1974) show that swell trains have well-defined maxima in group
heights that would have phased arrival times measured in hours
over an array on the order of a hundred kilometers in length. A
more complete treatment of these array concepts and the results
obtained are contained in *Seymour and Higgins* (1978C).

SUMMARY AND CONCLUSIONS

A reliable system for sampling the coastal wave climate has
been demonstrated. A compact representation of wave directionality is contained in the spectrum of the longshore component of
momentum flux obtained routinely with this system and this has been
shown to be useful in predicting longshore sediment transport.
Wave statistics obtained from this system have been used for the
design of harbor entrance improvements, shoreline erosion protection structures and for a large number of scientific investigations,
including large-scale island shadowing experiments and providing
ground truth for evaluating techniques for the remote sensing of
waves. The effectiveness and usefulness of the system has led to
plans to expand it to cover the entire coastline of California.

REFERENCES

Borgman, L. E. 1978. Directional wave spectra from wave sensors,
    In: *Ocean Wave Climate,* M. D. Earle and A. Malahoff (Eds.),
    Plenum, N.Y., (this volume).

Higgins, A. L., R. J. Seymour, and S. S. Pawka. 1978. A compact
    representation of ocean wave directionality. (In press.)

Longuet-Higgins, M. S. 1970. Longshore currents generated by
    obliquely incident sea waves, In: *J. Geophys. Res., 75,*
    6778-6801.

Longuet-Higgins, M. S., D. E. Cartwright, and N. D. Smith. 1963.
    Observations of the directional spectrum of sea waves using the
    motions of a floating buoy, In: *Ocean Wave Spectra,* Proceedings
    of a Conference, Prentice-Hall, Inc., Englewood Cliffs, N.J.

Seymour, R. J. and A. L. Higgins. 1978A. A slope array for
    estimating wave direction, In: *Proceedings of a Workshop on
    Coastal Processes Instrumentation,* June 16, 1977, La Jolla, CA,
    University of California, San Diego, Institute of Marine Re-
    sources Ref. No. 78-102, Sea Grant Pub. No. 62.

Seymour, R. J. and A. L. Higgins. 1978B. Continuous estimation of
    lonshore sand transport, In: *Coastal Zone 78 Symposium on
    Technical Environmental, Socioeconomic and Regulatory Aspects
    of Coastal Zone Management,* American Society of Civil Engineers
    Symposium, San Francisco, California, March.

Seymour, R. J. and A. L. Higgins. 1978C. Deepwater wave direction
    from an intensity array, In: *Proceedings, Sixteenth International
    Conference on Coastal Engineering,* Hamburg, Germany, Sept.
    (In press.)

Seymour, R. J., A. L. Higgins, S. L. Wald, and A. E. Woods. 1978.
    California coastal engineering data network second annual report,
    January 1977 through December 1977, California Department of
    Navigation and Ocean Development, Sacramento, California.

Seymour, R. J. and M. H. Sessions. 1976. A regional network for
    coastal engineering data, In: *Proceedings, Fifteenth Inter-
    national Conference on Coastal Engineering,* Honolulu, Hawaii,
    July.

Seymour, R. J., M. H. Sessions, S. L. Wald, and A. E. Woods. 1976.
    Coastal engineering data network first semi-annual report,
    December 1975 to June 1976, Institute of Marine Resources,
    University of California, IMR Ref. 76-11, Sea Grant Pub. No. 50.

Seymour, R. J., M. H. Sessions, S. L. Wald, and A. E. Woods. 1977.
    Coastal engineering data network second semi-annual report,
    July 1976 to December 1976, Institute of Marine Resources,
    University of California, IMR Ref. 77-103, Sea Grant Pub. No. 56.

Thompson, W. C. and R. C. Smith, Jr. 1974. Wave groups in ocean
    swell, In: *Proceedings of the International Symposium on Ocean
    Wave Measurements and Analysis,* American Society of Civil Engi-
    neers, New Orleans, Louisiana, Vol. I, Sept., 338-351.

A Note on the Wave Climatology of UK Waters

Laurence Draper

*Marine Information and Advisory Service,
Institute of Oceanographic Sciences, UK*

ABSTRACT

Wave measurement started in earnest during the second world war. Subsequently, fundamental studies were initiated to try to understand the processes at play and to evolve effective wave prediction methods. As a consequence of this work, large amounts of instrumentally-measured wave climate data have been accumulated and found to be valuable for engineering purposes. An international network is being set up to collect and disseminate such information.

INTRODUCTION

The first major impetus for wave research came during the second world war when the need arose to predict wave conditions at landing beaches. The task was facilitated by the developments in electronics which enabled reasonably good measurements to be made, and these data, together with visual observations of waves, resulted in the evolution of empirical wave prediction methods. By about 1950 it was realized that, because wave recorders then in existence could only operate quite near to coasts and in shallow water, the data then available referred only to waves which had been severely affected by the seabed; it became apparent that in order to measure unmodified waves it would be necessary to devise an instrument which would operate on either a buoy or a ship. The problem was first solved in 1952 with the Shipborne Wave Recorder (*Tucker*, 1956). By 1960, two UK weather ships carried these instruments, as did some French and Dutch vessels. *Darbyshire* (1955) used the UK ship data to derive a wave prediction formula

for oceanic areas. After another instrument had been installed,
first in the Morecambe Bay Light Vessel in the Irish Sea and
then in Smith's Knoll Light Vessel in the southern North Sea, he
was able to extend his wave prediction formula to shallower waters
(*Darbyshire*, 1963). Following the success of these two Light
Vessel installations, further instruments were used and their
coverage gradually spread over most UK waters so that good measured
wave climate information for duration of at least a year exists for
about twenty offshore locations (these are listed in *MIAS* and *Waves*).

As well as the open-water sites, data have been collected and
analyzed for an appreciable number of locations very near to the
coast for specific projects, mainly using cable-connected underwater
pressure sensors. In the early days, voltage or current reporting
sensors were used, but because of cable insulation problems, which
cause gradual and often undetected deterioration in the transmitted
data, these have long been superseded by frequency modulation
systems. These, in general, either work well or fail catastroph-
ically, so that recorded data are usually reliable.

EXTREME WAVE CONDITIONS

The increase in the number of existing and proposed semi-
permanent offshore installations, mainly for hydrocarbon and wave
power purposes, has demanded considerable expenditure of effort
into devising techniques of estimating extreme wave conditions.
The UK Department of Energy publishes maps, an example of which is
shown on Figure 1 (*HMSO*, 1977), presenting the most probable values
of the heights of the highest waves expected to occur in a fifty
year duration. It is only five years since it has been possible
to produce such maps (*Draper*, 1972), but although many different
techniques have been investigated and much more data obtained, the
subsequent changes needed have been small, especially in the
critical areas.

Studies of accumulations of wave data, now running into many
years at a small number of stations, continue in an attempt to
understand the distributions of wave heights under severe conditions. For example, an analysis of five years' data from Sevenstones Light Vessel off Land's End is presented in IOS Report No.
39 (*Fortnum and Tann*, 1977), and a similar study is about to be
published giving five years' data from the North Sea at 57°30'N 3°E.

WAVE MEASUREMENT PLANNING

In its early stages the UK wave climate programme was not a
planned operation; it just grew as opportunities appeared. Now that

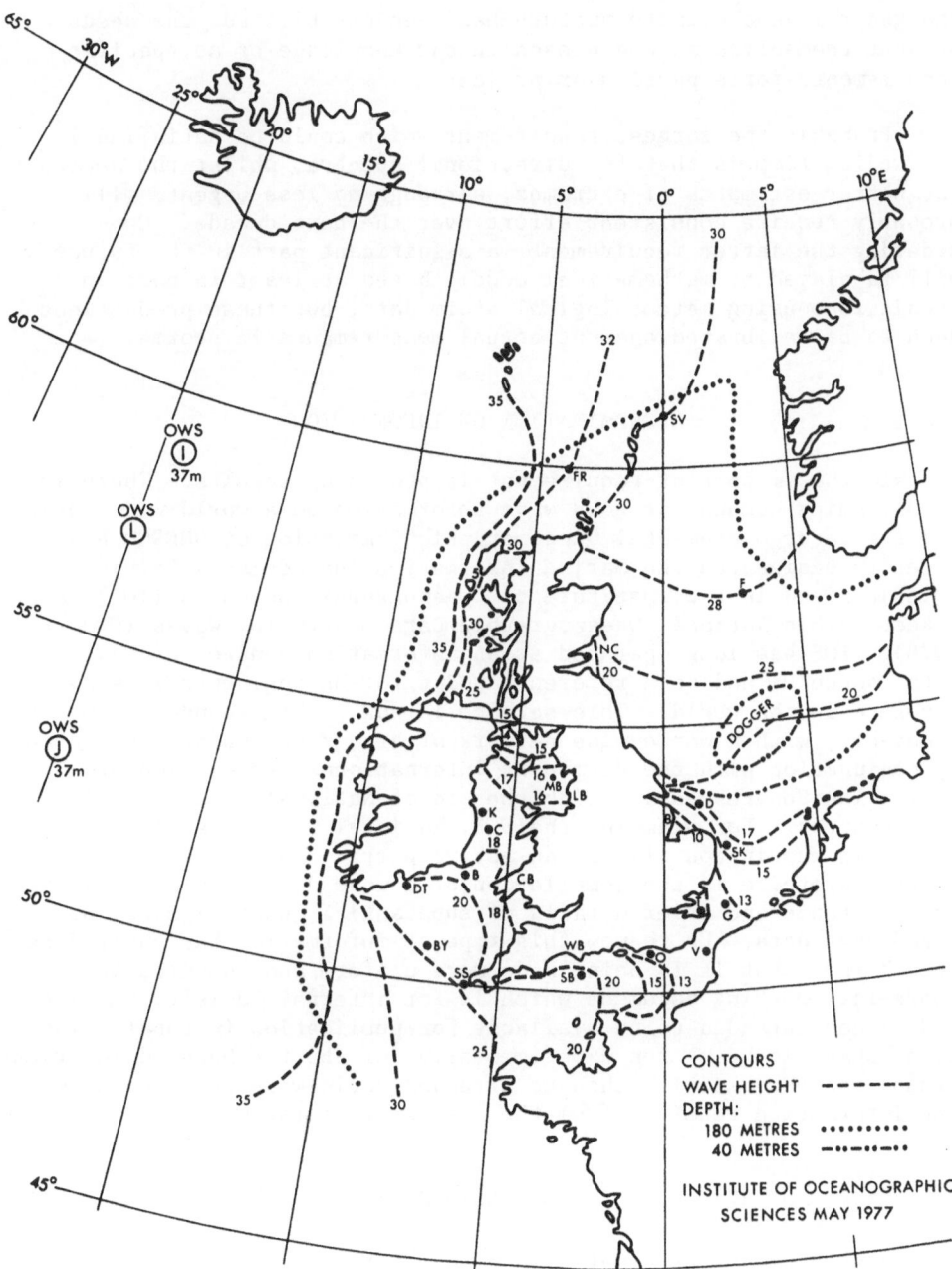

Fig. 1.   The 50-year extreme wave height around the British
Isles (from *HMSO*, 1977).   Wave heights are given in meters.   These
results are based largely on data collected or archived by the
Marine Information and Advisory Service.

the general wave climate picture has been established, the needs
present themselves as weaknesses in our knowledge or as specific
requirements for a particular project.

Probably the largest requirement which could be satisfied in
the medium term is that for directional spectra, whilst the demand
for better estimates of extremes, although no less urgent, will
probably require consistent effort over the next decade. Considering the latter requirement, a significant part in the future
will be played by mathematical models based at least in part on
predictions using meteorological storm data, but these predictions
need to be calibrated against actual measurements in storms.

DISSEMINATION OF INFORMATION

Another important requirement is providing results. There is
a continuing demand for good wave information on a world-wide basis
and the Intergovernmental Oceanographic Commission of UNESCO has
recently designated the Marine Information and Advisory Service
(MIAS), based in the Institute of Oceanographic Sciences (IOS), as
a Responsible National Oceanographic Data Centre for Waves (*Draper*,
1978). IOS has long operated as an information centre for wave
data in the UK and as a reference point for UK engineers working
anywhere in the world. This service is now to be expanded considerably, with a world-wide network of area representatives chosen
in conjunction with the Permanent International Association of
Navigation Congresses and with the aid of National Oceanographic
Data Centres. For example, the U.S. National Oceanographic Data
Center in Washington will be an active participant. The first phase
in this exercise is the compilation of a wave data catalogue listing
the existence and brief details of substantial quantities of analyzed wave data. Up to now this type of information has existed as
a card index but it is being made more uniform and compiled into a
loose-leaf working document which is for internal advisory service
and is not intended to be available for publication in total. The
first phase is being run experimentally for about a hundred locations
around the UK. It will then be extended world-wide in response to
the information received from the area representatives.

CONCLUSIONS

Over a period of twenty years the UK Wave Climate programme
has evolved from an almost unintentional start to an active coordinated programme capable of planning in response to specific needs.
It is continuing to provide data for projects; gradually reducing
the average distance between data stations by installing and maintaining equipment for at least a year; and building up long-term

data series at specific stations to enable better extreme predictions to be made. Results are studied in conjunction with
mathematical models, resulting in a degree of understanding which
was almost undreamt of a quarter of a century ago.

REFERENCES

Darbyshire, J. 1955. An investigation of storm waves in the
     North Atlantic Ocean. In: *Proceedings of the Royal Society,
     A 230,* 560-569.
Darbyshire, J. 1963. The one-dimensional wave spectrum in the
     Atlantic Ocean and in coastal waters. *Ocean Wave Spectra.*
     Prentice-Hall, 27-31.
Draper, L. 1972. Extreme wave conditions in British and adjacent
     waters. In: *Proceedings of the 13th Conference on Coastal
     Engineering,* Vancouver, Vol. 1, Chapter 6.
Draper, L. 1978. A world wave data centre. Accepted for the
     16th Conference on Coastal Engineering, Hamburg.
Fortnum, B. C. H. and H. M. Tann. 1977. Waves at Sevenstones
     Light Vessel, Jan. 1968 to June 1974. Institute of Oceanographic
     Sciences IOS Report No. 39.
HMSO (Her Majesty's Stationery Office). 1977. Offshore installa-
     tion: Guidance on the design and construction.
MIAS (Marine Information and Advisory Service) and Waves.
     Occasional publication. Institute of Oceanographic Sciences, UK.
Tucker, M. J. 1956. A shipborne wave recorder. *Transactions of
     the Institute of Naval Architects,* London, *98,* 236-250.

III

RECOMMENDATIONS OF
SYMPOSIUM WORKING GROUPS

COASTAL AND OCEAN ENGINEERING APPLICATIONS FOR
WAVE DATA WITH EMPHASIS ON STRUCTURE DESIGN

Major applications of wave data include: offshore platform
design, breakwater and jetty design, site selection criteria,
operational planning, military operations, transportation studies,
environmental impact studies, real-time input to wave forecasting,
search and rescue, and pollution clean up.

Wave statistics should be available in all U.S. coastal waters
particularly outer continental shelf lease areas. This requires
both observations and wave hindcasting. Emphasis should be placed
on observations at relatively deep water sites to minimize local
bottom effects. Shallow water measurements should be made for
specific applications or purposes. For coastal and offshore structure design, extreme event data is important and, above a selected
threshold value, wave conditions should be monitored continuously.
Long-term measurements are needed for offshore platform design.
However, since platform designers are faced with costly economic
and engineering decisions in the immediate future, design requirements must be based on hindcasting. Wave models must be verified
with good data from storms and hurricanes. Many users are capable
of working with raw wave data (with appropriate calibrations and
some quality control) and these users would like to obtain the
wave data quickly before it is archived. Some users would be able
to make use of real-time data. Because real-time data from
National Oceanic and Atmospheric Administration (NOAA) wave programs will be provided to the National Weather Service, it should
be possible to provide these data in real-time to other users. The
measurement of directional wave spectra should be an important goal
but the difficulty in making directional measurements should not
preclude making wave height measurements with available state-of-
the-art equipment.

Although cost savings due to better wave data would occur for
the above applications, the working group did not wish to provide
estimates. For offshore structure design, cost savings calculations are complicated and involve risk analysis.

As a comprehensive wave monitoring program is implemented, the
program should concentrate on measurements at one or a few locations

335

with high reliability equipment to assure a high data return probability. Specifically, one fixed platform should be instrumented to measure directional spectra reliably to test directional spectra representations and to verify wave forecasting and hindcasting models. A developmental program to measure directional spectra should parallel measurements by available techniques. The group strongly emphasizes the importance of setting up the data processing, quality control, and information dissemination system before the measurements are made.

Members of Working Group:
T. Fallon, Chairman
L. E. Borgman
L. Draper
E. Escowitz
D. L. Harris
L. LeBlanc
C. S. Niederman
K. G. Nolte
K. Steele
R. W. Whalin

APPLICATIONS OF WAVE DATA TO SHIP DESIGN
AND SHIP ROUTING

Among the most important applications of wave data in the marine field are the specification of ship responses, ship design, and ship operation planning.

Ship Characteristics - There is little foundation other than tradition for specification of the seakeeping qualities which the designer of a ship is expected or required to achieve. A critical missing factor is a wave climatology which would permit rational analysis of requirements as a function of mission and expected operating environment.

Ship Design - Data is needed for many aspects of design, including development of ship geometry (form, freeboard, flare), structure (main hull, girder, local strength), and mission equipment (sensor systems, helicopter handling systems, weapon systems, securing cargo).

Ship Operations - Historical data as well as highly reliable forecasts and descriptions of present conditions are needed for ship routing, planning of tactical operations, and ship handling in severe storms.

A wide variety of data are needed. Some of the questions which arise relate to the shape of the spectra as a function of frequency and the directional bandwidth, duration of storm conditions, duration of wave groups, magnitude of swells as well as wind seas, wind direction and speed, extreme conditions for a wide variety of responding vessels, and accuracy of forecasts and hindcasts.

Hence, one of the principal goals for both open ocean and nearshore ship applications is development of a capability to analyze and predict the spectral distribution of energy at the sea surface. In order to accomplish this objective, measured wave data are required as data input and quality control for numerical wave spectra prediction models. Owing to economic and time constraints imposed on even the largest computers, the

definition of the computation grid and parameter specification
(direction bandwidth and frequency bandwidths) can never toally
represent the ocean surface. Therefore, high quality observed
wave data, available in real-time, would provide information
required for adjusting and updating wave spectral forecasts and
allow synoptic modification to compensate for environmental variations and numerical design constraints.

For ship design and ship routing wave data are needed world-
wide. For example, shipping around the South African Coast
encounters hazardous conditions under some combinations of waves
and currents. Wave data are needed on the Great Lakes for "now-
casting" and warnings during bad weather. It has also been
suggested that more use be made of shipboard radar screens as an
aid in providing better estimates of wave direction at night.

In addition to the need for a climatology of wave parameters,
there is a growing request for joint distributions of wind and
wave parameters. Joint distributions of wind speed, wind direction,
wave height, wave period, and wave direction would enable a tre-
mendous improvement in the development of limiting operational
conditions of aircraft from ships at sea.

Cost will be heavily influenced by adequate wave data, but
there are so many unknowns that it is impossible to make any
precise statement with respect to ship design. However, in ship
routing real cost savings have been identified. For example,
current government per diem costs for a typical contract vessel
are approximately $20,000 per day. By applying ship routing in
conjunction with wave forecasts and ship responses, minimum transit
time can be realized resulting in savings of 12 to 36 hours. This
also means an additional savings in fuel costs.

A wave monitoring program to address these needs would
include:

(a) Continuation and gradual expansion of environmental data
buoys deployed by the NOAA Data Buoy Office including addition of
a directional wave capability. Directional bandwidth resolutions
of about 15 degrees would be very useful for many ship requirements
but are probably not possible with buoys. At present, wave spectral
density accuracies will probably be based on intercomparison with
Waverider type buoys simply because we have no other better standard for open ocean application.

(b) Drifting buoys which provide real-time data on wave
conditions would be enormously valuable for upgrading forecasts
and ship routing. The first five Fourier coefficients of the
directional spread of wave energy can be obtained from a floating
buoy, and would be most helpful.

(c) Less sophisticated disposable wave buoys to be dropped from cooperating ships would offer a much wider spread of wave information which would supplement the drifting buoy measurements.

(d) A major effort should be made to upgrade the surface wind data which is the input to forecast and hindcast models.

A central goal of the wave monitoring program should be wave model validation. However, before any wave models can be compared, the wind fields used to drive these models must be verified.

Members of Working Group:
P. DeLeonibus, Chairman
S. L. Bales
W. A. Cleary
G. Christoph
W. E. Cummins
D. Hoffman
F. Middleton
N. Stevenson
D. A. Walden

WAVE DATA REQUIREMENTS FOR COASTAL PROCESSES STUDIES

Good measured wave data are needed for studies and applications involving sediment-structure interaction, shoreline erosion, design of navigation channels, beach nourishment, coastal hydrodynamics, coastal zone planning, coastal operations, wave related biological losses, harbor surging, dredging, recreation, and marine facilities design. The following table qualitatively summarizes data requirements for a wave monitoring program.

There is no question that a comprehensive wave monitoring program will result in substantial cost savings for coastal applications. Maintenance dredging presently costs about $500 million annually and better wave data could result in savings of $150 million annually. Property loss due to erosion is on the order of $1 billion annually and better wave data could reduce losses by up to $500 million. These data would also reduce the excess cost both for construction and maintenance of coastal structures. Currently, about $500 million annually is spent for coastal structures and savings of up to $250 million could occur. Cost savings would occur for other coastal applications but these savings would be small in comparison to the large numbers that we have just provided. These other applications include increased operational efficiency of vessels, better safety and advisory services, reduced cost of dredged material disposal, and harbor design optimization including effects of long period waves. Overall, better wave data suitable for coastal processs studies could result in cost savings on the order of $1 billion annually. Although this figure is an estimate, cost savings would, without doubt, be many times the cost of a program to collect good measured wave data.

Members of Working Group:
O. T. Magoon, Chairman
W. L. Edge
V. Goldsmith
J. G. Housley
A. Malahoff
J. W. Maresca, Jr.
R. P. Savage

R. J. Seymour
C. L. Vincent
W. E. Woodward

Wave data requirements for coastal processes studies.

Studies and Applications	Geographic Coverage	Data Time Duration	Extreme Event Data	Deep and/or Shallow Measurements	Directional Spectra	Accuracy
Sediment Transport	-local -site specific -high density measurements	Unknown but very long term	Very Important	Need shallow data (near-shore and breaker zone).	Yes	High for wave height, period, and direction (very difficult)
Wave Limited Operation	Same	-real time for safety -monthly seasonal yearly, etc. for planning (i.e. wave persistence)	-not very important for operations (work is not possible if wave conditions exceed given values)	Deep and shallow data needed.	Yes	Moderate
Coastal Hydrodynamics	Same	Long-Term	Very Important	Deep and shallow data needed for shallow water transformations.	Yes	High

USE OF WAVE DATA FOR DEVELOPING AND VERIFYING
WAVE FORECASTING AND HINDCASTING MODELS

If hindcast models are to be further developed, improvements in monitoring the sea surface are required in both deep and shallow water. There exists a critical need for a measurement network that can provide continuous long-term measurements in terms of frequency spectra generally, and directional spectra at some sites. The system could consist of a network of buoys on the edge of the U.S. continental shelf including the East Coast, Gulf of Mexico, Gulf of Alaska and the Bering Sea.

With the exception of the Bering Sea, the present configuration of the NOAA Data Buoy Office (NDBO) buoy network would provide adequate coverage and could serve as the basis of the deep ocean wave monitoring system provided:

(a) the accuracy of the wave spectrum measurement system is established firmly through definitive studies to determine buoy transfer function, and overall system capability,

(b) that the systems are continuously upgraded until satisfactory accuracies are attained,

(c) a system of quality control is established that includes regular *in situ* checks on system calibration, and

(d) data is available every 3 hours, but hourly in tropical cyclones.

The deep ocean network should include in each broad region (e.g. North Atlantic, South Atlantic, Gulf of Mexico, West Coast, Gulf of Alaska, Bering Sea) at least one directional spectrum measurement system. Wave measurement programs should include meteorological measurement programs that provide at least: wind speed and direction, suitably averaged; sea level pressure; air temperature; sea surface temperature. Such measurements are already available from the NDBO buoys. The long-term deep ocean network needs to be augmented at specific regions and for limited time periods by a more intensive measurement network. The intensive network would

consist of several buoys spaced 100-200 km apart, including a long
term deep water site. The network should be deployed for a period
of time (e.g. 1 year) in a given region and then moved to another
region. The intensive network should include at least one site
that obtains directional spectra. All measurement sites in the
intensive deep water network should obtain at least one dimensional
spectra and the meteorological measurements described above. Near
shore intensive wave measurement programs that operate for shorter
time spaces (e.g. weeks to months) and on smaller space scales (e.g.
10's of kilometers) should be planned to complement existing offshore programs and should be coordinated with the intensive deep
water network. For the U.S. East Coast, the working group recommends that the near shore shallow intensive network be located
between roughly Cape Hatteras and Montauk Point.

It is recommended that, whenever appropriate to the needs of the
monitoring program, existing offshore platforms be utilized as
sites for wind and wave sensors. This could be done on a cooperative basis as is the case in Europe (North Sea) and Canada. It is
further recommended that joint industry, government and institutional
measurement and analysis programs be explored and encouraged.

A comprehensive wave monitoring program should include a program
supportive of hindcast and forecast model development and intercomparison. It is recommended that after a period of model development and wave data acquisition from the intensive networks, the
program coordinate a thoroughly objective test of contemporary
models, including at least one model from the following classes:

(a) significant wave

(b) discrete spectral Pierson-Tick-Baer type source function

(c) discrete spectral nonlinear term in source function

(d) parametric

(e) hybrid (parametric plus discrete swell)

The tests should be run with the same wind fields, evaluated
against independent and withheld wave data representative of a
wide range of conditions (e.g. tropical cyclones, extratropical
cyclones, limited fetch, swell, etc.). The program should strive
to assemble and archive wave data and meteorological data in a
form accessed by modelers, particularly where there is joint
government, institutional, and industry participation in a measurement program. The program should also archive routine meteorological
analysis and forecast fields from the NOAA National Weather Service
operational models, in grid form for all marine regions.

The working group recognizes the value of airborne and satellite wind and wave remote sensing systems, particularly during the intensive measurement phases. Such sources of data should be utilized, especially to fulfill measurement objectives not easily attained by platform mounted sensors; e.g. measurement of wave spectra in the right quadrant of fast moving intense hurricanes after recurvature.

Members of Working Group:
V. J. Cardone, Chairman
J. E. Overland
D. B. Ross
E. G. Ward

RADAR WAVE MEASUREMENTS

To be suitable for operational applications, a radar technique must be able (after development) to provide wave data at a given point or points, periodically during a day, and to operate relatively continuously for months or years. Thus, some techniques such as aircraft measurements and synthetic aperture coastal radars are suitable for research but not operational use. Synthetic aperture coastal radars could be operated on a regular basis but wave periods greater than 7 seconds cannot be measured due to interference with broadcast frequency bands.

Radar methods have several advantages over *in situ* measurement methods. Radar methods provide more continuous coverage both in space and time, are logistically simpler since ocean operations are avoided, and the data can be collected and processed in real-time. A disadvantage of radar methods is that time series and point geographic observations are often unavailable. Instead, the observations are in terms of statistical representations typically over 1/2 hour time periods and over approximately 1 km spatial regions.

There is a need to better understand the physics of both the radar and *in situ* measurements. For example, how does one compare and make use of area wave field measurements? The working group also recommends work to determine accuracies of both radar and *in situ* techniques, to locate radar error sources, and to reduce these errors. In addition, careful *in situ* measurements with known accuracies are needed for comparison to radar measurements.

Members of Working Group:
D. E. Barrick, Chairman
D. Enabnit
W. Evans
D. L. Harris
B. J. Lipa
J. W. Maresca, Jr.

M. Mattie
C. Parsons
C. C. Teague
W. E. Woodward

WAVE MEASUREMENTS FROM BUOYS

The working group stressed the need for buoys with a capability
of determining the direction of wave propagation. Nevertheless,
there is an extensive requirement for nondirectional wave measuring
buoys. Two such buoys are commercially available. One, the Waverider, is manufactured by Datawell and the other is made by Coastal
Data Service. The latter buoy has recently (1978) become part of
the product line of Environmental Devices Corporation (ENDECO).
The Datawell Waverider has been extensively used in the field over
the last eight years. The cost of a Waverider system (including
buoy, shore receiving station, parts, and hardware) is about $15,000.
The Coastal Data Service buoy with telemetry capabilities has not
been regularly used at sea. A disadvantage of small telemetering
buoys such as these is the limited transmission range, approximately
twenty miles.

The group recognized that the NOAA Data Buoy Office (NDBO) is
the only organization presently deploying buoys with wave direction
measurement capabilities. These buoys cost on the order of $100,000
with annual maintenance costs of about $50,000-$75,000. It was
recommended that NDBO deploy five to ten such systems as soon as
possible in deep water at locations to be recommended by other
working groups. In undertaking this work, NDBO must establish and
document the accuracy of the wave recording systems and, when the
buoys have been deployed, service and calibrate the systems at
regular intervals. The matter of buoy calibration, including small
Waverider type buoys, was emphasized. All buoys should be calibrated
in real situations and there is a need for calibrations at the low
frequency end of the wave spectrum.

Future wave program work should include the development of
instrumentation to resolve wave directions with a resolution of
15°. Recommendations for the wave monitoring field work include:
the measurement of wind speed and direction, whenever possible,
along with wave measurements, collection of data at hourly intervals (rather than three hour intervals) to obtain better time
histories during storms and hurricanes, and quality control of the
data from the time of measurement through all analysis procedures.

Comments during symposium discussion of recommendations.
A question was asked about mooring effects on Waverider measurements.
W. F. Baird stated that J. Nath of Oregon State University studied
mooring effects for the Canadian Wave Climate Study and that there
is not a major problem.

Members of Working Group:
W. F. Baird, Chairman
V. J. Cardone
P. DeLeonibus
A. Johnson, Jr.
L. LeBlanc
F. Middleton
K. Steele

WAVE MEASUREMENTS FROM OFFSHORE PLATFORMS

The probability of maximum data return with minimum investment
and development is possible by utilizing existing and proposed
public and private offshore structures. The working group recommends that a selected set of these structures be instrumented as
part of a wave monitoring program. In addition to meteorological
and other oceanographic parameters, directional spectra should be
measured. The time period of measurements at a given location
should be variable and should be based on the applications of the
collected data. In addition, the use of alternate more cost
effective measurement systems, such as radar or laser methods,
should be considered on a regular basis when these systems are
proven for regular field use. While wave monitoring is conducted
with presently available methods, the monitoring program should
support the development of advanced wave measurement systems. A
major concern of non-government wave data users is the rapid
availability of data with appropriate calibrations. These data
should be provided to all users who request it within approximately
thirty days after data collection.

Two projects should be immediately implemented to help satisfy
the above recommendations. First, the U.S. Navy's Air Combat
Maneuvering Range Tower Number 3 off Cape Hatteras should be instrumented. This tower represents an ideal base for proof-of-concept
of wave measurement instrumentation which should consist of: a
minimum of three wave staffs, up to six electromagnetic current
meters, a redundant set of meteorological instruments, and appropriate on-board recording or telemetering equipment. The instrumented tower should serve as a calibration standard for testing
Waverider buoys, other buoys, and different buoy moorings. The
tower offers the following advantage: U.S. east coast exposure
to deep water waves, early data return with high probability of
success, high exposure to a variety of meteorological and oceanographic conditions including hurricanes and severe extratropical
storms, good location for ground truth for a variety of experiments,
and good location for long-term measurements for application to
coastal and offshore projects.

A second project is government cooperation with the private
sector and other public agencies by furnishing automatic and

portable meteorological/oceanographic monitoring and recording systems to be placed on fixed and floating structures. These cooperative effects could be modeled after the highly successful Canadian Department of the Environment wave measurement program. The working group noted that there are automatic, NOAA furnished, meteorological stations installed on twelve petroleum industry structures off the Louisiana Coast. An ideal first step would be to upgrade some of these systems by the addition of wave measurements.

Comments during symposium discussion of recommendations.
Some participants were concerned about platform interference with wave measurements. R. A. Stacy said that the offshore oil industry had studied interference and that problems can be largely avoided by careful planning and placement of wave staffs including multiple staffs on a single platform.

Members of Working Group:
R. A. Stacy, Chairman
M. D. Earle
E. Escowitz
H. R. Frey
K. G. Nolte
R. W. Whalin

NEARSHORE WAVE MEASUREMENTS

The working group loosely defined "nearshore" as between the
surf zone and the location where waves are first significantly
affected by refraction. Three major classes of wave measuring
instruments were identified. These are: wave staffs, pressure
sensors, and current meters. Table 1 is a comparison of the
advantages and disadvantages of these instruments. A fourth
type, which is dependent on wave forces and which include tilting
spars and force instrumented spheres, are under development or have
had limited use. Two other types, radar and instrumented buoys,
were determined to be of primary importance for nearshore wave
measurements but are being covered by other working groups. Wave
staffs, pressure sensors, and current meters could be used for nearshore measurement without major development. An important task,
however, is the calibration of instruments against an accepted
standard and cross calibration between instrument types.

Estimates of a nearshore directional spectrum are possible
with state-of-the-art instruments. Approximate resolution capabilities are: surface following buoys and wave slope arrays $\pm15°$
to $\pm90°$, space arrays $\pm5°$ to $\pm90°$, wave staff and orbital velocity
meter arrays $\pm10°$ to $\pm15°$. Resolution limits could be improved by
careful assumptions of a wave direction model and by data adaptive
analysis techniques. A wave slope array is capable, however, of
defining a mean angle for each frequency band with a resolution of
about $\pm3°$. Table 2 summarizes several advantages and disadvantages
of space arrays and single point measurement systems for wave
direction measurements.

Costs of the various approaches are highly dependent upon site
specific factors. However, costs have been estimated. A slope
array of pressure sensors, near a pier, and part of a large data
collection system would average about $5,000 annually. An offshore
pitch, roll, and heavy buoy which is part of a large data collection
system might cost about $10,000 to $25,000 annually. An offshore
platform instrumented with wave staffs and current meters would
provide wave data at a cost of about $100,000 annually (several
symposium participants believe that use of existing platforms would
result in costs less than $25,000 annually).

350

   Comments during symposium discussion of recommendations.
During the presentation of these recommendations, it was pointed
out that coastal processes studies would greatly benefit if wave
directions are determined to within ±1°. Some participants considered this requirement too stringent. Several participants,
including T. Saville, felt that measurements of wave heights and
periods with estimates of wave directions would still be beneficial
for many applications. D. B. Ross noted that one should keep in
mind the difference between the need for fine scale directional
resolution of the directional spectrum and accurate measurements
of mean wave direction. O. T. Magoon stated that one program
should collect data on all types of waves including: wind
generated waves, long-period waves which produce harbor surging,
tides, and tsunamis.

Members of Working Group:
R. J. Seymour, Chairman
L. E. Borgman
W. L. Edge
V. Goldsmith
D. H. Harris
O. T. Magoon
D. T. Resio
R. P. Savage

TABLE 1.   Comparison of measuring instruments
for nearshore wave measurements.

General Instrument Group	Disadvantages	Advantages
Surface Piercing Staffs	1. Tides increase dynamic range 2. Requires a structure for support 3. May be shielded by structure 4. Subject to fouling 5. Subject to vandalism (on piers) 6. Subject to icing at high latitudes	1. Cheap 2. Linear and direct measurement of sea surface 3. Reliable, available
Pressure Sensors	1. Nonlinear indirect measurement of sea surface 2. Hard to service, install 3. Subject to scour and burial 4. Low pass filter 5. Useful in limited depth range 6. Damaged by trawls	1. Requires no platform 2. No interferences to navigation 3. Not susceptible to vandalism 4. Cheap 5. More direct estimate of bottom velocities
Current Meters	1. Subject to fouling 2. Nonlinear and indirect measurement of surface elevation 3. Expensive 4. Unreliable compared to staffs and pressure sensors	1. Direct measurement of velocity components at a point

TABLE 2. Comparison of major direction
measuring systems.

Type		Disadvantages		Advantages
Space Array	1.	Suffers from the spatial inhomogeneity of wave field	1.	Yields an estimate of a directional spectrum
	2.	Difficult and expensive to install	2.	Self checking between elements
	3.	Shielding and interference of support structures (pier or platform)	3.	Expandable
	4.	Narrow band in frequency and direction for afford-able arrays		
Point Measure-ment Systems	1.	Limited resolution	1.	Cheaper to install
	2.	More susceptible to small scale anomalies in wave field from fixed objects		

SUMMARIES OF PRESENTATIONS BY THE

U.S. ARMY CORPS OF ENGINEERS

The U.S. Army Corps of Engineers provided three presentations addressing Corps' wave activities and wave data needs. These presentations are briefly summarized below.

CALIFORNIA COASTAL DATA NEEDS, Richard Connell (Brig. Gen.), *U.S. Army Engineer Division, South Pacific*

Because of immediate problems related to coastal erosion, coastal processes and harbor design along the California coast, the South Pacific Division is initiating a comprehensive field data collection effort, the California Coastal Data Program. This program, which will begin in fiscal year 1979 and will extend over four years, will obtain wave measurements at approximately 100 sites along the California coast. Offshore accelerometer buoys and near-shore pressure sensors will monitor real-time wave data and coastal processes data such as beach profiles will be collected. General Connell's talk was accompanied by numerous slides illustrating critical wave-related coastal problems along the California coast.

U.S. ARMY CORPS OF ENGINEERS WAVE HINDCAST CLIMATOLOGY STUDIES, Donald T. Resio, *U.S. Army Engineer Waterways Experiment Station*

In 1974, a project was initiated at the U.S. Army Engineer Waterways Experiment Station to obtain a climatology of extreme wave heights for U.S. shorelines in the Great Lakes. A method involving a velocity- and stability-dependent transformation of over-land winds to over-lake winds was developed and calibrated with ship observations on each lake. This method of estimation of over-lake winds converged toward an rms error in wind speed of about 4 knots for over-lake velocities above 20 knots. A numerical scheme for surface waves in an arbitrary geometry was formulated and appeared to work well even under conditions of changing wind fields. The RMS error in approximately 300 comparisons between model outputs and wave gage data provided by the Canadian Department of the

Environment was under 1.5 ft. In a blind comparison between model
outputs and wave gage data from the U.S. Army Engineer Coastal
Engineering Research Center, the rms error was approximately 1.2 ft.
The same wave model has been used, unmodified, for verifications
on all of the Great Lakes, with the wind fields determined independently from the objective transformation technique mentioned
above. Reports on design wave conditions have been published for
the Great Lakes.

Reliable wave information is still lacking in many coastal areas
of the United States. The Corps of Engineers needs these data for
planning, design, operation and maintenance of navigation channels,
breakwaters, jetties, inlets, harbors and beaches. In reponse to
these needs, a Wave Information Study was begun to obtain a climatology of waves along the Atlantic, Pacific, and Gulf coasts. Not
only will the study provide a basic set of data and condensed tables
of information, but it will also systematize the data into a com-
puter-based information system containing state-of-the-art programs
to access wave information; calculate nearshore wave transformations;
and perform statistical analyses for specific locations. By fall,
1978, wind and wave models will be evaluated for the U.S. Atlantic
Coast. These models will then be used to provide a comprehensive
directional spectra data base. An additional phase of the program
will determine wave transformations close to the coast. High
quality measured wave data will provide an invaluable check,
particularly in shallow water, on the data generated by the Corps'
hindcast effort.

WAVE DATA NEEDS FOR MODEL STUDIES AND CONSTRUCTION PROJECTS,
Robert W. Whalen, *U.S. Army Engineer Waterways Experiment Station*

Wave data form the basic input for practically every planning,
design, operation, and maintenance problem concerned with coastal
projects of the Corps of Engineers. Extreme waves and water
levels are used for determining the size of stone or concrete
armor units and the crown elevation for coastal structures (breakwaters and jetties), the elevation of hurricane surge barriers
and/or control structures, and for flood plain management. Daily
climatologies are needed for optimal scheduling of dredging
operations which are constrained to operate in relatively mild
climates, to determine the required capacity of sand bypassing
systems, to evaluate the effect of proposed projects on adjacent
shorelines, to plan and design beach-fill projects, to evaluate
alternative methods of mitigating beach erosion problems, and to
determine the adequacy of alternative borrow sites for beach
nourishment.

The Waterways Experiment Station uses wave data and wave statistics
for a variety of model studies which are listed in the following

table. As just described by D. Resio, a major effort is underway
to numerically hindcast a directional wave spectra climatology
with known confidence limits for the east, west, and Gulf coasts
of the continental United States and Hawaii. Although the Waterways Experiment Station does not normally collect its own wave
data, it utilizes wave data both to determine model inputs and to
verify the numerical techniques used in developing the directional
wave spectra climatology. While existing wave measurements are
providing data and planned programs will provide additional data,
these programs of other organizations do not emphasize long period
(25 second to 15 minutes) waves which are important for harbor
oscillation studies. Hopefully, such data will be collected in
the future. In summary, the Waterways Experiment Station uses
physical and numerical models for coastal engineering and coastal
processes studies and is developing an extensive hindcast wave
climatology for input to these models.

Wave model studies at the U.S. Army Engineer Waterways Experiment Station

Application	Model Type	End Results	Wave Data Needed
Harbor design (sea and swell)	Physical	Breakwater and jetty characteristics (length, orientation, elevation, etc.) to assure low wave heights in navigation channels and mooring areas.	Wave climatology Extreme wave data
Harbor oscillations	Physical Numerical (finite element)	Knowledge of harbor oscillations which may affect moored ships and design character- istics to prevent sig- nificant oscillations.	Directions of long period waves (25 seconds to 15 minutes) incident to the harbor
Inlets	Physical	Jetty characteristics (length, orientation, elevation, etc.) to maintain safe naviga- tion channels through inlets. Percentage of time that dredges can operate.	Wave climatology including wave directions and joint probability of waves and water levels.

Wave model studies at the U.S. Army Engineer
Waterways Experiment Station (continued)

Application	Model Type	End Results	Wave Data Needed
Beach erosion	Physical	Determination of structural designs (such as submerged or semi-submerged breakwaters and groins) to mitigate beach erosion.	Wave climatology including wave directions and joint probability distribution of of waves and water levels.
Breakwater jetty design	Physical	Optimum design (including initial costs and probable maintenance costs) for extreme wave conditions.	Extreme wave data and joint probability distribution of waves and water levels.

OTHER GOVERNMENT WAVE PROGRAMS

A purpose of the *Ocean Wave Climate Symposium* was to increase knowledge about important wave programs of various government agencies. These programs are sometimes not well known but the data and results of these programs are most useful for scientific studies and practical applications. In addition to the contributed papers included in this volume, the symposium included discussions of several other wave related activities of government agencies. Here, these other activities are briefly noted.

Several wave programs are located within the National Oceanic and Atmospheric Administration (NOAA). The Great Lakes Environmental Research Laboratory is using Waverider accelerometer buoys to collect wave data in Lake Michigan. These data are being used to develop and validate a Lake Michigan spectral wave model. The Pacific Marine Environmental Laboratory is concentrating on site specific wind field models and shallow water wave models for application to coastal areas near the Gulf of Alaska and Seattle where tanker shipments and offshore activities are planned. A real-time wave data monitoring system has been developed by the National Ocean Survey for use off the Atlantic coast. In this system, a central minicomputer controls phone-line data acquisition from several shore stations which receive telemetered data from Waverider accelerometer buoys. Satellite collected wave data are used by the National Environmental Satellite Service, the Pacific Marine Environmental Laboratory, the Atlantic Oceanographic and Meteorological Laboratory and the Wave Propagation Laboratory. Significant wave heights are currently being obtained from radar altimeter measurements with GEOS-3. SEASAT, a satellite dedicated to oceanographic measurements, was successfully launched by the National Aeronautics and Space Administration (NASA) in June 1978 and collected approximately three months of data before its failure in October 1978. Wave research studies by NOAA, NASA, and contractors will evaluate and utilize its radar altimeter for significant wave height measurements, its synthetic aperture radar for wave length and wave direction measurements, its microwave scatterometer for wind speed measurements, and its microwave radiometer for high wind speed measurements. Marine forecasts, including wave forecasts, are provided by the National Weather Service. At the present time, waves are forecast with a significant wave height model but consideration is being given to utilizing spectral wave models and parametric hurricane models. The National Oceanographic Data Center has archives of measured wave data including data collected by NOAA data buoys and some data collected by the offshore oil industry. These archives are being expanded to include wave measurements made by other organizations and government agencies.

Within the U.S. Army Corps of Engineers, major wave projects are located at the Waterways Experiment Station, the Coastal Engineering Research Center, and the South Pacific Division which is responsible for coastal protection and coastal engineering projects along the California coast. The work of the Waterways Experiment Station and South Pacific Division has already been noted. The Coastal Engineering Research Center has built a large research pier north of Cape Hatteras. This pier is instrumented with several wave staffs and two tide gages and will be used to study coastal processes. The Coastal Engineering Research Center has also begun a wave transformation study to better relate deep water waves to shallow water waves near coasts. This work will be used with wave hindcast results developed by the Waterways Experiment Station.

The U.S. Navy Fleet Numerical Weather Central is operationally forecasting directional wave spectra for the Northern Hemisphere on a twice-daily basis and is involved in a 20-year deep water wave hindcast study for the Northern Hemisphere to provide a wave spectra climatology for ship design applications. The Navy will also utilize SEASAT wind and wave data to verify and improve wind and wave forecasting models.

Finally, note should be made of a Canadian program, the Canadian Wave Climate Study, which has been operated by the Canadian Department of the Environment since 1968. Since that time, wave data has been collected at about 125 locations in marine and inland waters throughout Canada. Waverider accelerometer buoys are generally used and, at any time, measurements are usually being made at about 20 locations.

LIST OF CONTRIBUTORS AND PARTICIPANTS

GERALD F. APPELL, National Ocean Survey, NOAA, Rockville,
   Maryland 20852
FRANK AUGUSTINE, Aerospace Corporation, P.O. Box 92957,
   Los Angeles, California 90009
LEDOLPH BAER, NOAA, 6010 Executive Blvd., Rockville, Maryland 20852
WILLIAM F. BAIRD, Marine Directorate, Department of Public Works,
   Ottawa K1AOM2, Canada
SUSAN L. BALES, Naval Ship Research & Development Center, Code 1568
   Bethesda, Maryland 20084
WILLIAM BARBEE, Office of Ocean Engineering, NOAA, Rockville,
   Maryland 20852
DONALD E. BARRICK, Environmental Research Laboratories, NOAA,
   Boulder, Colorado 80302
CELSO BARRIENTOS, National Weather Service, NOAA, Gramax Building,
   Silver Spring, Maryland 20910
JOHN BENNETT, Great Lakes Environmental Research Laboratory, NOAA,
   2300 Washtenaw Avenue, Ann Arbor, Michigan 48104
JOSEPH M. BISHOP, Environmental Data Service, CEDDA, NOAA,
   3300 Whitehaven Street, Page Building 2,
   Washington, D.C. 20235
LEON E. BORGMAN, Statistics Department, University of Wyoming,
   Laramie, Wyoming 82071
VINCENT J. CARDONE, Oceanweather, Inc., 170 Hamilton Avenue,
   White Plains, New York 10601
G. CHRISTOPH, Sun Shipping, Chester, Pennsylvania 18016
WILLIAM A. CLEARY, JR., Office of Marine Safety, USCG Headquarters,
   Washington, D.C. 20591
RICHARD M. CONNELL, Division Engineer, U.S. Army Engineer Division,
   South Pacific, 630 Sansome Street, San Francisco, California,
   94111
W. E. CUMMINS, Naval Ship Research & Development Center, Ship
   Performance Department, Bethesda, Maryland 20084
PAT DELEONIBUS, NOAA, 6010 Executive Blvd., Rockville, Maryland
   20852
JOHN M. DIAMANTE, National Ocean Survey, NOAA, Rockville,
   Maryland 20852
LAURENCE DRAPER, Marine Information Advisory Service, Institute
   of Oceanographic Services, Wormley, Godalming, Surrey,
   United Kingdom
MARSHALL D. EARLE, National Ocean Survey, NOAA, Rockville,
   Maryland 20852
WILLIAM L. EDGE, Civil Engineering Department, Clemson University,
   Clemson, South Carolina 29631

DAVID ENABNIT, National Ocean Survey, NOAA, Rockville, Maryland
    20852
EDWARD ESCOWITZ, Naval Facilities Engineering Command, Ocean
    Engineering Construction Project Office, Washington, D.C.
    20374
WILLIAM EVANS, Atlantic Richfield, P.O. Box 2819, Dallas,
    Texas 75221
TONY FALLON, Chevron Oil Field Research Co., P.O. Box 446,
    La Habra, California 90631
HENRY R. FREY, National Ocean Survey, NOAA, Rockville, Maryland
    20852
VICTOR GOLDSMITH, Virginia Institute of Marine Sciences,
    Gloucester Point, Virginia 23062
JOHN GREGORY, U.S. Geological Survey, Reston, Virginia 22092
D. LEE HARRIS, Coastal Engineering Research Center, Kingman
    Building, Ft. Belvoir, Virginia 22060
JOHN HEIDEMAN, EXXON Production Research Co., P.O. Box 2189,
    Houston, Texas 77001
DANIEL HOFFMAN, Hoffman Maritime Consultants, 9 Glen Head Rd.,
    Glen Head, New York 11545
JOHN G. HOUSLEY, U.S. Army Corps of Engineers, Washington, D.C.
    20314
ANDREW JOHNSON, JR., NOAA Data Buoy Office, National Space
    Technology Laboratories, Bay St. Louis, Mississippi 39520
CHARLES KEARSE, Engineering Development Laboratory, National
    Ocean Survey, NOAA, Riverdale, Maryland 20840
EDWARD KERUT, NOAA Data Buoy Office, National Space Technology
    Laboratories, Bay St. Louis, Mississippi 39520
LESTER LE BLANC, University of Rhode Island, Kingston, Rhode
    Island 02882
GORDON G. LILL, National Ocean Survey, NOAA, Rockville, Maryland
    20852
BELINDA J. LIPA, Center for Radar Astronomy, Stanford University,
    Stanford, California 94305
PAUL C. LIU, Great Lakes Environmental Research Laboratory, NOAA,
    2300 Washtenaw Avenue, Ann Arbor, Michigan 48104
ORVILLE T. MAGOON, U.S. Army Engineer Division, South Pacific,
    P.O. Box 26062, San Francisco, California 94126
ALEXANDER MALAHOFF, National Ocean Survey, NOAA, Rockville,
    Maryland 20852
JOSEPH W. MARESCA, JR., Stanford Research Institute International,
    Menlo Park, California 94025
MICHAEL MATTIE, Coastal Engineering Research Center, Kingman
    Building, Ft. Belvoir, Virginia 22060
M. E. McCORMICK, Naval Systems Engineering Department, U.S. Naval
    Academy, Annapolis, Maryland 21402
FOSTER MIDDLETON, University of Rhode Island, Kingston, Rhode
    Island 02882

MAX W. MULL, National Weather Service, NOAA, Silver Spring,
    Maryland 20910
C. S. NIEDERMAN, Office of R&D, U.S. Coast Guard, Washington, D.C.
    20590
K. G. NOLTE, Amoco Production Co., P.O. Box 591, Tulsa, Oklahoma
    74102
JAMES E. OVERLAND, Pacific Marine Environmental Laboratory, NOAA,
    3711 15th Avenue, N.E., Seattle, Washington 98105
ROY OVERSTREET, Environmental Research Laboratories, NOAA,
    Boulder, Colorado 80503
CHESTER PARSONS, Wallops Flight Center, NASA, Wallops Island,
    Virginia 23337
DONALD T. RESIO, U.S. Army Engineer Waterways Experiment Station,
    P.O. Box 631, Vicksburg, Mississippi 39180
RICHARD L. RIBE, National Ocean Survey, NOAA, Rockville, Maryland
    20852
M. E. RINGENBACH, Engineering Development Laboratory, National
    Ocean Survey, NOAA, Riverdale, Maryland 20840
DUNCAN B. ROSS, Atlantic Oceanographic & Meteorological
    Laboratories, NOAA, 15 Rickenbacker Causeway, Virginia Key,
    Miami, Florida 33149
EUGENE RUSSIN, National Ocean Survey, NOAA, Rockville, Maryland
    20852
R. P. SAVAGE, U.S. Army Corps of Engineers, Coastal Engineering
    Research Center, Ft. Belvoir, Virginia 22060
THORNDIKE SAVILLE, JR., Coastal Engineering Research Center,
    Kingman Building, Ft. Belvoir, Virginia 22060
RICHARD J. SEYMOUR, Scripps Institute of Oceanography, La Jolla,
    California 92093
OMAR SHEMDIN, Jet Propulsion Laboratory, 4800 Oak Grove Drive,
    Pasadena, California 91103
R. A. STACY, Mobil Research & Development Corporation, Offshore
    Engineering, P.O. Box 900, Dallas, Texas 75221
KENNETH STEELE, NOAA Data Buoy Office, National Space and
    Technology Laboratory, Bay St. Louis, Mississippi 39520
NORMAN STEVENSON, Fleet Numerical Weather Central, Monterey,
    California 98940
CALVIN C. TEAGUE, Center for Radar Astronomy, Stanford
    University, Stanford, California 94305
C. L. VINCENT, U.S. Army Engineer Waterways Experiment Station,
    P.O. Box 631, Vicksburg, Mississippi 39180
DAVID A. WALDEN, Office of R&D, U.S. Coast Guard, Washington, D.C.
    20590
E. G. WARD, Shell Development Co., P.O. Box 481, Houston, Texas
    77001
WELLINGTON WATERS, Environmental Data Service, NOAA, Washington,
    D.C. 20235

ROBERT W. WHALIN, U.S. Army Engineer Waterways Experiment Station, P.O. Box 631, Vicksburg, Mississippi 39180

J. R. WILSON, Marine Environmental Data Service, Dept. of Fisheries & Environment, 580 Booth Street, Ottawa, Canada

WILLIAM E. WOODWARD, Office of Ocean Engineering, NOAA, Rockville, Maryland 20852